SPATIAL PLANNING AND RESILIENCE FOLLOWING DISASTERS

International and comparative perspectives

Edited by Stefan Greiving, Michio Ubaura and Jaroslav Tešliar

First published in Great Britain in 2018 by

Policy Press
University of Bristol
1-9 Old Park Hill
Bristol
BS2 8BB
UK
t: +44 (0)117 954 5940
pp-info@bristol.ac.uk
www.policypress.co.uk

North America office:
Policy Press
c/o The University of Chicago Press
1427 East 60th Street
Chicago, IL 60637, USA
t: +1 773 702 7700
f: +1 773-702-9756
sales@press.uchicago.edu
www.press.uchicago.edu

British Library Cataloguing in Publication Data
A catalogue record for this book is available from the British Library

Library of Congress Cataloging-in-Publication Data
A catalog record for this book has been requested

ISBN 978-1-4473-2359-4 paperback
ISBN 978-1-4473-2362-4 ePub
ISBN 978-1-4473-2363-1 Mobi

Cover design by Hayes Design
Front cover image: Nadine Mägdefrau
Printed and bound in Great Britain by CMP, Poole
Policy Press uses environmentally responsible print partners

Contents

List of figures and tables

Figures

Acknowledgements

This handbook is a result of the research project 'Increasing resilience of urban planning' (URBIPROOF), which was funded through the CONCERT–Japan framework by the Federal Ministry of Education and Research (BMBF) and the Japan Science and Technology Agency (JST). The aim of URBIPROOF was to help cities to enhance their resilience by improving institutional and societal capacities and urban development strategies.

The editors would like to sincerely thank Ms Nadine Mägdefrau for her hard work and continuous effort in the editing of this book. We extend our thanks and appreciation for her assistance in coordination of the various chapters and content. Without the efforts of Ms Mägdefrau, realising the outcome of this project would not have been possible.

The editors would also like to thank and acknowledge the contributors of this publication for their joint effort and fruitful collaboration.

Abbreviations

AICP	American Institute of Certified Planner
BAKORNAS	National Board for Disaster Management
BAPPENAS	National Development Planning Agency
BNBP	National Agency for Disaster Management
BNOB	Bring New Orleans Back
BNPB	National Disaster Management Authority
BPBD	Local Disaster Management Authority
BRR	Rehabilitation and Reconstruction Agency
CAFRA	Coastal Area Facility Review Act 1973
CDBG	Community Development Block Grant
CDBG-DR	Community Development Block Grant Disaster Recovery
CLP	Regulation on the Classification, Labelling and Packaging
CNDA	Comprehensive National Development Act 1950
DCA	Department of Community Affairs
DCBA	Disaster Countermeasures Basic Act
DCP	Department of City Planning
DHS	Department of Homeland Security
DOB	Department of Buildings
DOT	Department of Transportation
DRR	disaster risk reduction
EDC	Economic Development Corporation
EEA	European Environment Agency
EIA	Environmental Impact Assessment Act
ESCR	East Side Coastal Resiliency Project
EU	European Union
FAR	floor area ratio
FBC	Federal Building Code
FEMA	Federal Emergency Management Agency
FMA	Flood Mitigation Assistance
GDP	gross domestic product
GEJE	Great East Japan Earthquake
GORR	Governor's Office of Recovery and Rebuilding
GOSR	Governor's Office of Storm Recovery
HFA	Hyogo Framework for Action
HMA	Hazard Mitigation Assistance
HMGP	Hazard Mitigation Grant Program

HPD	Department of Housing Preservation and Development
HRO	Housing Recovery Office
HUD	Department of Housing and Urban Development
IAIA	International Association for Impact Assessment
IPCC	Inter-governmental Panel on Climate Change
IRGC	International Risk Governance Council
LPS	Office of Local Planning Services
LTCR	Long-Term Community Recovery
MLIT	Ministry of Land, Infrastructure, Transport and Tourism
MoE SR	Ministry of Environment of the Slovak Republic
MPO	Metropolitan Planning Organization
MTA	Metropolitan Transportation Authority
NDRF	National Disaster Recovery Framework
NFIP	National Flood Insurance Program
NGOs	non-governmental organisations
NORA	New Orleans Recovery Agency
NRP	National Response Plan
NSPA	National Spatial Planning Act 2008
NYCHA	New York City Housing Authority
NYMTC	New York Metropolitan Transportation Council
NYRCR	New York Rising Community Reconstruction Program
NYRTTT	New York Rising to the Top
OEM	Office of Emergency Management
ORR	Office of Recovery and Resiliency
PDM	Pre-disaster Mitigation Program
PHA	Project Home Again
PP	New Jersey Board of Professional Planners
PSPAG	Post Sandy Planning Assistance Grant Program
RBD	Rebuild by Design
ROG	Federal Spatial Planning Act
RPA	Regional Plan Association
SBS	Department of Small Business Services
SEA	Strategic Environmental Assessment
SIRR	Special Initiative for Rebuilding and Resiliency
UNISDR	United Nations International Strategy for Disaster Reduction
UNOP	United New Orleans Plan
WHG	Federal Water Act

Notes on contributors

Editors

Prof. Dr. Stefan Greiving (Germany) is the Executive Director of the Institute of Spatial Planning at the Technical University of Dortmund, Germany. In parallel, he runs the consultancy company plan + risk consult. His areas of expertise include: disaster risk assessment and management, climate vulnerability assessment and adaptation planning, and urban and regional planning. He has led 21 international research projects and has been involved in nearly 50 additional projects within the last 15 years. From these projects, nearly 180 publications have been produced. Prof Greiving holds a Diploma in Spatial Planning, a PhD in urban planning and a habilitation (post-doctoral thesis) in planning and administration. He is also a full member of the German Academy for Spatial Planning and Research (ARL) and a member of the Architects' Chamber.
Email: stefan.greiving@tu-dortmund.de

Prof. Dr. Michio Ubaura (Japan) is an associate professor at Tohoku University, Department of Architecture and Building Science in Graduate School of Engineering and concurrently at the International Research Institute of Disaster Science. He specialises in urban planning and land use in Japan and Germany, focusing particularly on sustainable urban form and post-disaster planning. Since 2011, he has been extensively involved in recovery planning, both as a member of an advisory committee and a planner in Tohoku region. Some of his recent key book publications, among many, include: *Community making forefront in the recovery from the Great East Japan Earthquake* (2013) and *Land use planning in society with population decline* (2010) (both in Japanese and co-authored). He holds Bachelor of Laws, Masters and Doctorate degrees in urban engineering, all from the University of Tokyo. He has taught in Osaka City University in his earlier career.
Email: ubaura@archi.tohoku.ac.jp

Dr. Jaroslav Tešliar (Slovakia) is a director of the Agency for the Support of Regional Development Kosice, Slovakia (ASRD). In addition to managing the organisation, since 2007 he has cooperated on development of more than 40 international development and research projects dealing with balanced development of urban

and rural regions, integrated management of natural resources, job creation and human capacity development. He has also been involved in several projects related to risk management topics. Dr. Tešliar studied at the Forestry Faculty of the Technical University in Zvolen, where he received a PhD in Agro-forestry Sciences and he holds a post-doctorate Certificate in International and Regional Development from University of Washington, Seattle, USA.
Email: jaroslav.tesliar@arr.sk

Authors (in order of appearance)

Prof. Kanako Iuchi (Japan) has joined the International Research Institute of Disaster Science (IRIDeS), Tohoku University in 2013 as an associate professor. Her areas of expertise include disaster management planning, urban and regional planning, and community development in international settings. Prior to joining IRIDeS, she worked as an urban specialist in the World Bank and as an international development planner and researcher with bilateral and multilateral organisations; national, regional, and local governments; and communities in more than ten countries across East and South Asia, South America, and Eastern Europe. Her research interests are post-disaster relocation, planning decisions in recovery processes, and policy influences on communities. She holds a BS from Tsukuba University, an MRP from Cornell University, and a PhD from the University of Illinois, Urbana-Champaign, in urban and regional planning.
Email: iuchi@irides.tohoku.ac.jp

Prof. Tamiyo Kondo, PhD (Japan) is an associate professor at Kobe University, Department of Architecture in Graduate School of Engineering. Her areas of expertise include planning for built environment, community-based planning, housing policy and post-disaster housing recovery. She has been continuing housing rebuilding field surveys in New Orleans after Hurricane Katrina since 2005. After the Great East Japan Earthquake, she is especially interested in transformation of the built environment by individual-based housing relocation, community-driven resettlement, and volunteer-based housing renovation, all of which represent people's resilience after disaster. Some of her recent key book publications include: 'Clustered and Community-driven Housing Recovery: Lesson learned from Hurricane Katrina and Great East Japan Earthquake' in *Cities and Disasters* (CRC Press, 2015) and 'Compensation or assistance? Law and policy

for post-disaster housing recovery in the US and Japan' in *Asian Law in Disasters: Toward a Human-Centered Recovery* (Routledge, 2016).
Email: tamiyok@people.kobe-u.ac.jp

Prof. Yuka Karatani, PhD (Japan) is a Professor at the Graduate School of Urban Science, Meijo University. She teaches urban planning, disaster management planning and post-disaster recovery planning. Her areas of expertise include community-based disaster reduction planning for multi-hazards with stakeholders. She has been studying housing recovery and community assessment due to relocation after the Great East Japan Earthquake (2011). She lived with victims in the shelter and the temporary housing immediately after the EQ to investigate the resilience for disaster reduction.
Email: karatani@meijo-u.ac.jp

Togu Santoso Pardede (Indonesia) has been an affiliated researcher from April 2013 and a PhD candidate at the International Development and Regional Planning Unit, Department of Urban Engineering, University of Tokyo. He is a government officer at BAPPENAS (State Ministry of Development Planning/National Development Planning Agency) in the Republic of Indonesia. He started to work at BAPPENAS from 1996 as a young planner at Secretariat of National Spatial Planning Coordination Board (BKTRN). In 1999 he moved to Directorate of Housing, Settlement, Urban Development, Land and Spatial Planning. Since 2003, he has been working at the Directorate of Special Areas Development, responsible for hazard prone areas. In this section, he has been directly involved in the recovery planning of two mega-disasters; Aceh Tsunami (2004) and Yogyakarta Earthquake (2006). He has been responsible for conducting Damage and Need Assessment as well as developing Aceh's Blue Print Plan and Yogyakarta's Action Plan. At this moment, he is concurrently serving as the Head of Knowledge Management and Handling Complaints Unit, Secretariat for the Indonesia Multi Donor Fund Facilities for Disaster Recovery (IMDFFDR). He holds a BE from Bandung Institute of Technology in urban and regional planning, a Master of International Development Studies from National Graduate Institute for Policies Studies (GRIPS) Tokyo, Japan.
Email: togupar@bappenas.go.id

Gita Chandrika Munandar (Indonesia) is an urban and regional development planner with areas of expertise in policy and institutionalisation analysis, urban development and spatial planning,

and strategic planning for local development, and more than 20 years' working experiences. She was the Program Director of Urban and Regional Development Institute (URDI) from 2007–10. She earned her Masters degree in Community Planning from the School of Planning, College of Design, Architecture, Art and Planning, University of Cincinnati in 1990 with recognition (receiving Graduate Student Award) and received her Bachelor degree in Urban and Regional Planning from Bandung Institute of Technology in 1987. Before she joined URDI in December 2005, she had been working as independent consultant in the National Development Planning Board (BAPPENAS), and in the Directorate General of Human Settlements, Department of Public Works, as well as working with donor agencies United Nations Development Programme (UNDP) and Canadian International Development Agency (CIDA).
Email: gitanapitupulu@yahoo.com

Gusti Ayu Ketut Surtiari (Indonesia) is a PhD researcher at the United Nations University, Institute for Environment and Human Security (UNU-EHS) in the section on Vulnerability Assessment, Risk Management and Adaptive Planning (VARMAP). She is also a researcher at the Research Center for Population, Indonesian Institute of Sciences (PPK-LIPI) and doing various studies on the issue of human and environment linkages. Since 2008, she has been conducting research on urban development, climate change impact, and adaptation of communities and government, especially in urban areas and disaster risk reduction. Currently, she is doing her PhD research on the dynamic of vulnerability and adaptation in Jakarta, Indonesia.
Email: surtiari@ehs.unu.edu

Dr. Neysa Jacqueline Setiadi (Indonesia) is a Research Associate at the United Nations University, Institute for Environment and Human Security (UNU-EHS), in the section on Vulnerability Assessment, Risk Management and Adaptive Planning (VARMAP). Neysa holds a PhD from the University of Bonn, Institute for Geodesy and Geoinformation, with a thematic focus on tsunami early-warning response capability and related considerations for urban planning in Indonesia. Previously, she graduated as an Environmental Engineer at the Institut Teknologi Bandung, Indonesia, and obtained her Master's degree in Water Resources Engineering and Management at the University of Stuttgart, Germany. Her research interest lies in the field of disaster risk reduction, vulnerability, emergency and urban planning.
Email: neysajacqueline@yahoo.com

Dr. Matthias Garschagen (Germany) is an Associate Academic Officer at the United Nations University, Institute for Environment and Human Security (UNU-EHS), in the section on Vulnerability Assessment, Risk Management and Adaptive Planning (VARMAP). Matthias holds a PhD from the University of Cologne, Department of Geography, with a thematic focus on dynamics in vulnerability and adaptation to natural hazards in urban Vietnam under conditions of transformation. He serves as a contributing author to the IPCC's *Fifth Assessment Report* (AR5) in WG II, chapter 24 on Asia. Matthias further was a background author to UN-ISDR's *Global Assessment Report 2013*. He has received scholarships from the German National Academic Foundation and the German Academic Exchange Service. He is currently an active member of the working group on Southeast Asian Studies within the German Geographical Association as well as the International Expert Working Group on Measuring Vulnerability, organised by UNU-EHS.
Email: garschagen@ehs.unu.edu

Prof. Dr. Joern Birkmann (Germany) is the Director of the Institute of Spatial and Regional Planning (IREUS) at the University of Stuttgart. He holds a PhD in Spatial Planning from the Dortmund University and a post-doctoral degree (*Habilitation*) in Geography from the University of Bonn, Department of Geography. He has authored or co-authored more than 100 scientific publications, including monographs such as *Measuring Vulnerability to Natural Hazards: Towards Disaster Resilient Societies.* He has expertise in the field of spatial planning, risk and vulnerability, sustainable development and environmental assessment. He specialises in development of assessment methodologies and indicators to estimate and evaluate different socio-economic trends and environmental degradation at the sub-national, local and household scale. Joern Birkmann has been a lead author of the Intergovernmental Panel on Climate Change (IPCC) Special Report *Managing the Risk of Extreme Events and Disasters to Advance Climate Change Adaptation.* He is also lead author of the chapter on 'Emergent Risks and Key-Vulnerabilities' in the Fifth Assessment Report of the IPCC. He is member of the Scientific Committee of the International Council for Science (ICSU)-supported programme on Integrated Research on Disaster Risk (IRDR).
Email: joern.birkmann@ireus.uni-stuttgart.de

Dr. Riyanti Djalante (Germany/Indonesia) has taken several positions related to disaster risk reduction (DRR) and climate change adaptation (CCA) research and policy. Supported by the Alexander

von Humboldt Foundation's Scholarship for Experienced Researchers, she is currently a visiting scientist at the United Nations University – Institute of Environment and Human Security (UNU-EHS) in Bonn, Germany. She has a doctorate in the field of Human Geography related to Resilience to Disasters and Climate Change in Indonesia. She has been actively involved in the following research networks: Earth System Governance (ESG) Research Fellow; World Social Science (WSSC) Fellow of the International Social Science Council (ISSC), Urbanisation and Global Environmental Change (UGEC); Enhancing scientific capacity to inspire informed action on global environmental change (START) Fellow; Integrated Research on Disaster Risk (IRDR) Fellow. In addition to this, she is a Future Earth Young Scholar Member. In Indonesia, Dr Djalante has worked for the local government in Indonesia accumulatively for 10 years where she has been involved in development planning and disaster management. She has been also been appointed as an honorary lecturer at the Halu Oleo University Indonesia, teaching on the subjects of disaster risk reduction, climate change adaptation, environmental planning and sustainable development.
Email: djalante@ehs.unu.edu

Prof. Dr. Yekti Maunati (Indonesia) is now a research professor at the Research Center for Regional Resources at the Indonesian Institute of Sciences (PSDR-LIPI). Over the past several years, Yekti has been conducting research on tourism, border issues, and the lives of indigenous people/minorities living in the Southeast Asian countries. One of her most recent article is 'Networking the Pan-Dayak', in Wendy Mee and Joel S. Kahn (eds) *Questioning Modernity in Indonesia and Malaysia*, Singapore and Japan: Nus Press in association with Kyoto University Press, 2013 (Kyoto Cseas Series on Asian Studies 5). Since 2011 she has been involved in research on urban resilience in North Jakarta and Bangkok. Yekti obtained her PhD from the Department of Sociology and Anthropology, La Trobe University, Melbourne and obtained her Master's from the Department of Anthropology and Sociology, Monash University, Melbourne. Her Bachelor's degree was from the Department of Sociology, Faculty of Social and Political Studies, University of Gadjah Mada, Yogyakarta.
Email: yektim@yahoo.com

Donovan Finn, PhD (USA) is Assistant Professor of Environmental Design, Policy and Planning in the Sustainability Studies Program and School of Marine and Atmospheric Sciences at Stony Brook University in New York. His research focuses on sustainability and resilience as

policy goals, and the role of public participation in the planning process. His research has been funded by the National Science Foundation, the Natural Hazards Center and the Lincoln Institute of Land Policy. He holds a Master's degree in Urban Planning and a PhD in Regional Planning, both from the University of Illinois at Urbana-Champaign.
Email: donovan.finn@stonybrook.edu

Divya Chandrasekhar (USA) is Assistant Professor in the City and Metropolitan Planning program at the University of Utah. Her research interest is in how communities plan for recovery and resilience after major disaster events. Her studies have examined recovery and reconstruction planning processes after disasters, community participation these processes, emergence of new institutions, post-disaster displacement, and community capacity to recover from major disasters in the US and Asia. Her research has been funded by the National Science Foundation, the Natural Hazards Center at Boulder, CO, and the Mid-America Earthquake Center, and her work has been published in national and international journals. She holds a doctoral degree in urban and regional planning from the University of Illinois at Urbana-Champaign.
Email: d.chandrasekhar@utah.edu

Yu Xiao (USA) is an associate professor in the Department of Landscape Architecture and Urban Planning and a Faculty Fellow with the Hazard Reduction and Recovery Center at Texas A&M University.
Email: yuxiao@email.tamu.edu

Alena Kučeravcová (Slovakia) is a project manager at the Agency for the Support of Regional Development (ASRD), Kosice, where she has coordinated a number of projects that focus on regional and urban planning, rural development, and sustainable tourism since 2011. Before joining ASRD, Ms. Kuceravcova worked for Slovak Environmental Agency as a landscape planning specialist, where her main tasks were focused on preparation of strategic documents on sustainable development, preparation of regional/local ecological stability systems and development of spatial/territorial plans. She holds a Master's degree in Environmental Science and Geography from Faculty of Natural Sciences, Constantine the Philosopher University in Nitra.
Email: alena.kuceravcova@arr.sk

Jan Dzurdženík (Slovakia) has been working as a project manager at the Agency for the Support of Regional Development (ASRD), Košice, Slovakia since 2008, where he focuses on

regional planning, rural development and sustainable natural source utilisation. During period 2003–08 he has worked for Slovak Environmental Agency as a specialist. He holds a Master's degree in Environmental Ecology from the P.J. Šafárik University in Košice.
Email: jan.dzurdzenik@arr.sk

Jozef Šuľak (Slovakia) has been working as a project manager for the Agency for the Support of Regional Development Košice, Slovakia since 2005. He focuses on development and implementation of different projects in various areas including spatial planning, environmental protection and flood management. Part of his work also includes the building of partnerships among different stakeholders at local or regional level, as well as providing training to support the regional and community development processes. Mr. Sulak graduated from University of Economics in Bratislava, Faculty of Business Economy in Košice.
Email: jozef.sulak@arr.sk

Nadine Mägdefrau (Germany) is a lecturer and coordinator of research at the Institute of Spatial Planning at the Technical University of Dortmund, Germany. She holds a Diploma in Spatial Planning from the Technical University of Dortmund. Ms. Mägdefrau has worked in the research project 'Increasing resilience of urban planning' (URBIPROOF) and is currently writing her dissertation on 'Increasing urban resilience after a disaster through spatial planning'. The focus of her research is the Tohoku Region in Japan.
Email: nadine.maegdefrau@tu-dortmund.de

Teresa Sprague (USA) is a lecturer and researcher at the Institute of Spatial Planning at the Technical University of Dortmund, Germany. She received her MSc in Water Science, Policy and Management from the University of Oxford and a BA in Political Science/International Relations and Environmental Systems from the University of California, San Diego. She was granted a Marie Skłodowska-Curie Research Fellowship working within the EU Seventh Framework Programme project 'CHANGES – Changing Hydro-meteorological Risks – as Analyzed by a New Generation of European Scientists', and is currently completing her PhD on 'Understanding the intangible: improving "good" risk governance for water-related extremes through EU policy and local spatial context'.
Email: teresa.sprague@tu-dortmund.de

Disaster response and spatial planning – key challenges and strategies

Stefan Greiving

Introduction

Worldwide, the urban development and disaster management arena finds itself at a critical crossroads. This is driven by rapid urbanisation (and de-urbanisation), as well as a growing volume of damage caused by natural (and unnatural) disasters, which are increasingly affecting urban and rural inhabitants. Bearing this in mind, experiences from disaster management and especially from disaster recovery have led to advances in the field and an increase in the importance of the role of spatial planning.

Priority 2 of the new *Sendai framework for disaster risk reduction* (UNISDR, 2015, p 17) points at 'Strengthening disaster risk governance to manage disaster risk' and states that 'Clear vision, plans, competence, guidance and coordination within and across sectors as well as participation of relevant stakeholders are needed'. This enlightens the key role of spatial planning as a comprehensive, overarching sectoral actor. More explicitly, land-use planning is regarded as an important actor for Priority 4: 'Enhancing disaster preparedness for effective response and to "Build Back Better" in recovery, rehabilitation and reconstruction' (UNISDR, 2015, p 21). While spatial planning is traditionally seen as a key actor for preventive measures (see, eg, Greiving et al, 2006), the *Sendai framework* underlines its importance for the recovery phase: 'use opportunities during the recovery phase to develop capacities that reduce disaster risk in the short, medium and long term, including through the development of measures such as land-use planning' (UNISDR, 2015, p 21f).

That is why this book brings together the experiences and knowledge of spatial planning after significant disasters and highlights ongoing efforts to improve spatial resilience across the globe. One of the main goals is to understand the influence of significant disasters on spatial planning and spatial resilience under different legal-administrative and cultural framework conditions. That is why selected contributions from different continents (Europe, Asia and Northern America) were chosen. The other continents (Africa, Australia and Southern America) are not presented by country reports, but considered in Part B of this book.

The further chapters in Part A present different prominent authors highlighting the specific relationship in various countries: authors from Japan, Indonesia, the US, Slovakia and Germany write about their experiences and efforts to rebuild their communities in a more resilient manner after major disasters and thus give an overview of the state of the art. These countries represent high- (US, Germany, Japan), medium- (Slovakia) and low-income (Indonesia) countries and, at the same time, Western and Eastern cultural traditions. Moreover, all of them faced (different) extreme events in the past and came up with response strategies for which spatial planning played a considerable role. At the beginning of each country section, key information on the selected country and its hazard profile is presented.

Part B gives a cross-country analysis of five important topics: the transformation of spatial planning after significant disasters; efforts in building spatial resilience after disasters; coordination in building spatial resilience after disasters; participation in rebuilding space more resilient; and spatial planning and uncertainties. This part further identifies key factors that can be shared throughout the countries. A particular focus will be on how disasters have influenced spatial planning and have potentially improved the resilience of space. Key findings are illustrated by useful examples from the country reports in Part A, as well as further examples from third countries derived from the literature.

The book arises out of the project 'URBIPROOF – Improving the resilience of urban planning', funded by the joint European–Japanese initiative 'CONCERT Japan'.

Spatial planning and natural hazards

In the following, the connections between spatial planning and disaster risks will be discussed. Space can be defined as the area where human beings and their artefacts are threatened by spatially relevant hazards. Every hazard has a spatial dimension. Spatial effects might occur if a hazard turns into a disaster or by spatial planning response.

Generally speaking, natural hazards pose a risk to human beings and their properties; nature and its ecosystems have always adapted to natural hazards. Natural hazards can be strengthened by human activities, however, and can then pose a threat to ecosystems, at least from a human perspective. Risk is a function of a hazard (or hazards) and vulnerability. The key challenge is thus to control or influence the main driving forces behind risk. Economic growth and the concentration of population in threatened areas lead to increasing vulnerability, even if hazards do not occur more often than before. This combination of natural and manmade factors is the main cause behind the rapidly increasing losses caused by natural hazards in Europe (Schmidt-Thomé and Greiving, 2008). Here, spatial planning comes into play because spatial planning takes decisions for society as to whether and how space is used. In consequence, spatial planning can influence both the (flood) hazard and the vulnerability to floods. Of course, every hazard has a spatial dimension (disasters take place somewhere). However, the occurrence of spatially relevant hazards is limited to a certain disaster area that is regularly or irregularly prone to hazards (eg river flooding, storm surges, volcanic eruptions). Spatially non-relevant hazards occur more or less anywhere (ie pandemics or asteroid impacts). These hazards cannot be managed by spatial planning.

It is due to authors like Burby (1998) or Godschalk et al (1999) that the need for spatial planning and the important role it has to play in the whole disaster management cycle was highlighted. In recent years, however, this not only has been accepted by planners and policymakers, but also corresponds with latest research initiatives in which the potential role of spatial planning in risk management has been stressed (eg EC, 2007).

In spatial planning, hazards are generally determined in accordance with the order of optimisation. This means that its concerns must be considered within the planning process but can be outweighed by other reasons that are more important for the case. However, to a great extent, risk assessment and management is taken on by sectoral planning (eg river floods by water management authorities).

Spatial development as key driver for disaster risks

Urbanisation is a long-term process that has currently accelerated to the point that we can state that one of the most fundamental and most radical changes recently experienced by humanity is global urbanisation. At the global scale, we observe a rapid growth in the urban population and its concentration in few mega-cities or super-cities. In 1980, less than 3% of the world's population lived in cities. At the beginning of the 21st century, humanity entered the 'century of cities' (Kofi Annan), with the majority of people worldwide living in urban areas. This percentage is likely to rise further. Moreover, it will be crammed into a few large conurbations – often located in developing and economically vulnerable countries – with more than 20 million inhabitants, such as Tokyo, Chongqing, Seoul, Mexico City, Delhi, Mumbai, New York City and São Paulo:

> Between 2007 and 2050, the world population is expected to increase by 2.5 billion, passing from 6.7 billion to 9.2 billion (United Nations 2008). … At the same time, the population living in urban areas is projected to gain 3.1 billion, passing from 3.3 billion in 2007 to 6.4 billion in 2050. Thus, the urban areas of the world are expected to absorb all the population growth expected over the next four decades while at the same time drawing in some of the rural population. (United Nations, 2008, p 1)

More specifically, by 2050, the urban population will amount to 75% of the total population (UN DESA, 2014).

In Europe, for instance, approximately 70% of the population live in urban areas. Yet, the majority of cities in Europe are medium and small. One half of the population of Europe lives in small towns of 1,000–50,000 people, one quarter in medium-sized towns of 50,000–250,000 people and one quarter in cities of more than 250,000 people (UNCHS, 2001). The largest urban hubs are restricted to a few conurbations with populations between 6 million and 8 million inhabitants each, distributed over several municipalities that together form a metropolitan area. Moreover, most of the urbanised areas in Europe are subject to processes of both expansion (of peripheral, peri-urban areas) and shrinking (of historic neighbourhoods, city centres and isolated marginal settlements). Expansion, in particular, assumes the form of urban sprawl, that is, a more or less planned incremental urban development, characterised by a low density mix of land uses on the urban fringe: 'European cities were more compact and less sprawled in the mid 1950s than they are today, and urban sprawl is now a common phenomenon throughout Europe. Moreover, there is no apparent slowing in these trends' (EEA, 2006, p 5).

In addition, the socio-demographic composition of urban populations is increasingly complex. Globalisation, integration, easier transport and the high attractive power exerted by larger cities in terms of opportunities and assets lead to an increasing level of heterogeneity in the socio-demographic status of citizens, especially in terms of racial groups, educational attainment, self-declared religious and political affiliation, income and labour status, family composition, housing condition, and the possession of goods. This heterogeneity is often recognisable in the physical organisation of the urban space, with parts of the city frequented only by rich people and professionals (gentrification, gated communities) and other parts hosting immigrants, poor families or other unwelcome communities (ghettos, marginalised peripheral areas).

A considerable proportion (32%) of city dwellers worldwide are therefore forced to live in congested and substandard low-income housing areas (UN, 2003) Thus, many parts of contemporary cities have become unhealthy, unsafe and unpleasant places to live. Frequently, multiple hazards are created that result in excessive environmental risks. Climate change risks are higher in these areas, but also in cities in general. Since a risk is the product of the probability of an event and the vulnerability of a site, it follows that cities, as the most densely inhabited areas, are the places where risks are the highest. Thus, spatial planning, being responsible for decisions as to whether and how space is used, is a producer of risk.

From sustainability to resilience

The concept of 'sustainable development' has been developed and applied for dealing with such complex development issues as urbanisation. Accordingly, the preferable development path of a society (or city) would be one that 'meets the needs of the present without compromising the ability of future generations to meet their own needs' (Brundtland et al, 1987, p 8). This concept has been translated into strategies and monitoring indicators that would ensure long-term sustainability by balancing ecological, economic and social development goals. As for the European Union (EU), sustainable growth, that is, promoting a more resource-efficient, greener and more competitive economy, is one of three priorities of the recently launched 'Europe 2020' strategy (EC, 2010). However, such a long-term-oriented development model (or sustainable systems more generally) works best under stable conditions but might be threatened by external shocks or sudden stresses, such as natural hazards, global economic crises and so on. Such events might seriously derail an otherwise sustainable development. Therefore, the concept of sustainability is now often complemented by the concept of resilience. A massive shock or stress, for example, caused by an extreme event like a tsunami, results in a sudden drop of development. It is only due to countervailing forces that this drop can be stopped and a further collapse be prevented. Given the right conditions and/or interventions, the system can recover and continue its previous development path (albeit from a lower starting point).

Figure 1: The concept of resilience

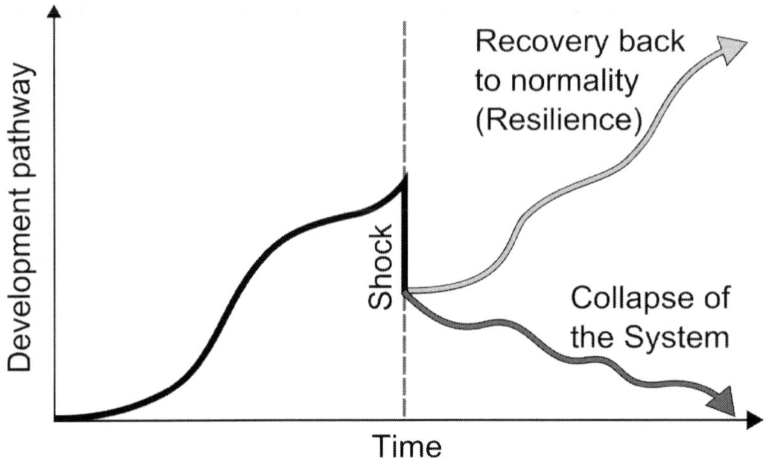

Source: Own figure

'Resilience' is a term that has been used in a variety of contexts. Henstra et al (2004) show that resilience has been defined in different ways since the 1970s. They suggest the following definition in the context of disaster-resilient cities, which, however, can also be transferred to other future changes: resilience is thus 'the capacity to adapt to stress from hazards and the ability to recover quickly from their impacts' (Henstra et al, 2004, p 5). These characteristics are categories for which indicators will be developed to assess the resilience of urban structures.

The United Nations International Strategy for Disaster Reduction (UNISDR, 2009, p 24) defined resilience as follows:

> The ability of a system, community or society exposed to hazards to resist, absorb, accommodate to and recover from the effects of a hazard in a timely and efficient manner, including through the preservation and restoration of its essential basic structures and functions.

With regard to cities, Godschalk (2002, p 2) characterised the concept of resilience as follows:

> Such [resilient] cities are capable of withstanding severe shock without either immediate chaos or permanent deformation or rupture. Designed in advance to anticipate, weather, and recover from the impacts of natural or

> technological hazards, resilient cities are based on principles derived from past experience with disasters in urban areas. While they may bend from hazard forces, they do not break.

Resilience is also a way to cope with uncertainty because future changes can rarely be predicted, and because the vulnerability of community systems cannot be fully known in advance. Thus, cities must be designed in a 'change-proof' way, that is, to have the flexibility to accommodate extremes without failure and the robustness to rebound quickly from undesired impacts (Henstra et al, 2004, p 8). Projected climate change-related effects on temperature and precipitation lead to new uncertainties because past events might not be representative anymore. Similarly, other changes in the catchments (eg deforestation, melting of glaciers, surface sealing through settlement development, surface modification by infrastructure, etc) will lead to high uncertainties. Here, the perspective changes from probabilities to just possibilities. With public decision-making not having any precise information at hand, restrictions for private property rights are probably not legally justifiable anymore. Hereby, the justification of actions and consensus about thresholds for acceptable risks and response actions becomes more important and calls for a 'change-proof' planning.

With respect to the impact of such a concept on development discourse, the concept of resilience matches with that of 'sustainable recovery' declared, for instance, in the strategy of 'Europe 2020' (EC, 2010, p 7):

> Europe is left with clear yet challenging choices. Either we face up collectively to the immediate challenge of the recovery and to long-term challenges – globalization, pressure on resources, ageing – so as to make up for the recent losses, regain competitiveness, boost productivity and put the EU on an upward path of prosperity ('sustainable recovery'). Or we continue at a slow and largely uncoordinated pace of reforms, and we risk ending up with a permanent loss in wealth, a sluggish growth rate ('sluggish recovery') possibly leading to high levels of unemployment and social distress, and a relative decline on the world scene ('lost decade').

According to Godschalk (2002), current guidelines do not sufficiently accommodate the particular vulnerabilities of 'cities under stress'. In order to create resilient cities, Godschalk (2002, p 5) derives the characteristics – or principles – of resilient systems that should be taken into account for the design and management of cities:

- *Redundancy*: systems designed with multiple nodes to ensure that failure of one component does not cause the entire system to fail;
- *Diversity*: multiple nodes versus a central node in order to protect against a site-specific threat;
- *Efficiency*: positive ratio of energy supplied to energy delivered by a dynamic system;
- *Autonomy*: capability to operate independent of outside control;
- *Strength*: power to resist a shock or attack;
- *Interdependence*: integrated system components to support each other;
- *Adaptability*: capacity to learn from experience and the flexibility to change;
- *Collaboration*: multiple opportunities and incentives for broad stakeholder participation.

A key strategy in this context is 'no regret', which can be traced back to the 4th Assessment Report of the Inter-governmental Panel on Climate Change (IPCC, 2007): 'A policy that would generate net social and/or economic benefits irrespective of whether or not anthropogenic climate change occurs'. This can be adapted to questions of any change process. However, what is completely missing in the scientific discussion is concrete advice on what is really meant by 'no regret' and how to integrate this principle into decision-making practices. For a successful communication of an adaptation strategy that is based on the no-regret principle, synergies with other development interests have to be optimised and conflicts have to be avoided as much as possible.

These principles have to be operationalised for practice, and they guide the further discussions in Part B of this book. No regret and other strategies for dealing with uncertainty are addressed in more detail in Chapter Five.

Vulnerability, adaptive capacity and resilience

As already pointed out, resilient cities are capable of withstanding severe shock without either immediate chaos or permanent deformation or rupture. Cities that lack resilience are vulnerable in manifold ways. For '*vulnerability*', several definitions can be found in the literature (Cutter, 1996, p 531). When referring to vulnerability, it is important to distinguish between the origins of vulnerability: in general, vulnerability is defined as a potential for loss, but, in most cases, it is not clearly defined what type of loss and whose loss is meant. The following types of losses – and therefore origins of vulnerability – can be distinguished (Cutter, 1996, p 530): (a) *individual vulnerability* – individual potential for and sensitivity to losses, occurring in spatial and non-spatial domains; (b) *social vulnerability* – the susceptibility of social groups or society at large to potential structural and non-structural losses from hazardous events and disasters occurring in distinct spatial outcomes or patterns and variations over time; and (c) *biophysical vulnerability* – potential for loss derived from the interaction of society with biophysical conditions that affect the resilience of the environment to respond to the hazard or disaster. They also influence the adaptation of society to such changing conditions, occurring also in explicit spatial outcomes.

Today, the concept of vulnerability is no longer restricted in its use to the analysis of the functionality of specific ecosystems and as a parameter in the determination of risks (Birkmann, 2006; Schmidt-Thomé and Greiving, 2008; Greiving et al, 2012), but rather refers to a wide spectrum of social, economic, institutional and ecological hazards, as well as their interactions, and is even used to describe a so-called 'generic vulnerability of social groups and societies' (Young, 2010; Christmann et al, 2011).

However, there is also a clear difference in the understanding of vulnerability between the disaster risk and climate change community. According to the IPCC (2007):

> Vulnerability is the degree to which a system is susceptible to, and unable to cope with, adverse effects of *climate change*, including *climate variability* and extremes. Vulnerability is a function of the character, magnitude, and rate of climate change and variation to which a system is exposed, its *sensitivity*, and its adaptive capacity [emphases added].

Here, vulnerability is understood as the final outcome of the assessment. On the contrary, UNISDR (2009, p 30) defines vulnerability as: 'The characteristics and circumstances of a community, system or asset that make it susceptible to the damaging effects of a hazard'. Thus, vulnerability is hazard/context-dependent and just one element of risk.

When making a risk assessment, a specific hazard, which is expressed in statistically determinable correlations of frequency and magnitude, is contrasted with a vulnerability component, which, on account of its indicators, resembles the concept of sensitivity as applied to the assessment of vulnerability (see also Costa and Kropp, 2012). The outcome of the assessment is the risk (see Figure 2).

Figure 2: The disaster risk concept

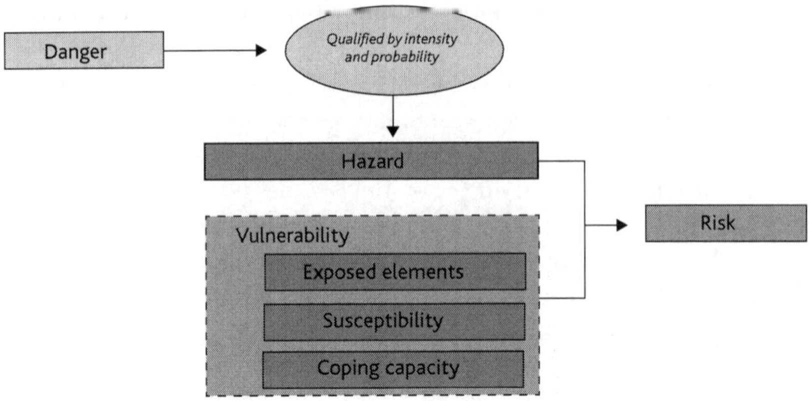

Source: Own figure

Recently, the term 'disaster risk' was defined by the IPCC as follows:

> The likelihood over a specified time period of severe alterations in the normal functioning of a community or a society due to hazardous physical events interacting with vulnerable social conditions, leading to widespread adverse human, material, economic, or environmental effects that require immediate emergency response to satisfy critical human needs and that may require external support for recovery. (IPCC, 2012, p 558)

The special report *Managing the risks of extreme events and disasters to advance climate change adaptation* (Field et al, 2012) throws light on the connections between climate change and extreme events, as well as the complementarity/overlap of risk management and adaptation to climate change. Both approaches aim at reducing vulnerability and exposure to disaster risks triggered by weather and climate events. However, disaster risk management addresses many more events that are not influenced by climate change, and adaptation strategies also deal with creeping changes in temperature, precipitation and related impacts on the environment that are not necessarily related with extreme events.

Many scientists disagree that risk is just a mathematic term; instead, arguing that it is socially constructed (IRGC, 2005; Löfsted, 2005). How risk policy is carried out is a matter of the definition of risk. Moreover, defining risk is an expression of power. Slovic (1999) thereby argues that whoever controls the definition of risk, controls risk policy. In addition, the term 'risk' is purely negatively defined.

Positive changes – which are possible due to climate change – cannot be expressed when using the concept of risk.

An important aspect in the context of vulnerability concepts is the emerging concept of *adaptive capacity*. Developing and implementing flexible strategies to cope with disaster risks require institutions that are capable of enhancing the adaptive capacity of societies. For cities to be able to anticipate and rapidly respond to disaster risks, they will need to have appropriate institutions. This implies that the institutions that are governing cities should be able to support social actors and enable them to respond to challenges proactively. Societal and institutional adaptive capacity building is therefore a critical part of 'change-proof' or resilient cities. Adaptive capacity refers to the ability of a system or individual to adapt to climate change, but it can also be used in the context of disaster risk (IPCC 2012, p 73) as 'The ability of a system to adjust to climate change … to moderate potential damages, to take advantage of opportunities, or to cope with the consequences'. Applying this definition to institutions, Gupta et al (2010, p 461) define adaptive capacity as 'the inherent characteristics of institutions that empower social actors to respond to short and long-term impacts either through planned measures or through allowing and encouraging creative responses from society both ex-ante and ex-post'.

Institutional setting

The legal framework and the political-administrative system significantly determine how strategies and measures for disaster risk response are designed and by which institutions they are implemented (Greiving and Fleischhauer, 2012). As an example, the setting of legally binding and spatially specific objectives (eg to keep an area free of further settlement development) presumes that there are laws enabling the enactment and enforcement of such spatial objectives. Thus, the differences in the political-administrative systems were taken into consideration for the selection of the country studies discussed in Part A of this book. Furthermore, there is a need to know about the characteristics of the political-administrative systems of the assessed countries before the transferability of main findings to other countries can be estimated. Therefore, each chapter in Part A contains some basic information about the institutional background of the respective country.

In some of the countries, for instance, in Germany, a new development is legally allowed when it conforms to the land use as laid down in the legally binding regional plan. This so-called regulatory function of spatial planning is known under the term 'conforming planning' in the international discourse on planning theory (Rivolin, 2008, pp 167ff). In most of the countries and thus also in the majority of country studies covered by this book, the so-called development function dominates at the regional level, which is discussed under the term 'performing planning'. This planning type is characterised by legally non-binding programmatic and/or strategic statements. Potential projects are then evaluated against the question of whether they support the implementation of the programme or strategy. Furthermore, there are – if at all – only partially binding effects for the subordinated local level. At the local level, in contrast, the similarities between the planning systems of the countries are much higher compared to the regional level. Throughout Europe and abroad – with the well-known exception of the UK – there are two-level planning systems at the local level, consisting of a legally binding zoning of the urban or municipal area ('conforming planning').

In developing countries, even though they may have an institutionalised planning system, the reality is different. Disasters in developed countries are normally a cause of enormous economic damage, but only a few people die. Disasters in developing countries kill people – mainly living in informal settlements. Many of the informal settlements are located in areas that are out of the scope of formal development because they are demarcated as hazard zones (steep slopes, river channels, etc). Neither planning nor building rights can be granted. In consequence, these areas attract migrants. People have no risk awareness and/or no alternative for settling. They are highly exposed and susceptible to the weak construction of their homes. Granting land ownership is potentially risky as people may just sell their land to new migrants and/or investors. Relocation projects normally fail as the new settlements are often not accessible. There is no one-size-fits-all solution, but comprehensive development strategies are most likely the best option. Investments in coping capacity and risk awareness may help.

Finally, disaster risk management in general and the different phases of the disaster risk cycle are complex tasks that call for the sound coordination of activities between the actors involved. First, horizontal coordination is required between spatial planning and the different sectoral planning divisions, such as geological surveys or water management authorities. This coordination is relevant for the development of the necessary evidence base for decisions about altering or coping with risks, but also for the decision-making about tolerating or altering risks. Second, vertical coordination between the central state government and its regional and local entities is indispensable for effective and, in the end, acceptable disaster risk management (Young, 2010; Greiving et al, 2012). Involvement of the public and relevant stakeholders is required in order to be able to take the perception and estimation of risk by the affected people into account. Here, risk governance concepts come into play, meaning the identification, assessment, management and communication of risks in a broad context (IRGC, 2005): 'The risk governance framework offers a systematic way to help situate … judgments about disaster management, risk reduction, and risk transfer within [a] broader context' (IPCC, 2012, p 56). This is also underlined by Priority 2 of the *Sendai framework for disaster risk reduction* (UNISDR, 2015).

In consequence, the following chapters of Part A of this book will explicitly address these requirements for coordination and stakeholder involvement.

References

Birkmann, J. 2006, *Measuring vulnerability to natural hazards: Towards disaster resilient societies*. Tokyo: United Nations Press.

Brundtland, G., et al 1987, *Report of the World Commission on Environment and Development: Our common future*. Transmitted to the General Assembly as an Annex to document A/42/427 - Development and International Co-operation: Environment.

Burby, R.J. 1998, *Cooperating with nature: Confronting natural hazards with land-use planning for sustainable communities*. Washington, DC: Joseph Henry Press.

Christmann, G., Ibert, O., Kilper, H., Moss, T. u.a. 2011, *Vulnerabilität und Resilienz in sozio-räumlicher Perspektive. Begriffliche Klärungen und theoretischer Rahmen*. Working Paper, Erkner, Leibniz-Institut für Regionalentwicklung und Strukturplanung.

Costa, L. and Kropp, J. 2012, *Linking components of vulnerability in theoretic frameworks and case studies*, Sustainable Sciences DOI 10.1007/s11625-012-0158-4.

Cutter, S. L. 1996, 'Vulnerability to environmental hazards', in *Progress in Human Geography*, Vol. 20, No. 2, pp. 529-539.

EC (European Commission) 2007, *Territorial agenda of the European Union: Towards a more competitive and sustainable Europe of diverse regions*. Berlin.

EC 2010, *Europe 2020: A Strategy for Smart, Sustainable and Inclusive Growth*, Communication from the Commission.

EEA (European Environment Agency) 2006, *Urban Sprawl in Europe: The ignored challenge*, EEA Report, n.10/2006, Luxembourg: Office for Official Publications of the European Communities.

Field, C.B., V. Barros, T.F. Stocker, D. Qin, D.J. Dokken, K.L. Ebi, M.D. Mastrandrea, K.J. Mach, G.-K. Plattner, S.K. Allen, M. Tignor and P.M. Midgley 2012, *Managing the Risks of Extreme Events and Disasters to Advance Climate Change Adaptation, A Special Report of Working Groups I and II of the Intergovernmental Panel on Climate Change*, Cambridge University Press, Cambridge, UK and New York, NY, USA.

Godschalk, D. R. 2002, *Urban hazard mitigation: Creating resilient cities*, Plenary paper presented at the Urban Hazards Forum, John Jay College, City University of New York, January 22-24, 2002 (July 10, 2007); http://www.arch.columbia.edu/Studio/Spring2003/UP/Accra/links/GodshalkResilientCities.doc.

Godschalk, D. R., Beatley, T., Berke, P., Brower, D. J. and Kaiser, E. J. (1999): *Natural hazard mitigation: Recasting disaster policy and planning.* Washington, DC: Island Press.

Greiving, S. & Fleischhauer, M. 2012, 'National climate change adaptation strategies of European states from a spatial planning and development perspective', in *European Planning Studies,* Vol. 20, No. 1, January 2012, pp. 27-47.

Greiving, S., Fleischhauer, M., Wanczura, S. (2006): European Management of Natural Hazards: The Role of Spatial Planning in selected Member States, *Journal of Environmental Planning and Management.* Vol. 49, No. 5, September 2006, S. 739 – 757.

Greiving, S., Pratzler-Wanczura, S., Sapountzaki, K., Ferri, F., Grifoni, P., Firus, K.and Xanthopoulos, G. 2012, 'Linking the actors and policies throughout the disaster management cycle by "Agreement on Objectives" – a new output-oriented management approach', *Nat. Hazards Earth Syst. Sci.*, 12, pp. 1085-1107, doi:10.5194/nhess-12-1085-2012

Gupta, J., Termeer, K. Klostermann, J., Meijerink, S:, van den Brink, M., Jong, P., Nooteboom, S. & Bergsma, E. 2010, 'The Adaptive Capacity Wheel: A Method to Assess the Inherent Characteristics of Institutions to Enable the Adaptive Capacity of Society' in *Environmental Science & Policy*, Vol. 13, No. 6, pp. 459-471.

Henstra, D., Kovacs, P., McBean, G. & Sweeting, R. 2004, Background paper on disaster resilient cities, Institute for Catastrophic Loss Reduction, Toronto/London, (Online assessed on July 10, 2007), http://www.dmrg.org/resources/Henstra.et.al-Background paper on disaster resilient cities.pdf.

IPCC (Inter-governmental Panel on Climate Change) 2007, 'Climate Change 2007: Impacts, Adaptation and Vulnerability', Contribution of Working Group II to the Fourth Assessment Report of the Intergovernmental Panel on Climate Change.

IPCC 2012, *Managing the risks of extreme events and disasters to advance climate change adaptation.* A Special Report of Working Groups I and II of the Intergovernmental Panel on Climate Change [Field, C.B., V. Barros, T.F. Stocker,D. Qin, D.J. Dokken, K.L. Ebi, M.D. Mastrandrea, K.J. Mach, G.-K. Plattner, S.K. Allen,M. Tignor, and P.M. Midgley (eds.)]. Cambridge University Press, Cambridge, UK, and New York, NY, USA.

IRGC 2005, *Risk governance towards an integrative approach*, Geneva.

Löfstedt, R. 2005, *Risk management in post-trust societies*, Houndmills, Basingstoke, Hampshire, New York.

Rivolin, U. J. 2008, 'Conforming and Performing Planning Systems in Europe: An Unbearable Cohabitation', in *Planning, Practice & Research*, Vol. 23, No 2, pp. 167-186.

Schmidt-Thomé, P. and Greiving, S. 2008, 'Response to natural hazards and climate change in Europe', in: Faludi, A. (ed.) *European Spatial Planning and Research.* Cambridge, MA: Lincoln Institute for Land Policy, pp. 141-167

Slovic, P. 1999, 'Trust, emotion, sex, politics, and science: Surveying the risk-assessment battlefield', *Risk Analysis*, Vol 19, No 4, pp. 689–701.

UN (United Nations) 2003, *The challenge of slums. Global report on human settlements 2003.* London and Sterling, VA: Earthscan Publications

UN 2008, *State of the world's cities 2008/2009: Harmonious cities.* New York: Un Habitat.

UNCHS (United Nations Centre for Human Settlements) 2001, *Global Report on Human Settlements 2001: Cities in a Globalizing World*, Earthscan, London.

UN DESA 2014, 'World Urbanization Prospects: The 2014 Revision', Highlights, New York.

UN-ISDR (International Strategy for Disaster Reduction) 2009, *Global Assessment Report on Disaster Risk Reduction.* Geneva, United Nations.

UNISDR 2014, 'UN Resilient Cities Campaign'. http://www.unisdr.org/campaign/resilientcities/ (assessed at 14.08.2014).

UN-ISDR 2015, *Sendai Framework for Disaster Risk Reduction 2015-2030*, http://www.wcdrr.org/uploads/Sendai_Framework_for_Disaster_Risk_Reduction_2015-2030.pdf

Young, O. R. 2010, 'Institutional dynamics: Resilience, vulnerability and adaptation in environmental and resource regimes', in *Global Environmental Change*, 20, pp. 378-385.

Part A

A1. Japan

A2. Indonesia

A3. USA

A4. Slovakia

A5. Germany

A6. Cross-case analysis

CHAPTER A1a

Disaster risk management and its relationship to land use geographies vulnerable to water-related disasters: an analysis of the Japanese legislative system

Kanako Iuchi

Japan	
Population: 128.8 million (2015)	**Government:** Centralized political structure
Area: 377,962 km²	**Administrative structure:** Japan's national government oversees 47 prefectures and other municipalities, including cities, towns, and villages (there are 790 cities, 745 towns, and 183 villages (total: 1,718 municipalities) in 2014). Additionally, 20 ordinance-designated cities – regions with populations over 500 thousand – hold semi-autonomous power and operate independently from the prefectures.
Hazard profile: Volcanic eruptions Earthquakes Tsunamis Typhoons Flood and inland inundation Storm surge Land slide Heavy snow (Japan sea side)	**Authorities in charge of risk assessment and management:** Cabinet office is responsible for coordinating related governmental organizations to collaborate in disaster management efforts. All levels of government, of national (mainly the Ministry of Land, Infrastructure, Transportation and Tourism (MILT)), prefectural and municipal, are responsible for assessing risk with their spatial scale.
Analysed event: The Great East Japan Earthquake and Tsunami (2011) Estimated return period: 1:1,000 Estimated damages: JPY 16 to 25 trillion (US$192 to 299 billion) Casualties: 15,890 death and 2,589 missing (March 2015)	**Role of spatial planning:** Responsible for long-term mitigation and post-event recovery. Actual spatial control tends to occur only after major disasters.

Japan's geography and assets vulnerable to natural disasters

Due to geographic, topographic and meteorological conditions, Japan's landscape is prone to various natural events, including earthquakes, volcanic eruptions, flooding, typhoons, torrential rain and heavy snowfall. Recent impacts of climate change are further aggravating the magnitude and frequency of disasters.

Japan's share of seismic activity globally represents its vulnerability to earthquakes and associated disasters like tsunamis. Between 2000 and 2009, 1,036 seismic activities larger than $M_W6.0$ were recorded worldwide. Of these, 212 events, or approximately 20% of the world's total, were recorded just in Japan (Cabinet Office, Government of Japan, 2012). More recently, in 2011, the Great East Japan Earthquake (GEJE) that ruptured off the Pacific coast of north-eastern Honshu recorded a magnitude of $M_W9.0$ and was the fourth-largest earthquake recorded worldwide since 1900 (USGS, 2015). It generated recorded tsunami waves of up to 9.3 metres and run-up heights of 35 metres that inundated 560 square kilometres of the Tohoku region (Central Disaster Prevention Council, Technical Investigation Committee, 2011) (for an image of the coastal inundation, see Figure 3).

Japan's geography is also prone to water-related disasters, including floods and storm surge. For instance, rivers in Tokyo are at a higher elevation than in other urban areas, unlike some major rivers in Europe (see Figure 4). These rivers are steep and narrow, thus increasing the potential for flash floods during torrential rainfall. This is significant considering that 50% of the total population and 75% of total assets are agglomerated in these flood-prone[1] areas: 5.4 million people live below sea level; 4 million of which are within Japan's three largest bays of Osaka, Ise and Tokyo (MLIT, 2007a). These geographic conditions, as well as exposure to hazards – that is, the agglomeration of population and assets in vulnerable areas – have made the country highly vulnerable to water-related disasters. The effects of climate change, such as sea-level rise and more frequent typhoons, also further expose people and assets to risk.

Figure 3: Lands inundated by tsunami following the Great East Japan Earthquake

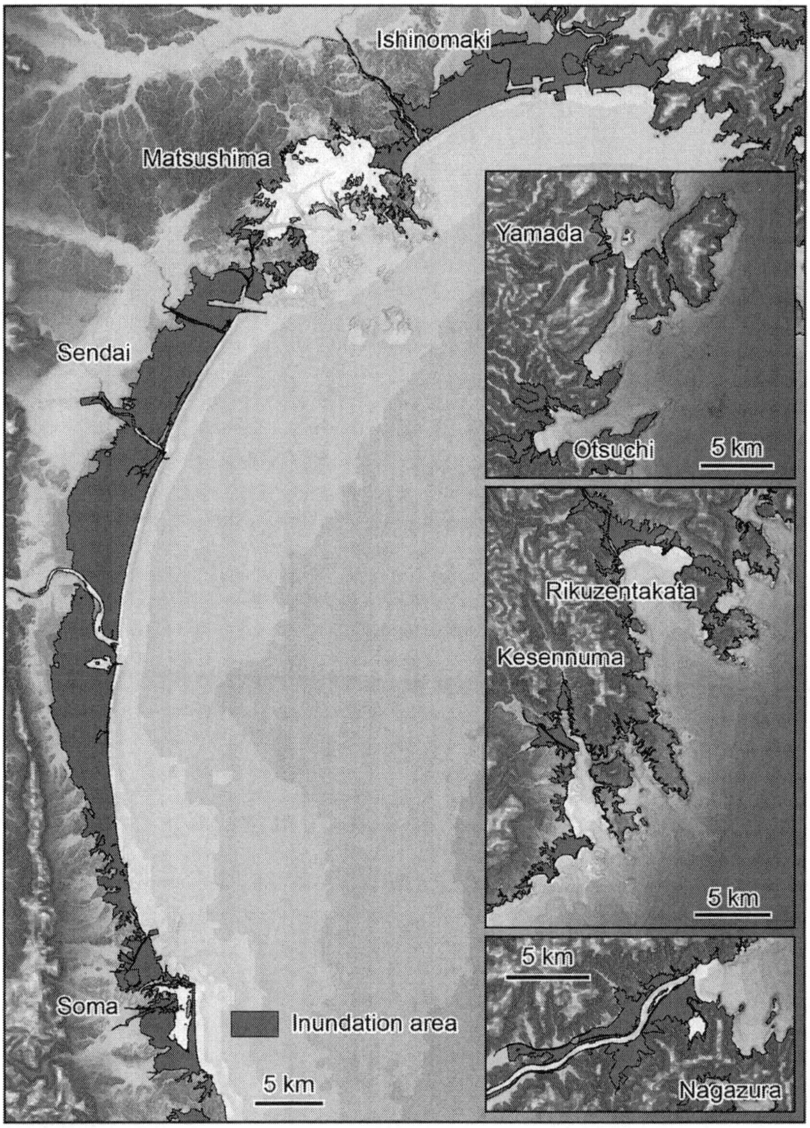

Map courtesy of Daisuke Sugawara, tsunami inundation data provided by the Japan Society of Geoinformatics

Figure 4: Conceptual cross-sections showing elevations of rivers and urban areas in four international cities

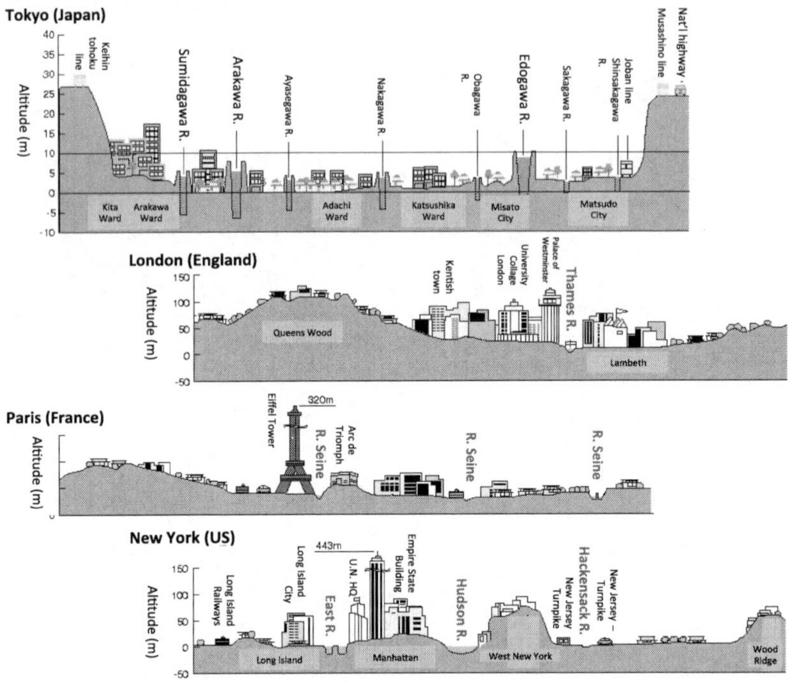

Source: Ministry of Land, Infrastructure, Transport and Tourism (MLIT) (2007b)

Institutional and legislative frameworks relevant to spatial risk management

Due to the recurrent nature of natural disasters in Japan, the government has developed institutional and legislative frameworks for natural disaster management. The earliest disaster-related Act came into effect in the 1880s. Since then, related laws and regulations have been passed and modified upon needs after each significant event, to better mitigate risk at various levels.

The current institutional and legal structure to manage spatial risk began to take shape in the 1960s and 1970s, with several acts adopted regarding disaster management and land use. At the time, the national economy was booming and strong forces pushed development, with little thought to spatial risk and sustainability. As a result, vulnerable urban structures were formed and expanded, and natural events, including typhoons and floods, exacerbated damage and loss once it occurred. To counter such problems, the Disaster Countermeasures Basic Act (DCBA) was adopted in 1961 and established the central role of the current national disaster management plan. Similarly, to reduce vulnerable development and to increase the appropriate use of urban land, the National Land Use Planning Act was endorsed in 1974 aside from the DCBA. Each discrete Act was enacted and enforced individually, though both were designed to include an aspect of risk reduction via spatial control when needed.

In accordance with the disaster management and land use acts, multiple levels of government bear certain responsibilities and powers to attain their goals. Principally, Japan has a centralised governing structure in which the national government oversees 47 prefectures and other local governments, including cities, towns and villages.[2] Local governments exercise their power based on a national delegation, and the prefectural governments play an intermediary role between the national and local governments. Additionally, 20 ordinance-designated cities[3] – regions with populations over 500,000 – hold semi-autonomous power and operate independently from the prefectures. Consequently, upon exercising the DCBA, local governments are given the lead role in all phases of disaster management and the national government continues to provide funding and oversee local implementation. When it comes to the appropriate use of land, the National Land Use Planning Act delegates plan creation and implementation to individual prefectures and local governments, whereas the national government sets the overall land use structure.

Institutional and legislative frameworks for disaster risk management

Three key disaster management Acts

There are three key disaster management acts central to water-related risk management: the DCBA of 1961, the Flood Control Act of 1949 and the River Act of 1964. The DCBA specifies the current disaster management framework. It was enacted in response to the Isewan typhoon of 1959 that devastated the Isewan Bay area, which extends across the Aichi and Mie prefectures. The DCBA aims to protect the lives, livelihoods and assets of citizens from natural disasters, and to safeguard social order, security and public welfare. It was designed to make disaster management function at the national, prefectural and local levels to prepare for and respond to different types of disasters. Notably, it clearly identifies the role and responsibilities of multiple levels of government, private sector corporations and individual citizens, and provides guidelines for managing disasters. Since its adoption, the DCBA has been revised multiple times after significant disasters, including the Kobe earthquake of 1995 and the GEJE of 2011. The Flood Control Act of 1949 and the River Act of 1964 are two key acts that manage water-related disaster risks. The Flood Control Act addresses public safety by monitoring and protecting citizens and assets through mitigating damage from flooding, tsunamis and storm surges. The River Act aims to comprehensively manage rivers to prevent disasters from flooding, tsunamis and storm surges so that land will be protected and developed for public safety and welfare.

The role of disaster management plans in the Disaster Countermeasures Basic Act

The DCBA appoints a mix of government representatives and citizens to develop disaster management plans. These plans provide a framework to guide disaster management efforts through three different phases, including: (1) mitigation and preparation; (2) response; and (3) reconstruction and recovery. Three types of plan are recommended for development: the Basic Disaster Management Plan, the Disaster Management Operation Plan and the Local Disaster Management Plan. The Central Disaster Management Council in the Cabinet Office prepares the Basic Disaster Management Plan, and the prefectures and municipalities play a central role in implementing the disaster reduction efforts. The Disaster Management Operation Plan is developed by designated governmental organisations and public corporations. Lastly, the Local Disaster Management Plan is prepared by prefectures and municipalities, and mirror the disaster management plans prepared at the national level (for governments' hierarchical relationship and disaster management responsibilities, see Figure 5).

The Basic Disaster Management Plan was first developed and adopted in 1963 by the national government, and has since been the main guide for Japan's disaster management. Its contents are revised as needed, with major changes made after the Kobe earthquake (1995) and the recent GEJE (2011). Prior to the GEJE, the Basic Disaster Management Plan was organised by four types of disasters, including sections of earthquakes, wind and floods, volcanic eruptions, and snowstorm. Each plan was organised by the phases of disaster management and articulated the roles of the national, prefectural and local governments, as well as residents. Post-GEJE, a new section that solely focuses on tsunami measures was created; previously, tsunamis were treated only as a sub-phenomenon of earthquakes.

Among different types of disasters focused on in the Basic Disaster Management Plan, sections on wind and floods, as well as tsunamis,

Figure 5: Structure of the disaster management system

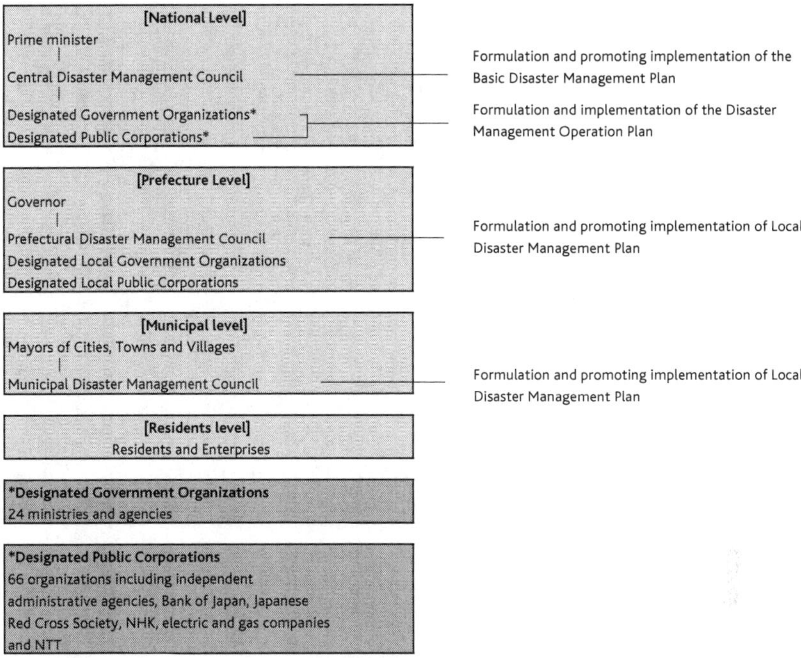

Source: Modified from Cabinet Office (2011 and 2015)

are relevant to spatial control in managing water-related disasters. For instance, upon preparing disaster management plans for floods and tsunami events, national and local governments, along with communities, are mandated to proactively promote mitigation efforts by preparing master plans. This includes the creation of publicly available hazard maps that indicate potential inundation. Another example is apparent in the newly created section on tsunamis, which includes a reference to the role of land use planning in making areas resilient and introduces a two-tiered approach for disaster reduction. Proposed land uses post-GEJE against tsunami hazards are strongly linked with the newly developed concepts of 'more frequent less height tsunamis (L1 tsunami)' and 'less frequent large height tsunamis (L2 tsunami)'. For instance, structural levee defences along the ocean are proposed as front-line measures against an L1 tsunami. In another case, a combination of structural and non-structural measures, including land use and evacuation routes, are recommended to protect against an L2 tsunami. Local governments are then primarily tasked with taking the lead on determining when to individually adopt and when to combine such different land use measures with current local land use. Detailed strategies include prohibiting residential use, providing conditions for types of permitted land uses and physically raising the height of land in areas of potential inundation from L2 tsunamis.

Land use system in brief: managing space and development

Independent from the disaster management scheme, a modern land use planning system in Japan took shape between the late 1960s to mid-1970s. The new City Planning Act was enacted in 1968, revising the old 1919 Act, to introduce new land management concepts and techniques. These included the detailed land use (*senbiki*) system, development permit (*kaihatsu kyoka*) system, and floor area ratio (FAR) to better handle hastened urbanisation, which exacerbated urban sprawl and degradation. Following the City Planning Act 1968, the National Land Use Planning Act was endorsed in 1974. This Act introduced a national land use planning system that organised the holistic use of land, and individual laws that regulated and systemised the use of urban, agricultural and forest land. The following section explains the national land use system and the city planning system to highlight the link between the land use system in urban areas and hazard management.

National land use

The National Spatial Plan, following the National Spatial Planning Act 2008 (NSPA), is the current primary national-level plan that aims to preserve healthy cultural livelihoods, as well as promote balanced development across the nation. Prior to this National Spatial Plan, five Comprehensive National Development Plans, following the Comprehensive National Development Act 1950 (CNDA), had led national spatial development for the latter half of the 20th century. Recently, CNDA was replaced by the NSPA because the concept of national development had to shift from 'expansion and growth' to 'smarter uses of land' as the country is no longer expanding but instead shrinking with respect to both population and economy. Unlike the former comprehensive national development plans that considered land use nationally, the new plan addresses development by region, yet continues to guide national spatial development. Consequently, the National Spatial Plan continues to be an anchor for more detailed land use plans developed at different levels of government, including prefectures and municipalities.

The National Land Use Planning Act 2005 created three governing levels for land use planning and management – a 'national plan', a 'prefectural plan' for the 47 prefectures and a 'municipal plan' for the local municipalities (which includes cities, towns and villages). It underscores the primacy of the 'national plan' by requiring 'prefectural plans' and 'municipal plans' to be based on it. Prefectures and municipalities must also incorporate local characteristics, including natural, economic and cultural conditions, for more accurate land use planning (see Figure 6). The three levels of plans collectively focus on future visions and ways to utilise land use strategies to achieve each jurisdiction's goals. However, progress on developing the land use plan at all governmental levels has not been smooth. For instance, the third national land use plan was adopted in 1996 and all prefectural governments developed their plans soon after following the revision. Yet, as of 2008, local governments had made minimal progress on their individual plan development, even after almost three decades of the concept being introduced (MLIT, 2008).

Figure 6: Various plan implementations of the National Land Use Law

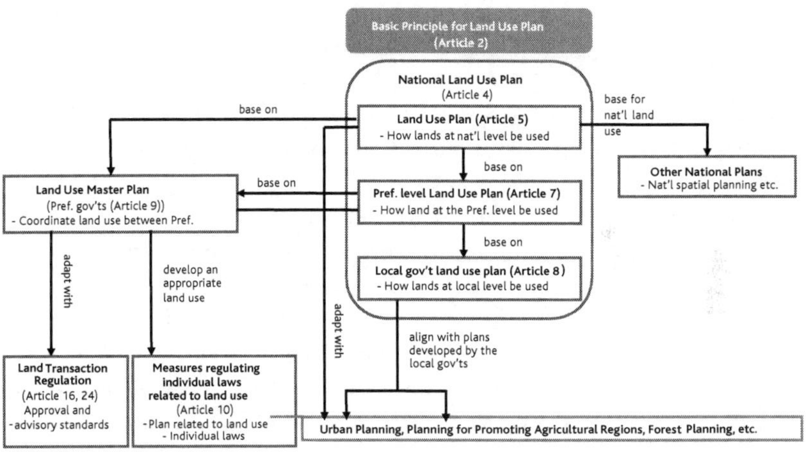

Source: Translated on the MLIT, N.A.

Related to the land use plans, the 47 prefectures are primarily responsible for creating land use basic plans to align with the corresponding national and prefectural plans. Prefectural governors make land use decisions under macro-developmental perspectives as they are responsible for guiding development in their jurisdictions. The land use basic plan follows the land classification of: urban, agricultural, forestry, national parks and natural preservation districts (see Figure 7).

Figure 7: Five land types in the Land Use Basic Plans

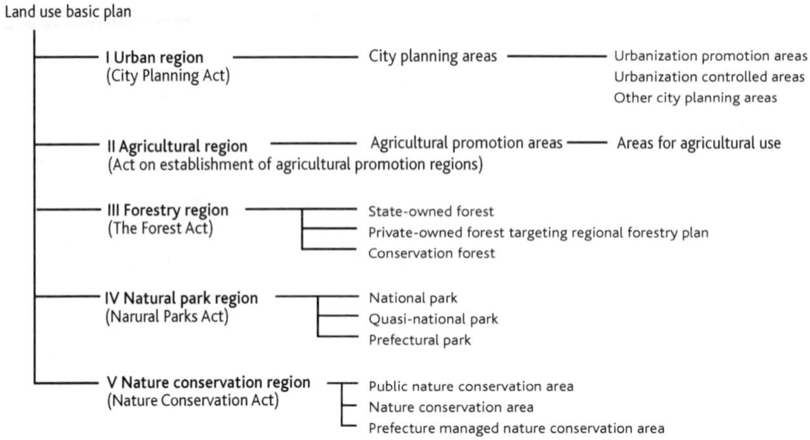

Source: Translated from Kato, eds (2006)

City planning areas

The City Planning Act 1968 aims to promote orderly spatial development by enforcing plans, regulations and projects in defined city planning areas. Designation of a city planning area is voluntary, however, meaning that prefectures can designate such areas within their territories and that its application level varies by prefecture. Ordinance-designated cities, on the other hand, are required to entitle city planning areas to achieve orderly spatial growth and management with higher population densities. As of 2011, there are 1,129 city planning areas in 1,354 local government jurisdictions. This accounts for 26.9% of land nationwide and encompasses 94.6% of the national population (MLIT, 2011). Once a city planning area is designated, the prefectural and local governments develop and implement master plans with a combination of land use regulations, urban facilities management and urban development projects (see Figure 8). Local governments typically handle all details, including granting permissions on land development and building construction to align with their land use, developing planned urban facilities, and initiating urban development projects listed in the local master plans.

In large and medium cities with populations of at least 100,000, prefectural governments are mandated to divide lands further to reflect specific land use details (*senbiki*, literally translated as an act of drawing a line to divide areas) (see Figure 9). Two examples of specific land use details are urbanisation promotion areas (*shigaika kuiki*) – which are accompanied with land use zoning – and urbanisation controlled areas (*shigaika chousei kuiki*) – where development should be avoided.

Figure 8: Basic structure of the city planning system

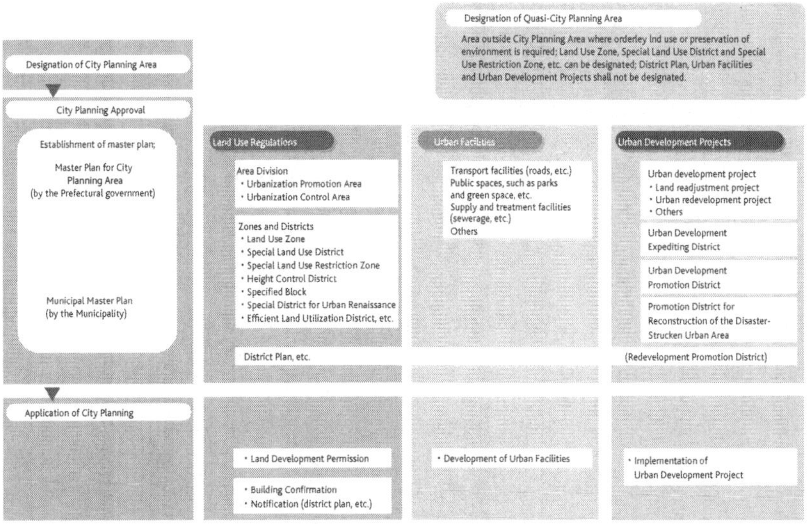

Source: MLIT Retrieved November 5, 2013 from http://www.mlit.go.jp/common/000234477.pdf

Figure 9: Land planning classification: conceptual diagram

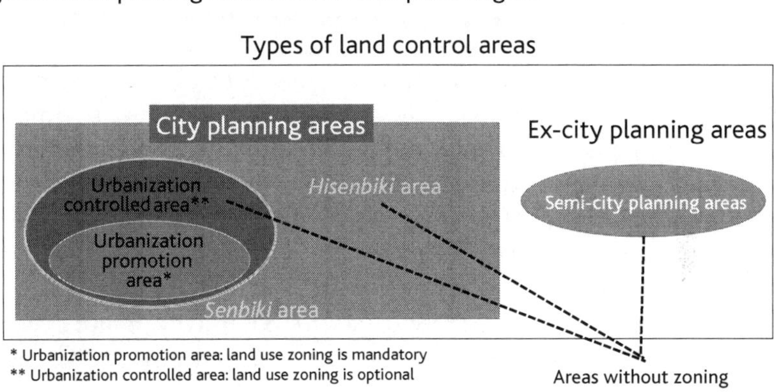

Source: Created from Ito et al (2011)

In medium- and smaller-sized cities, dividing an area into more detailed land use designations is voluntary. This leaves a significant proportion of city planning areas without assigned land uses (*hisenbiki*, literally translated as an act without drawing a line). Within these areas, land use is loosely regulated and zoning can be optionally included if the local governments determine it to be necessary. However, in practice, the majority of areas are left without any land control as most local governments do not enforce zoning in this land type.

Lastly, local governments can assign semi-city planning areas (*jun-toshikeikaku kuiki*) and introduce a zoning system outside identified city planning areas. This sub-designation was newly included when the City Planning Act was revised in 2000. However, this classification is voluntary and not strictly implemented. Under this current system operated in accordance with the City Planning Act, a vast amount of land has not been controlled for appropriate use and orderly development – instead, ad hoc land use is prevalent in areas without zoning, resulting in lower-density, sprawl-type development.

Managing development by land use to mitigate water-related disaster risk

The city planning system in Japan is designed to fully function in controlled urban areas by aligning with planned land use at the national level. At the same time, institutional and legislative frameworks on disaster management systems were developed to function at all national, regional and local levels in order to prepare for and respond to various disasters. Efforts have also been put in place to manage the use of hazardous lands through regulation, such as Article 39 of the Building Standards Law and Article 7 of the City Planning Act. These systems and regulations, however, do not necessarily interact seamlessly, because the planning system that controls land use and the disaster management system that functions to reduce disaster impacts were established independently. For example, human use should ideally be restricted in areas exposed to hazards in order to minimise loss and damage in the event of a disaster. However, areas prone to disasters are continuously developed and used due to various reasons, including locational superiority, a history of use and agglomerated assets. Other municipalities are concerned about hazards but usually do not have the financial power to purchase alternative lands for relocation. Additionally, even if there is a desire to avoid using high hazard areas, developable areas are limited in Japan, resulting in continuous development.

Land use control in hazardous areas

Enforcing regulations and providing financial incentives are two available approaches to restricting use in hazardous areas. In practice, regulations are more frequently used to restrict land use; incentivised approaches through tax reductions or funding allocations are as yet minimally available. There are two key regulations that can address water-related disaster risk: Article 39 of the Building Standards Law and Article 7 of the City Planning Act. However, their application has so far been limited.

Article 39 of the Building Standards Law

Article 39 of the Building Standards Law (enacted in 1950) is one of the key regulatory approaches to control land due to hazard potential. Local governments have the power to designate hazardous areas from events including landslides, surges, floods and tsunamis by enforcing ordinances and controlling building construction. For example, if existing residential buildings are in hazardous areas and potentially risky, the local government could determine that it is better to relocate, and designate the area hazardous to support residents relocate to safer areas. Once an area is designated as hazardous, the land will be restricted from any future residential use, meaning that current owners must relocate and the construction of new residential buildings will be permanently prohibited.

There are two programmes available for collective or individual relocation. The 'Collective relocation promoting programme for disaster prevention' (*Bousai no tameno shudan iten jigyou ni kakaru kuni no zaiseijou no tokubetsu sochi tou ni kansuru horitsu*) allows residents to collectively and proactively relocate to safer areas (MLIT, 2004b). This law was passed in 1972 after more than 400 people perished in a series of heavy rainstorms that affected multiple localities in early to mid-July of that year. It has often been used by communities in isolated, geographically severe areas, as relocating to safer locations – often designated in urban areas as locations with better access to infrastructure and services – is an attractive solution. The 'Relocating programme for hazardous residential buildings adjacent to steep slopes' (*Gakechi rinsetsu kiken jyutaku iten jigyo*) supports the proactive relocation of individual residents specifically living near collapsible slopes in order to prevent damage and loss from slope failure. Unlike the former collective relocation programme, households targeted could relocate individually instead collectively, since slope failures occur in spots rather than in areas. Since, by design, this programme does not require collective household decisions, it provides flexibility and is used much more frequently than the collective relocation programme.

In five decades between 1950 – the time when the Building Standards Law became effective – and 2004, 17,800 locations have been designated as hazardous, most of which resulted from spot and slope failures (MLIT, 2004a). However, local jurisdictions have not been proactive in designating hazardous lands as only 40 local governments had regulations related to Article 39 in 1978, and with no increases through 2013 (Kodama and Kubota, 2013). Additionally, the collective housing relocation programme has been sparingly utilised; only 35 local jurisdictions had used it to relocate a total of 1,834 housing units prior to the GEJE (Iwate Nippo, 2013).

Although intended to promote relocation prior to disasters, the Building Standards Law has primarily been used reactively after significant disasters. The types of disasters that triggered relocation varied and included earthquakes (Hokkaido in 1993; Chuetsu in 2004), volcanic eruptions (Nagasaki in 1991; Hokkaido in 2000), landslides (Niigata in 1980; Kumamoto in 1982), heavy rainfall (Aomori in 1981; Kagoshima in 1993), snow avalanches (Niigata in 1981) and typhoons (Hyogo and Tokushima in 1976). The median number of relocating households by each locality is 23, with a maximum of 329 housing units overall (MLIT, 2014). Prior to the GEJE, this programme had only been adopted once as a response to tsunami disaster – after an M_w7.8 earthquake ruptured off Hokkaido near Okishiri Island in 1993, to relocate 55 housing units away from the shore (Iwate Nippo, 2013).

Upon implementing a collective relocation programme, local governments are responsible for adopting and implementing these programmes in coordination with local communities and individuals. The cost needed for constructing relocation sites, however, is largely covered by the national government, traditionally at 75% of the total cost required. The land lots developed will then be rented or sold by the local governments to interested residents affected by the disaster, who are then responsible for rebuilding their residence on their own.

Article 7 of the City Planning Act and the guidelines of the national departments

Article 7 of the City Planning Act regulates the need to classify land between urbanisation promotion areas and urbanisation control areas in order to prevent unregulated development. In accordance with this Article, areas that may be damaged due to floods, tsunamis and surges, as well as areas that may be exacerbating such disasters by urbanisation, are excluded from the urbanisation promotions areas. For instance, in 1970, the national Ministry of Land, Infrastructure, Transport and Tourism (MLIT) announced that lands along rivers subject to flooding with precipitation of 50mm/hour, or potential inundation depths exceeding 50cm, should not be included in the urbanisation promotion areas unless they have appropriate flood control measures. The MLIT also states in their city planning management guideline that hazards must be taken into account upon defining urbanisation promotion and urbanisation control areas, so that within the urbanisation control areas, hazardous lands will be kept away from development.

In theory, hazard control via land use could be managed through this land classification system. However, in practice, land prices and the potential for speculation override the enforcement of hazard-controlled land use, inevitably overlooking the importance of such a designation. Even in Japan, where these systems are developed, controlling hazardous areas through land use has been difficult.

Efforts at mitigating water-related disaster risk through land use management

As of 2002, 17 districts totalling 7,060 hectares had been designated as hazardous due to the potential for water-related disasters. Of these, 90% are located in the Southern part of Nagoya City, identified in 1961 after the Isewan typhoon of 1959 (Mizutani, 2002). As these data reveal, controlling hazardous lands in a proactive manner has been a difficult task. Saito and Ubaura (2012) suggest that only 32 local governments across the nation have so far adopted Article 39 of the Building Standards Law in relation to managing water disasters, and 80% of local governments had done so only after advisories provided by the national and prefectural governments. Among these 32 local governments, only four had used the collective relocation programme in parallel as the majority only designated uninhabited land as hazardous (Saito and Ubaura, 2012).

In populous areas, the three cities of Nagoya (Aichi, designated 1961), Sapporo (Hokkaido, designated 1966) and Miyazaki (Miyazaki, designated 2005) have adopted Article 39 to limit development (see Table 1). Among these three areas, Nagoya is the only city that enforced Article 39 in the urbanisation promotion areas. Sapporo City adopted Article 40 of the Building Standards Law that gives certain conditions for the residential use in its urbanisation promotion areas, but did not entirely prohibit residence. Miyazaki City only designated hazardous areas in the urbanisation control areas. The limited number of Article 39 adoptions and prohibitions on residential land use in urbanisation promotion areas suggests that land use control for mitigating damage and losses is still minimally utilised.

To improve the current status of land use tied to water-related disaster risks, in 2006, the MLIT introduced the 'Programme on integrated disaster management of land use and water-related disasters' (*Tochiriyo ittaigata mizu bousai jigyo*). This programme aims to invest in

Table 1: Cities adopting Article 39 of Building Standards Law and its conditions

City	Triggering Disasters	Designating hazardous land in urbanization promotion areas
Nagoya, Aichi (Adopted 1961)	Isewan Typhoon (1959)	*Yes* Hazardous areas by Article 39 are designated in urbanization promotion areas
Sapporo, Hokkaido (Adopted 1966)	Recurrent floods with snow melting (Year n.a.)	*No* Article 39 is designated only in the urbanization controlling areas; Article 40 of the Building Standards Law is adopted in urbanization promotion areas to require building structure (e.g. raising) resistant to flooding
Miyazaki, Miyazaki (Adopted 2005)	Typhoon No. 14 (2005)	*No* Article 39 are designated only in the urbanization controlling areas

infrastructure to protect existing residential areas in flood-prone areas along the river. An example is the combination of constructing a ring levee (*wajutei*) and raising the level of land in the targeted communities in flood-prone areas, instead of constructing gigantic levees along rivers. Flood-controlling facilities, as well as water retention facilities, will be constructed in the targeted lands (Otsuka, 2006). In this way, minimum investment could maximize protecting people and assets.

A paradigm shift? Water-related disaster risk mitigation efforts after the Great East Japan Earthquake and tsunami

After the unprecedented GEJE and tsunami in 2011, perceptions of the use of land use control to mitigate against potential hazards are changing. Areas designated as hazardous after the GEJE are very different from previous ones. In March 2013, the MLIT announced that the total area designated hazardous due to a potential tsunami added up to 13,000 hectares (130km^2). In 2014, a confirmed 15,723 hectares (157.23km^2) were designated as hazardous in the three most affected prefectures of Miyagi, Iwate and Fukushima (Araki and Hokugo, 2014). This is extremely large compared to the areas designated as hazardous in the past; as mentioned earlier, the total area designated as hazardous prior to the GEJE was 7,060 hectares, or a little less than half the 2014 amount. Additionally, the designation of hazardous areas from non-water-related disasters, like those designated due to the Kobe earthquake in 1995 and the Chuetsu earthquake in 2004, are also minimal, adding only up to 1.95 and 27.97 hectares, respectively (see Table 2). These numbers suggest the significance of Article 39's application in post-GEJE recovery.

Table 2: Areas designated hazardous by Article 39

Area designated hazardous due by the water-related disasters		**Examples on the use of Article 39 from other types of disasters**
Pre-GEJE	***Post-GEJE***	
70.6 km^2	157.23 km^2 (As of Dec. 2014)	Kobe earthquake (1995): 0.0195 km^2 (1.95 ha) Chuetsu earthquake (2004): 0.2797 km^2 (27.97 ha)

The amount of local governments adopting collective relocation programmes in rebuilding from the GEJE is also notable. According to the MLIT, 337 districts have been approved to implement collective relocation programmes in the Tohoku region, whereas only three districts were adopted in the aftermath of the 2004 Chuetsu earthquake (Reconstruction Agency, 2014).

Designating vast areas along the shore as hazardous will create unprecedented challenges for the Tohoku region. First, controlling land by legally binding certain areas and then relocating people from or restricting them from rebuilding in the targeted areas requires a significant amount of time. A small number of public housing units in the relocation sites became available only early in the fourth year after the GEJE, and many other sites still lack housing (Iuchi et al, 2015). Second, there are concerns that dividing hazardous lands originally used for housing from new residential areas will influence the lifestyles and economies of communities and regions. As it is, placing residential, commercial and recreational uses in close proximity due to space restrictions as a result of the hazardous land designation is becoming difficult.

Alternative efforts to mitigate water-related disaster risks: sharing hazard information

So far, imposing higher costs to deter the use of hazardous lands has not been utilised. Although there are minimal differences in levying premiums for earthquake insurance by level of seismic activity across Japan, no such linkage exists between flood maps and assessed risk, which could control insurance premiums and thus land use (Mizutani, 2002). However, there are other recent efforts of note.

Since the Flood Control Act was revised in 2001, the MLIT and the prefectural governments have proactively publicised hazard maps that include different types of water-related disasters, including flood, inland inundation, storm surges and tsunamis (see Figure 10). This type of information sharing will ideally aid people living in hazardous areas to prepare for floods and evacuation procedures by understanding the phenomenon and its potential impacts.[4] The MLIT is responsible for integrating information from local governments and providing online links through their website so that users have easy access to this material. Via the website, users can zoom into specific regions, prefectures and local governments to examine detailed hazards maps.

Figure 10: Interactive flood hazard map

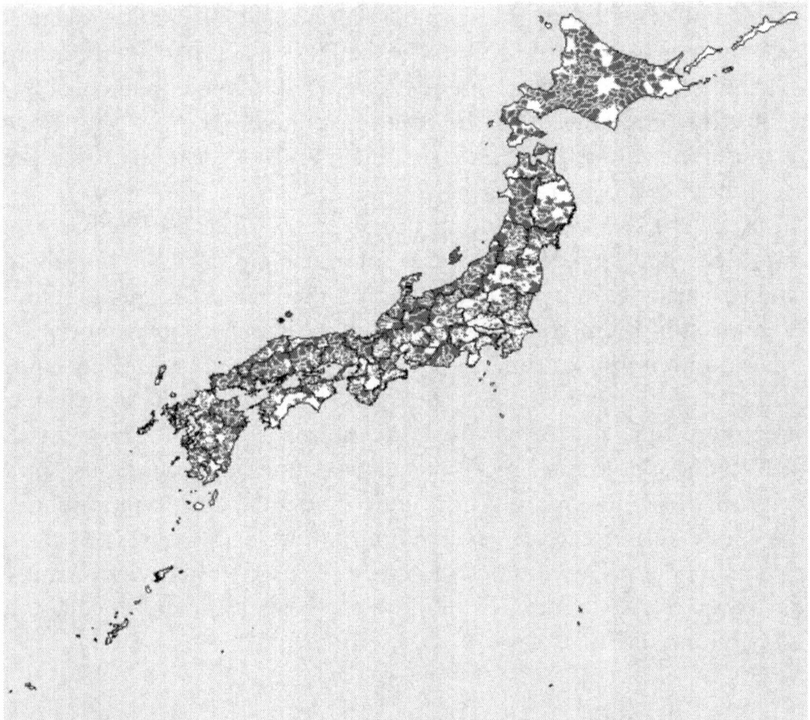

Source: Ministry of Land Infrastructure, Transport and Tourism at: http://disapotal.gsi.go.jp/

Until recently, efforts such as revealing hazard maps to the public have been a taboo in Japan. Policymakers were anxious about the potential negative impact of water-related disaster risk on real estate prices. However, unveiling such risk via hazard maps did not significantly affect land prices, and sharing information with the public has become mainstream. As of December 2005, inundation risks for 248 major and 940 medium to small river basins have been publicised (Cabinet office, Government of Japan, 2011). In addition, technology used to share information is currently being upgraded – for example, hazard (eg potential area for inundation) and exposure (eg roads and population) data can now be overlaid with geographic data.

Since 2010, the need to understand the level of public concern regarding hazards has been increasingly underscored. Often called 'coping capacity', policymakers and planners have begun to understand the ways in which hazard information should be shared and used in communities in order to act appropriately before, during and after disasters, in addition to raising awareness. The importance of these efforts became evident, for instance, after a recent national survey identified that only 30% of citizens polled had accessed any type of earthquake hazard maps available at the local government office (Cabinet Office, Government of Japan, 2010). To increase awareness, the Fire and Disaster Management Agency and the Cabinet Office websites both provide evaluation tools to measure the 'coping capacity' of communities.

The GEJE experience further highlighted the need to better link self, mutual and public support systems upon responding to mega-disasters, and this was legally reflected in the revision of the DCBA in June 2013. Similarly, the community disaster management planning system began operation with the publication of its guidelines by the Cabinet Office in April 2014. With this system in operation, communities are anticipated to better identify risk, capability and needs upon certain disasters, including water-related events.

Conclusion

Due to geographic, topographic and meteorological conditions, Japan has invested extensive efforts in promoting resiliency against natural events. Institutional and legislative frameworks to reduce disaster risk that incorporate spatial control have therefore been shaped for almost half a century. The city planning system has been implemented to manage the use of space wisely, and, in particular, the land classification system adopted with the City Planning Act provides the technical capacity to control hazardous lands to protect lives and assets from natural events.

However, up until the recent devastation of the Tohoku area from the GEJE and tsunami of 2011, only minimal efforts had been made to manage space for resilient development. Although regulations and rules for managing hazardous lands exist – including mechanisms to mitigate against water-related disasters, represented by Article 39 of the Building Standards Law and Article 7 of the City Planning Law – few were proactively used and strictly enforced by different levels of government. Controlling hazardous lands in order to mitigate water disasters was viewed as impractical due to limited available inhabitable land and the fact that the majority of the population and assets were already concentrated along flood-prone areas. Effectively, prohibiting the use of certain portions of land would negatively affect short-term financing and long-term economic activities tremendously.

Rebuilding from the GEJE and tsunami by restricting the use of potentially inundating areas may be a turning point in hazard-considered land use in Japan. National, prefectural and local governments are proactively controlling hazardous lands and supporting extensive collective relocations. However, this control has caused a separation of residential, commercial and recreational uses, which has created its own set of issues. For instance, new residential areas have been developed along hillsides to reduce future tsunami risk. For communities, they are distanced from commercial and urban facilities being reconstructed closer to the shore. Also, since transportation services connecting activity nodes of working, residing and recreating are limited in these reconstruction areas, typical daily movement requires considerable effort. All that said, areas being redeveloped based on the concept of hazard-considered land use must also incorporate the idea of better mobility, and mixed land use for better living. It is too early to conclude whether Japan has or has not effectively incorporated land use control into rebuilding resilient spaces, though there certainly is a nationwide momentum to link water-related risk management and land use planning. Further discussion to refine the definition of 'resilient spaces' is needed as hazard reduction by itself will not lead regions to a resilient and sustainable future.

Notes

1. Defined as areas located along rivers that are lower than the planned maximum water level (ie maximum water capacity of river dikes).

2. As of April 2013, there are 1,719 municipalities across the nation.

3. Sapporo, Sendai, Niigata, Saitama, Chiba, Kawasaki, Yokohama, Sagamihara, Shizuoka, Hamamatsu, Nagoya, Kyoto, Osaka, Sakai, Kobe, Okayama, Hiroshima, Kitakyushu, Fukuoka and Kumamoto.

4. See the hazard map portal site prepared by the MLIT, available at: http://disapotal.gsi.go.jp/

References

Araki, Y., & Hokugo, A. (2014). Analysis of relationship among the designation of disaster risk areas and the loss rate of lives, houses and the rate of inhabitable land in areas devastated by the Great East Japan Earthquake Tsunami. *Memoirs of the Graduate Schools of Engineering and System Informatics*, (6), 24-31.

Cabinet office, Government of Japan. (2010). *Summary on social poll on disaster management.* Tokyo: Cabinet office.

Cabinet office, Government of Japan. (2011). *Disaster management in Japan.* Tokyo: Cabinet Office.

Cabinet office, Government of Japan. (2012). *White paper on disaster management of Japan, 2012.* Tokyo: Seruko.

Cabinet office, Government of Japan. (2015). *Disaster management in Japan.* Tokyo: Cabinet Office.

Central Disaster Prevention Council, Technical investigation committee. (2011). Meeting by the technical investigation committee for earthquake and tsunami on lessons from Tohoku region pacific coast earthquake. Retrieved from http://www.bousai.go.jp/jishin/chubou/higashinihon/1/index.html

Ito, M., Kobayashi, I., Sawada, M., Nozawa, C., Mano, Y., & Yakamoto, T. (Eds.). (2011). *Book that explains city planning and community development* (1st ed.). Tokyo: Shokokusha.

Iuchi, K., Maly, E., & Johnson, L. A. (2015). Three years after a mega-disaster: Recovery policies, programs, and implementation after the great east japan earthquake. In V. Santiago-Gandino, Y. A. Kontar & Y. Kaneda (Eds.), *Post-tsunami hazard reconstruction and restoration* (pp. 29-45). London: Springer.

Iwate Nippo. (2013). Barrier on community rebuilding. Retrieved from http://www.iwate-np.co.jp/311shinsai/saiko/saiko130713.html

Kato, N., & Takeuchi, D. (Eds.). (2006). *New city planning theory* (2nd ed.). Tokyo: Kyoritsu shuppan.

Kodama, C., & Kubota, A. (2013). Study on the idea of land use regulation focusing on article 39 of the building standards law. *The City Planning Institute of Japan*, 48(3), 201-206.

Mizutani, T. (2002). *Natural disasters and disaster management science.* Tokyo: Tokyo University.

MLIT (Ministry of Land, Infrastructure, Transport and Tourism) (no date). Legal structure on national land use. Retrieved from http://www.mlit.go.jp/common/000022635.pdf

MLIT (2004a). Current condition on areas designated hazardous to disasters. Retrieved from www.mlit.go.jp/river/shinngikai_blog/.../s1.pdf

MLIT (2004b). *Small-scale residential district improvement project*. Tokyo: MILT.

MLIT (2007a). Summary on flood-related projects, 2007. Retrieved from http://www.mlit.go.jp/river/pamphlet_jirei/kasen/gaiyou/panf/gaiyou2007/

MLIT (2007b). Considering water-related disaster risk reduction: Rivers in Japan at a higher elevation than residential areas, 2007. Retrieved from http://www.mlit.go.jp/river/pamphlet_jirei/bousai/saigai/kiroku/suigai/suigai_3-3-1.html

MLIT (2008). *National land use plan (fourth revision)*. Tokyo: Ministry of Land, Infrastructure, Transport and Tourism.

MLIT (2011). *Urban planning survey 2011*. Tokyo: Ministry of Land, Infrastructure, Transport and Tourism.

MLIT (2014). Urban disaster management: Collective relocation promoting program for disaster prevention. Retrieved from http://www.mlit.go.jp/toshi/toshi_tobou_tk_000009.html

Otsuka, M. (2006). *Recent water-related disasters and its measures - focusing on flooding from middle to small sized river and urban-type flooding*. (No. 544). Tokyo: National Diet Library.

Reconstruction Agency (2014). *Situation of the recovery and current efforts from the Great East Japan Earthquake (August 2014)*. Tokyo: Reconstruction Agency.

Saito, S., & Ubaura, M. (2012). Problem and actual condition of regulation of building restrictions to reduce flood damage – Article 39 of the Building Standards Law. *The City Planning Institute of Japan, 47*(3), 445-450.

USGS (2015). Largest Earthquakes in the World since 1900. Retrieved from http://earthquake.usgs.gov/earthquakes/world/10_largest_world.php

CHAPTER A1b

Spatial planning for housing recovery after the Great East Japan Earthquake

Tamiyo Kondo and Yuka Karatani

Introduction

After a devastating disaster, the affected area loses all functions that are necessary for people to sustain their lives. One of the most important functions to recover is housing, which serves as a basis for human living and community. That is why research about housing recovery is especially important. Here, the term 'housing recovery' is used rather than 'house' or 'reconstruction'. 'House' is physical structure, but 'housing' includes the system of providing houses for people and the process of human living in the house within the surrounding community. 'Recovery' of housing contains the meaning not only of 'reconstruction' as an action, but also of 'rebuilding' as a decision-making and reconstruction process.

Smith and Wenger (2006, p 237) define sustainable disaster recovery as 'the differential process of restoring, rebuilding, and reshaping the physical, social, economic, and natural environment through pre-event planning and post-event actions'. Disaster recovery is a process – neither a goal nor consequence. Here, 'disaster recovery' is defined as the process of restoring survivors' living and enhancing the sustainability and resilience of the built environment. The former often tends to fall into the category of 'put "it" back in place', and the latter falls into 'build "it" back better'. 'It' includes human living, community, housing and jobs for people, and industry, economy and urban landscape for cities. The perspective of this chapter and expected means of housing recovery includes not only social welfare policy, such as providing housing for those who became homeless after disaster and encouraging people to reconstruct their housing, but also spatial planning for survivors' housing reconstruction and urban infrastructure. It includes two dimensions: one is people-centred (Office of Communications and Partnerships Bureau United Nations Development Program, 2011) and the other is the spatial planning perspective. Basically, the people-centred focus is on human settlement, but the perspective extends over the built environment. Planning for the built environment following disaster has to pay attention to the spatial gradualness and time axis in order to contribute not only to quick housing reconstruction for survivors, but also to a long-term sustainable community and urban built environment for future generations. The relation between these is not one of trade-off; rather, they have to be considered together because human settlement and the built environment do not exist independently, but are interrelated: human settlement forms the built environment and the latter influences human living.

This chapter intends to demonstrate the significance of spatial planning 'guidance' for individuals' housing reconstruction and relocation in order to avoid scattered housing reconstruction and urban sprawl, which is a socially, physically and economically unsustainable development pattern for people's living and urban management in a depopulated region. How is the progress of housing recovery by homeowners and local government after the Great East Japan Earthquake? What kind of spatial planning has been carried out in tsunami-affected coastal areas? What is the role of spatial planning for housing recovery following disaster?

This chapter is based on continuing field surveys of housing rebuilding and interviews with homeowners in the Miyagi and Iwate prefectures after the Great East Japan Earthquake. The chapter focuses on recovery issues in the prefectures of Iwate and Miyagi along coastal areas, and does not discuss the impact of the nuclear accident in the prefecture of Fukushima. This is because the challenges of an earthquake/tsunami disaster are quite different from those of a nuclear disaster.

The characteristics and challenges of housing recovery after the Great East Japan Earthquake

There are two major challenges ahead for housing recovery after the Great East Japan Earthquake. One is that a significant percentage of housing would be controlled by post-disaster urban recovery projects such as land readjustment projects, with land-raising and collective relocation to mountainside areas. As Iuchi points out in Chapter A1a, strict land-use control for tsunami risk has been enforced by the government sector and an unprecedented area is planned to be designated as hazardous, in which building is restricted. One of the strong planning projects promoted by the government sector is collective residential relocation to mountainside areas in order to reduce tsunami risk.

Figure 11 shows the collective relocation project site in a mountainside area overlooking the sea.

Figure 11: Collective relocation project site in the City of Ofunato, Iwate Prefecture (May 2015)

Source: Author

The national government planned five years (2011–15) as a period of intense urban planning project implementation; however, the expected progress percentage by the end of FY 2015 is 41% in Iwate Prefecture and 53% in Miyagi Prefecture. Many people participating in the project have been waiting to reconstruct their housing since 11 March 2011. Several local newspapers report that the local government is facing a difficult decision on re-examination of the plan as to how much of the lot has to be developed in the collective relocation project.

In addition to the collective relocation, local governments are planning to rebuild a high sea wall and embankment of land along coastal areas and riverbanks. Figure 12 shows the gigantic civil engineering public works in the city of Rikuzentakata to develop residential land in mountainside areas and carry mountains of dirt to embankment areas in order to develop new elevated land areas. It is true that these government-driven projects are effective for tsunami risk reduction, but one of the problems for survivors is that it requires time to finish developing land to start housing reconstruction.

The other main challenge is that the tsunami experience influenced survivors' decisions on whether to relocate or not. Even if the local government does not designate areas as hazardous and forbid residential use, people are considering where to live in order to achieve a feeling of safety from tsunamis. There are two levels of decision-making about where to live. The first is decided by the government sector, which utilises land-use control and enforces where residents cannot live, and the second is individuals' own decisions based on their terrifying experience of tsunami. This is totally different from the experience after the Great Hanshin-Awaji Earthquake of 1995, which was the

Figure 12: Land readjustment projects with embankment and collective relocation to mountainside areas in the City of Rikuzentakata, Iwate Prefecture (August 2014)

Source: Author

most destructive urban disaster in Japan, when people could reconstruct their housing on their own lots. The transformation of the urban spatial structure after 1995, it is argued, was mainly dictated by local governments' recovery projects, such as land readjustment projects and public housing estate development. In contrast to this, it is expected that housing rebuilding and the relocation of survivors will create an enormous impact on the transformation of the urban spatial structure after the Great East Japan Earthquake. The question of where to reconstruct housing by individual survivors leads to the transformation of the built environment.

Housing rebuilding assistance subsidy programme by local governments

Housing recovery is a major component of disaster recovery. If people do not come back and reconstruct their housing, cities would not recover. This is why the housing rebuilding assistance subsidy for survivors is considered and justified as an obvious public need in Japanese society. This idea was not easily carried out and was strongly resisted by the Ministry of Finance in Japan right after the Great Hanshin-Awaji earthquake because the government sector did not want to use taxpayers' money to contribute to the development of private property.

How has the design and philosophy of the housing rebuilding assistance subsidy changed after the Great East Japan Earthquake? Kondo (2015a) explains that the motivation and objectives of housing rebuilding assistance subsidy programmes after the Great East Japan Earthquake was to adjust gaps between survivors. This concept comes from the aim to achieve fairness by the Japanese government sector. Figure 13 shows the relationship between land-use control and the housing rebuilding assistance subsidy programmes. The main targets of the subsidy programme are survivors who live outside 'hazardous areas', in which building is restricted. The 'collective relocation project' and 'relocation of dangerous building in the near-cliff area project' are spatial planning programmes inside hazardous areas, both of which aim to decrease the exposure to hazard risks such as tsunamis and landslides. It is especially important to mention here that this project is expected by local governments to function as housing rebuilding assistance for homeowners after the Great East Japan Earthquake (Kondo, 2015b).

Even if this housing rebuilding assistance subsidy helps people's housing reconstruction, it might function to trigger urban sprawl. There is a compelling need for the conversion of agricultural land and forest into building lots. This is because of the shortage of land that was not flooded by the tsunami, and a great demand for non-flooded land by survivors who have a fear of tsunamis. The increase

Figure 13: Land use control and housing rebuilding assistance subsidies after the Great East Japan Earthquake in the City of Higashi-Matsushima, Miyagi Prefecture

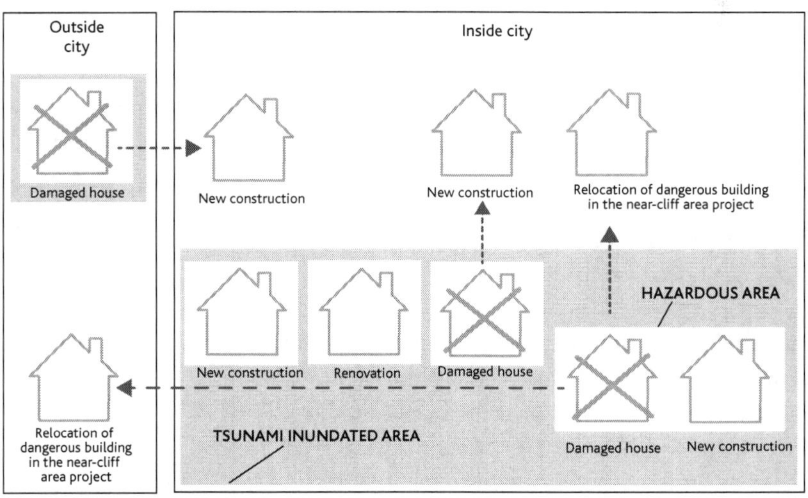

of real estate values is enormous in coastal areas in the Tohoku region, in contrast to other areas of Japan, where most land prices are continuing to slide. The subsidy to develop agriculture fields, forests and rice fields into residential land is critically important in order to reconstruct survivors' housing, although it cannot be denied that government subsidies for individual survivors, such as building lot development, road construction and water supply works, encourage people's housing reconstruction to occur in scattered bits and pieces all over cities, which leads to urban sprawl. It is clear that the housing rebuilding subsidy programme is not designed and consistent with the vision and grand design of post-disaster recovery planning and spatial planning. The focus is only people-centred, lacking long-term built environment perspectives.

Spatial distribution of individual self-help housing reconstruction with relocation

According to a housing rebuilding field survey, a questionnaire survey and interviews in the prefectures of Iwate and Miyagi since 2012, it becomes increasingly clear that individual voluntary relocation and housing reconstruction actions are increasing in coastal areas. Figures 14 and 16 show the spatial distribution of newly constructed buildings in the city of Rikuzentakata in Iwate Prefecture and the city of Ishinomaki in the Miyagi Prefecture. This map indicates newly constructed building after the Great East Japan Earthquake. Although all of these are not identified as consequences of individual relocation, it is a high possibility that this is the case because it is unlikely that the number of newcomers would increase in the disaster-stricken area. The methodology of the survey and mapping is as follows (Kondo and Karatani, 2016a). Nine municipalities along the coast were selected as case studies, all of which have a high percentage of housing and tsunami-inundated damage. The authors identify the newly constructed building after the tsunami by comparing the 'ZENRIN Residential Map' published by ZENRIN DataCom CO., Ltd, before and after the tsunami, which features the names of each building and residence, including the name of the household head. We converted this analogue information to digital information through the use of a geographical information system (ESRI ArcGIS ver.10). After identifying more than 4,000 newly constructed buildings in nine municipalities, in 2014, the authors distributed questionnaires on sites where new buildings are concentrated. Out of 988 questionnaires, 310 responded and the authors conducted additional interview surveys

Figure 14. Spatial distribution of newly constructed buildings and its images in the City of Rikuzentakata in Iwate Prefecture (Kondo and Karatani, 2016a)

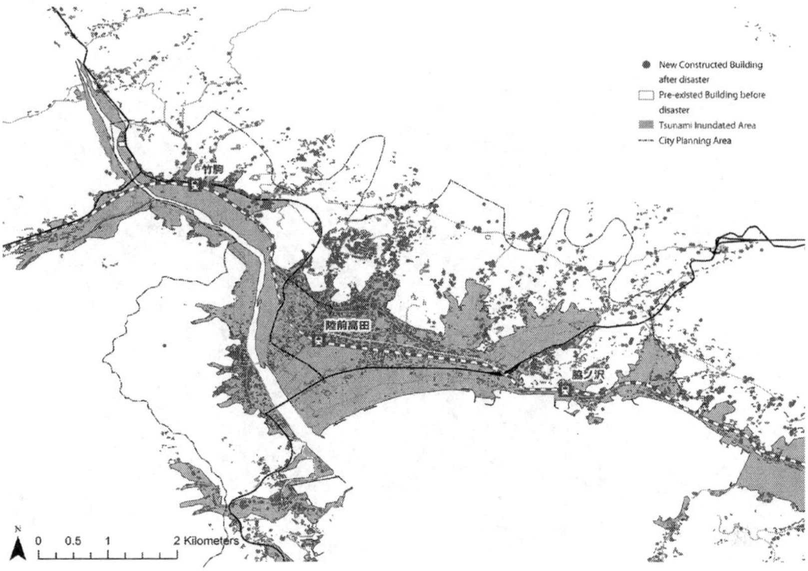

among the questionnaire respondents (N = 40 households) (Kondo and Karatani, 2016b).

Figure 14 shows the 'scattered urban sprawl' pattern in the city of Rikuzentakata, Iwate Prefecture. New buildings are dotted along the existing roads and the borderline of tsunami-inundated areas. New buildings are distributed in a low density and scattered citywide, and the footprint of the living area is expanding, which can be called urban sprawl. The populated area has expanded inland away from the coastline.

Figure 15 shows new detached housing that was developed by clearing trees and converting apple fields to residential land. The boundary line of the living area is expanding, which is urban sprawl. Individual survivors developed rice fields, agricultural land and forest as residential land to reconstruct their housing. Many landscapes like this can be found in the city of Rikuzentakata. There were many apple farmers in the city before the disaster; however, as many farmers did not have any successors, the disaster offered them a good chance to give up farming and sell their apple fields to private housing developers looking for land to satisfy the high demand for non-flooded areas.

Figure 14 also indicates that there are many new buildings located outside of the city planning area's boundary. There are two stakeholders who can control new residential development. One is the local

Figure 15. Landscape of new housing reconstruction in agriculture land and forest (August 2013)

Source: Author

government, which has the authority to permit development through the City Planning Act, which aims to avoid disorganised development. The other is the Agricultural Commission, which controls and gives permission for the conversion of agricultural land into residential land.

Both planning systems are not designed for post-disaster development, but are general land-use management procedures. It is unclear whether the criteria for permission differ for times of post-disaster recovery. It is argued that the local government and Agricultural Commission are faced with a conflict between people's quick housing reconstruction needs and a long-term land-use management perspective. They have to make decisions by weighing up speed and people's housing reconstruction against city-wide sustainability.

Figure 16 shows the 'infill development without urban sprawl' pattern in the city of Ishinomaki, Miyagi Prefecture. New building has been constructed by being inserted into pre-existing settlements, which is categorised as the infill development type as it does not expand the urban footprint. This is totally in contrast with Figure 14, in which new building is constructed in non-residential areas. New building construction is concentrated in areas where there used to be rice fields and agricultural land a decade ago before the earthquake struck. This area was developed by landowners associations through the land readjustment project: a method whereby an irregular pattern of agricultural land holdings is rearranged into regular building plots and equipped with urban infrastructure such as roads and open space. The existing vacant lots in pre-developed residential areas enabled the absorption of new individual housing reconstruction after the tsunami, when demand for residential land dramatically increased.

Figure 16. Spatial distribution of newly constructed buildings and its images in the City of Ishinomaki in Miyagi Prefecture (Kondo and Karatani, 2016a)

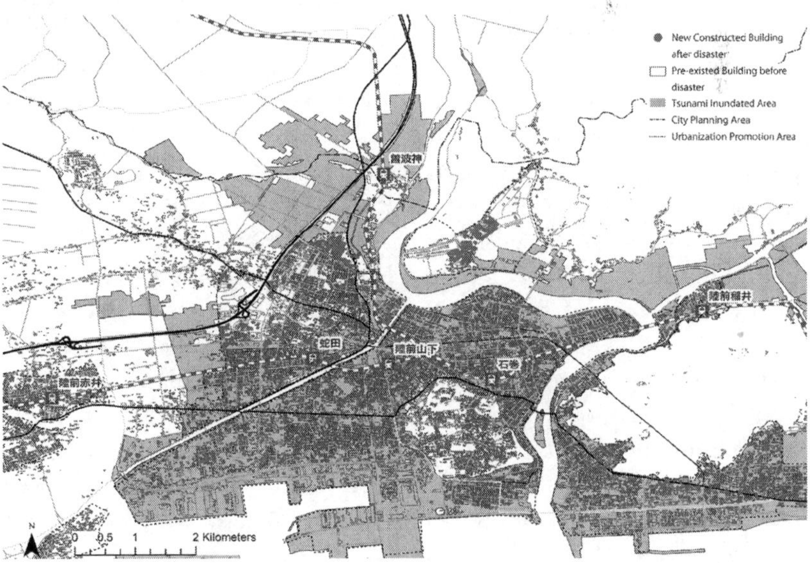

How do these individual self-help relocations result in changes of the spatial structure of urbanised areas? It is found that individuals' scattered housing rebuilding actions in the city of Rikuzentakata led to a physically unsustainable recovery in a depopulated society, which triggered urban sprawl and low-density development. Infill development housing recovery is considered to be an effective model of housing recovery from a physical sustainability perspective as it avoids low-density and scattered development.

Individual survivors' decision-making process for relocation

Even if this spatial distribution of housing reconstruction is a physically unsustainable pattern, these actions were the results of individuals' decision-making to sustain their living as quickly as possible and to avoid tsunami risk in order to achieve a feeling of security, which is considered socially sustainable and represents people's resilience. What kind of characteristics do individual relocated survivors have? Based on a 2014 questionnaire survey, more than 90% of their housing experienced severe damage by the tsunami. None of them are required to relocate by government land-use control. Approximately 60% of respondents' land was designated as in a hazardous zone, which restricted building; however, more than 20% can reconstruct their housing on their previous lot but decided not to because they are willing to relocate due to their tsunami experience (Kondo and Karatani, 2016b).

When and why did individual survivors decide to relocate? Almost 70% of the respondents decided to relocate before the local government had finalised the post-disaster recovery plan; 46% of the respondents had decided to relocate within six months; and 68% had decided to relocate within one year. This indicates that the prolongation and contents of government-driven planning projects did not influence 70% of the respondents. It is considered that the expectation of the project delay deepened survivors' resolve to prioritise speed in order to reconstruct their housing as early as possible (Kondo and Karatani, 2016b).

How did they decide where to relocate? There are various options for survivors about where to relocate. The top reason for selecting where to relocate was 'To decrease the risk of tsunami' (73.9%). The next reason was 'Found the land by chance' (31.6%). This indicates that they did not weigh up options for where to live in a hurried manner. The subsequent reasons of 'Near where they lived before the tsunami', 'Convenient for daily shopping' and 'Affordable land' made up almost 30%. This proves that they selected the land by affinity, convenience and affordability (Kondo and Karatani, 2016b).

How did they evaluate their own decision to relocate? Based on interviews with questionnaire respondents (40 households), what they positively evaluated were the speed of housing reconstruction, that they were happy to move out of prefabricated temporary housing, the individual- rather than group-based nature of the decision-making, the feeling of safety from tsunami, and so on (Karatani and Kondo, 2016). However, there were many negative aspects, especially for elderly people, who faced many challenges even if they had finished housing reconstruction. One is an increased financial risk to household budgets due to spending all their savings to buy new land and for housing reconstruction. The other is an inconvenient life due to relocating to mountainside areas and away from the town centre, for example, having to walk for half an hour to go to the supermarket because they do not have a car. It is demonstrated that transformation of built environment by individual relocation gives negative influence for peoples' living which is major component of social sustainability.

What kind of social contact within communities has decreased? Respondents answered this question as follows: 'Visit neighbour's house/each other' (41.6%), 'Lending of goods and sharing food' (32.7%) and 'Go shopping and enjoy leisure' (31.9%), which are deeper relationships. More than 40% of the respondents are 'very satisfied' and 'satisfied' with their livelihood, and more than 70% are willing to continue living in the new area. Survivors did finish constructing their housing but there are many people who feel isolated in old and new communities (Kondo and Karatani, 2016b). What people regret the most is the loss of connection between people and the destruction of their prior communities from before the tsunami.

Individual self-help housing recovery with relocation cannot avoid the destruction of communities and scattering. Is there any alternative that overcomes these challenges? Significant potential can be seen in 'community-driven collective relocation', in which residents form a group consisting of more than five households and take leadership in finding land and negotiating with landowners for buyouts. Community-driven collective relocation is not a reactive action based on choosing between options provided by government plans, such as where to live, but is a self-determining action among residents, on a group basis, about where to live and how to relocate.

Conclusion

Spatial planning for housing recovery and disaster risk reduction by the government sector after the Great East Japan Earthquake was too limited to collective relocation 'projects' and land-use 'control' for hazardous space. It has been shown by the questionnaire survey that the motivation for individual-based relocation is a fear of the next tsunami and an aggressive withdraw from the government's collective relocation projects due to their slow speed. This chapter emphasises the importance of spatial planning 'guidance' for individuals' housing reconstruction and relocation in order to avoid scattered housing reconstruction and urban sprawl, which is a socially, physically and economically unsustainable development pattern for people's living and urban management in a depopulated region.

Specific examples of 'guidance' include government subsidy incentives for individual survivors through suggesting areas to become residential areas and town centres defined by post-disaster recovery planning. 'Guidance' is a way of mobilising survivors' resilience to get back their home and is also essential for the private sector to develop subdivisions to accommodate the increased demand for residential land on higher ground. Spatial planning has to go beyond government-driven projects and control, and be expanded to guidance in order to pursue a sustainable and resilient society by promoting multiple stakeholders in housing recovery and resettlement.

However, this guidance and control for individuals' housing reconstruction and relocation is full of difficulties and hardships for local governments trying to balance the long-term sustainability of the region with survivors' short-term needs for housing reconstruction. Development permission as spatial control planning is effective in avoiding disorganised development; however, at the same time, if the restrictions on development are too tight, this will have a negative impact on survivors' housing reconstruction.

This chapter focused on the resiliency of individuals who relocated to mountainside areas, but, actually and ironically, this is one of the factors that has triggered the checkerboard housing recovery situation in low-lying areas, as Ubaura points out in Chapter A1c. When we focus on district-scale recovery by land use and housing reconstruction, we can find that there are survivors who select on-sight reconstruction but also survivors who decide to relocate. Checkerboard housing recovery can be seen at the city-wide level and also district level in coastal areas after the Great East Japan Earthquake. It is important to develop planning strategies and programmes to guide individual-based relocation before and right after disasters, but it also necessary to prepare the land-use management programme based on the premise of accepting individual-based relocation as a given phenomenon that represents human resiliency.

This chapter has implications for the ongoing process of recovery in Tohoku region. However, we need to enhance knowledge in order to promote pre-disaster housing recovery planning for future mega-disasters such as the Nankai Trough earthquake in Japan. One of the next research questions that needs to be answered is: 'What determines the pattern of individual-based housing reconstruction with relocation, such as infill development or sprawl and low-density development?' It is supposed that answers include: (1) the characteristics and vulnerability of cities and society; (2) the extent of damage; (3) government policies and programmes for recovery, such as the size of hazardous zones that restrict housing, collective relocation projects and land readjustment projects following disaster; (4) the development of residential areas by private developers who can find demand and markets for new land and housing; and (5) survivors' attachment and level of willingness to stay in their cities. It is unclear to date how land-use control patterns influence survivors' housing reconstruction. The next research task for the authors is to analyse the spatial distribution of individual relocation between these factors and clarify the mechanism of human resilient action and built environment transformation.

Acknowledgements

I would like to express my gratitude to a number of survivors who kindly accepted our questionnaire and interview survey on individual housing reconstruction with relocation in Miyagi and Iwate prefectures. Also, the housing relocation mapping was made possible by the enormous efforts of students in Kobe University and Meijo University. This research was supported by Japan Society for the Promotion of Science, Grants-in-Aid for Scientific Research Grant Numbers 25702021 (Principal: Tamiyo Kondo) and 24310123 (Principal: Yuka Karatani), and the Housing Research Foundation JUSOKEN.

References

Karatani Y. and Kondo T. 2016, 'The Merits and Demerits of Individual Self-Help Housing Reconstruction with Relocation' in H. Hashimoto and I. Hayashi (eds), *Succession and Creation for Disaster Culture*, Rinsen Book Co.

Kondo T. 2015a 'The Founding Background and Characteristics of Housing Rebuilding Assistance Subsidy Programs by Local Governments after Great East Japan Earthquake: Discussion on Housing Rebuilding Assistance after Mega Disaster', *Journal of Architecture and Planning* 80(707), pp. 135-144 (in Japanese).

Kondo T. 2015b 'The Utilization and Expected Role of the "Relocation of Dangerous Buildings in the Near-Cliff Area Project" in Post-Great East Japan Earthquake Recovery: Questionnaire and interview surveys for local governments and homeowners in Miyagi and Iwate Prefecture', *Journal of Architecture and Planning* 80(715), pp. 2043-2049 (in Japanese).

Kondo T. and Karatani Y. 2016a, 'The Transformation of Urban Built Environment and Spatial Characteristics of New Building Construction after the Great East Japan Earthquake: Focus on the self-help housing reconstruction with relocation in nine municipalities in Iwate and Miyagi Prefecture', *Journal of Architecture and Planning* 81(721), pp. 667-674 (in Japanese).

Kondo T. and Karatani Y. 2016b, 'The Decision-Making and Housing Reconstruction Action of Self-Help Relocated Survivors after the Great East Japan Earthquake: Case studies of nine municipalities in Iwate and Miyagi Prefecture', *Journal of Architecture and Planning* 81(719), pp. 117-124 (in Japanese).

Office of Communications/Partnerships Bureau United Nations Development Program. 2011, 'People-centered Development: Empowered lives. Resilient nations'

Smith G. and Wegner D. 2006, 'Sustainable Disaster Recovery', in H. Rodriguez, Quarantelli & R. Dynes (eds.), *Handbook of Disaster Research*, Springer.

CHAPTER A1c

Urban planning and reconstruction after the Great East Japan Earthquake

Michio Ubaura

Introduction

A paradigm shift in Japanese urban planning

Japan's overall population began to decrease as of around 2007, entering a phase of population decline that regional cities in the country had already begun to experience decades earlier. Not only has the population declined, but also population ageing continues apace at the same time. Moreover, as the debt of the government, including local administrations, has continued to increase, the impact on municipal finances has been severe.

Against the background of such a change in the social situation, Japan's urban planning is also experiencing a paradigm shift (see Figure 17). Namely, it has entered a period that allows Japan to envision a 'contemporary shrinkage model' of urban planning, replacing the 'modern expansion model' that has been followed until now. Specifically, from a land-use planning perspective, rather than orderly urban expansion, what has become a challenge is how to plan a pivot to a so-called 'smart shrink' model of land use that will shape a concentrated urban structure. From an urban planning project standpoint, this means that what is becoming important is spatial consolidation at the district level based on adjustments among a variety of business, rather than a silo style of business accumulation, as well as planning that focuses not only on the 'hard' aspects of projects, but also on 'community planning', which includes both 'hard' and 'soft' dimensions. In terms of decision-making mechanisms, what is called for is a move from government-centred 'top-down' decision-making to a 'bottom-up' model of decision-making that involves the participation and cooperation of local residents.

These modern and contemporary models of urban planning are not at odds with each other. Rather, there are more than a few elements that suggest a relationship whereby the contemporary model already exists in the foundations of the modern model. However, without adequately resolving the challenges faced by Japan's modern urban planning paradigm – for example, without developing the adequate tools for organising expansionary urban development – there will be a difficulty in a situation where it is necessary to resolve the challenges of contemporary urban planning.

Figure 17: Paradigm shift of spatial planning in Japan

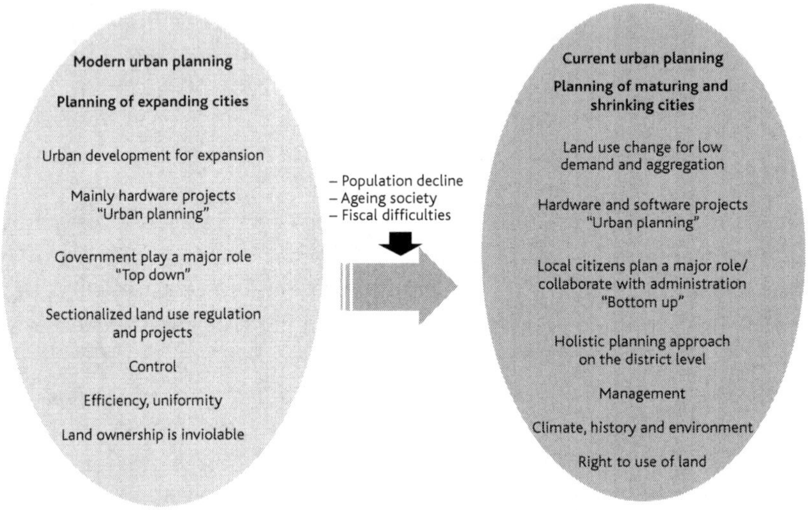

The Great East Japan Earthquake and reconstruction planning

It was in such a context that Eastern Japan was struck by a massive earthquake of magnitude 9.0 on 11 March 2011. The range of the tremor extended over an extremely wide area, approximately 500km to the north and south and 200km to the east and west. However, more than the earthquake itself, the majority of the damage was caused by the tsunami triggered by the earthquake. The tsunami, whose maximum run-up height reached 43.3m (Onagawa Town) and whose inundation area extended across 561km^2, led to the catastrophic death or disappearance of almost 22,000 people, and the partial or total destruction of approximately 400,000 buildings.

The principal affected areas were the prefectures of Iwate, Miyagi and Fukushima (see Figure 18). Except for in Fukushima Prefecture, where there were delays as a result of the nuclear disaster, reconstruction efforts commenced in the other areas immediately following the disaster.

Figure 18: Tohoku region and basic data of tsunami disaster

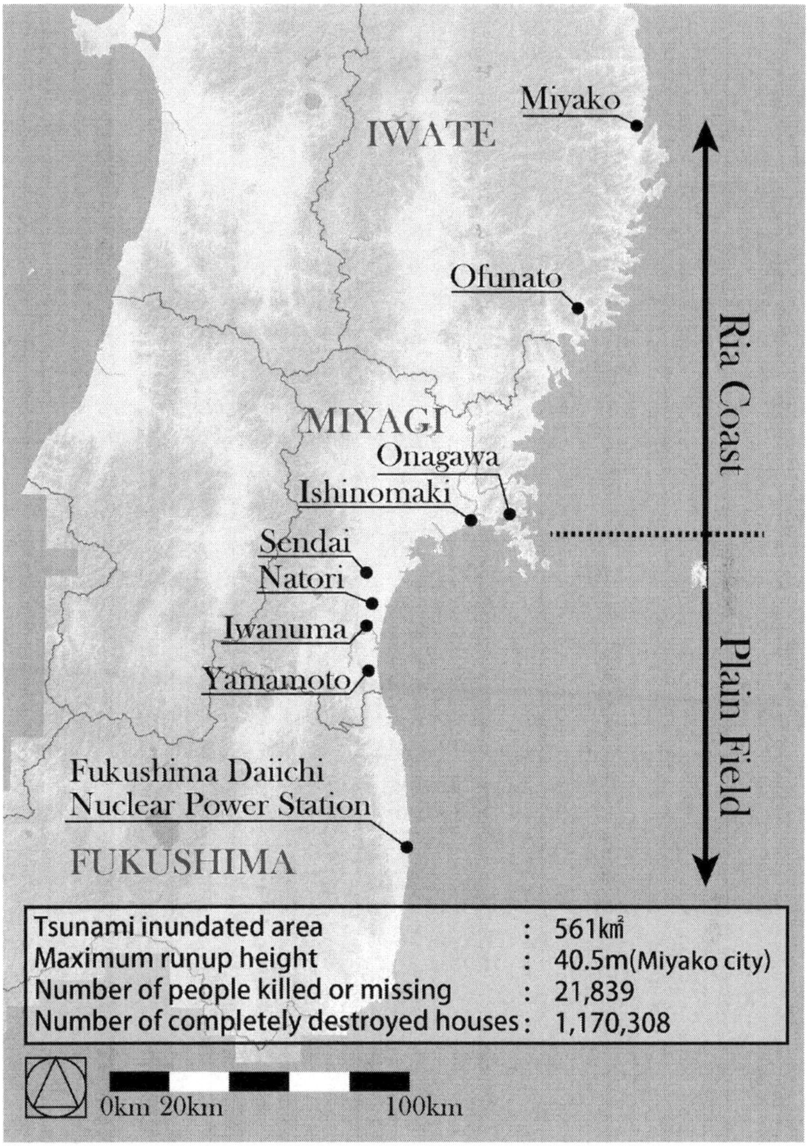

The details of urban planning in connection with reconstruction have varied by municipality since these have basically been formulated at the municipal level. Even so, they generally share the characteristic of being broadly divided into two core pillars.

The first of these is 'safe and secure community development' (*anzen anshin no machizukuri*). Originally, the opportunity for formulating reconstruction plans had generally been brought about by widespread damage caused by large-scale disasters. Also, the area affected by the disaster on this occasion was one that had previously been struck by tsunamis at a certain frequency and, in addition, was an area that was heavily damaged. Therefore, planning a community that will not be subjected to such damage ever again is now regarded as the primary goal of planning.

The other pillar is 'compact and sustainable community development' (*konpakuto de jizoku kanō na machizukuri*). Many of the affected areas were already acutely experiencing the aforementioned challenges confronting Japanese regional and rural communities, namely, population decline, population ageing and financial difficulty. Furthermore, in most municipalities, these problems were only aggravated by the disaster. In order to respond to these challenges, it will be important to build spatially compact communities. By doing so, it will become possible to reduce maintenance costs for infrastructure and social services, as well as to generate more liveable environments, such as where even elderly residents might be able to live without needing to drive, eventually leading to improvements in sustainability. Thus, compact community development initiatives to increase sustainability account for the second pillar.

In moving forward with these efforts, it will be vitally important to have an urban planning perspective that is not only 'modern' in the manner noted earlier, but also 'contemporary'. So, then, how can such reconstruction planning be drawn up and realised? What are its challenges? In the following, I would like to discuss these questions in specific detail.

The reconstruction land-use plan

In order to carry out the first pillar, namely, safe and secure community development, we are faced with the problem of how, and to what degree, to ensure tsunami disaster prevention capabilities. The document released by the Central Disaster Management Council (2011) contains standard criteria for the aforementioned purpose. The basic approach of the guidelines hypothesises two levels of tsunami: 'tsunamis that occur relatively frequently' (Level-1 [L1] tsunamis); and 'tsunamis that occur at an extremely low frequency, but which cause an enormous amount of damage' (Level-2 [L2] tsunamis). In the case of L1 tsunamis, the guidelines aim to 'prevent tsunamis from reaching the land using physical structures such as seawalls', while in the case of L2 tsunamis, importance is placed on 'developing countermeasures based on "disaster mitigation" that seeks to minimize damage'. Therefore, while 'reducing damage to the greatest extent possible using tangible countermeasures', the guidelines also call for 'intangible, evacuation-based countermeasures'. More specifically, municipalities should:

> implement countermeasures timely that correspond to local conditions by combining secondary levee construction that utilizes traffic infrastructure, land elevation, the provision of evacuation shelter, tsunami refuge buildings, evacuation routes, and emergency stairs, and land use and building regulations that consider inundation risk.

A significant number of municipalities have implemented land-use plans and regulations similar to those outlined in the following based on the approach outlined in the guidelines (although approaches differ slightly between municipalities).

First, seawalls are constructed to the height of L1 tsunamis. Then, following the construction of seawalls, the effects of an L2 tsunami are viewed through simulations. Areas in which the estimated inundation height generally exceeds 2m are designated as disaster hazard zones and the construction of residential buildings in these areas is, as a rule, prohibited. The reason for using a standard height of 2m is that the rate at which wooden houses are washed away increases when the inundation height exceeds 2m. Areas designated as disaster hazard zones are also defined as relocation promotion zones under the Disaster Prevention Group Relocation Promotion Project (see Figure 19), and residents living in these zones are relocated to housing complexes in elevated areas with the help of financial support from the Japanese

Figure 19: Conceptual diagram of the Disaster Prevention Group Relocation Promotion Project

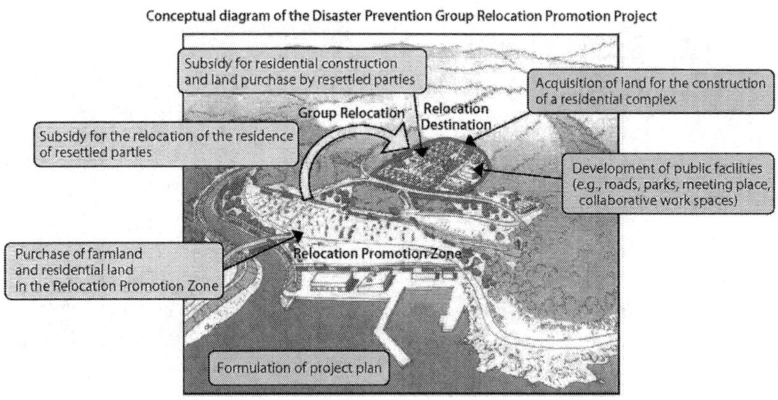

government. Furthermore, municipalities may decide whether to proceed under a disaster hazard zone designation or as a relocation promotion zone under the Disaster Prevention Group Relocation Promotion Project. Here, the former designation is intended for large urbanised areas located on low-lying plains, whereas the latter is intended for small fishing villages in Ria coastal regions.

In addition, it may not be necessary to relocate residents of relocation promotion zones en masse as it is also possible to carry out individualised reconstruction while obtaining substantially the same subsidy within the official framework of 'Projects to Relocate Residences Away from Cliffs and Other Hazardous Areas'.

Conversely, in principal, residential buildings in areas with an estimated inundation height of 2m or less are reconstructed in their original locations. However, in areas that were experiencing problems in terms of urban infrastructure – for example, areas that originally had a low road ratio or did not contain parks – municipalities have also capitalised on the opportunity to make improvements under land readjustment projects.

In addition, some municipalities are planning reconstruction initiatives under plans and schemes that differ from these so-called standard plans. For example, with regard to the specified range of disaster hazard zones, while the city of Ofunato designated as hazard zones all areas for which a simulation predicted inundation, due to the scarcity of available residential areas along the Ria coastline, the additional presence of a certain number of people who wished to rebuild in their current location and the necessity of making use of low-lying areas, three types of zonal divisions were created (namely, zones of types 1, 2A and 2B) for residential construction under certain conditions (see Figure 20). However, because the Disaster

Figure 20: Conceptual diagram of Disaster Hazard Zone designations in the city of Ofunato

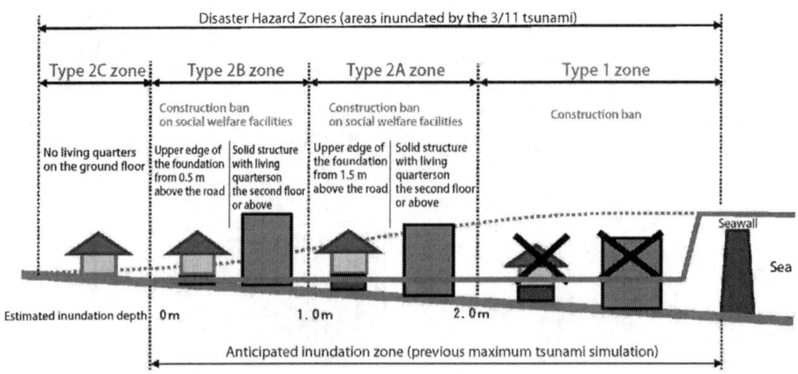

Prevention Group Relocation Promotion Project had been carried out beforehand, areas outside the predicted inundation zone were designated as relocation promotion zones, and since these were wholly included within the hazard zones, a supplementary type 2C zone was added to the designation based on the recent tsunami (see Figure 20). For this reason, type 2C zones are designated as hazard zones outside the inundation. Thus, one reason why such a broad designated range has been taken for the disaster hazard zones is economic; rather than an emphasis on safety by the city, the intention is to subsidise victims of the recent disaster who wish to relocate to higher areas on the basis of the Disaster Prevention Group Relocation Promotion Project.

Specific examples of reconstruction land-use plans

Now, let us turn to some specific examples of how plans are being drawn up. In the following, I take up the city of Ishinomaki (specifically, the downtown core) as a typical example of planning for a lowland plain, and the central district of the town of Onagawa as a typical example of a Ria coastal area in order to give a brief explanation.

Ishinomaki City (city centre)

Ishinomaki is the second-largest city in Miyagi Prefecture following Sendai, with a pre-disaster population of 163,000 people. The city's downtown core was struck by the tsunami, which advanced from Ishinomaki Bay on the city's southern flank, with the areas near the Port and Minamihama districts suffering devastating damage from a direct hit, while areas around the city centre suffered major damage from the tsunami's surge up the Kitakami River. Overall, the city suffered 3,700 dead or missing casualties, 2,500 of these from the downtown core (government office district) alone, making Ishinomaki one of the most seriously affected municipalities in the 2011 disaster.

Having suffered such damage, Ishinomaki City embarked on the formulation of plans to rebuild in the wake of the earthquake disaster, and in December 2011, unveiled the Ishinomaki City Basic Plan for Earthquake Disaster Reconstruction, which also included a determination for land-use planning in general. Since then, the city has sought to move forward with specific projects while amending the plan as necessary (see Figure 21).

Specifically, the city is erecting seawalls at a height of 7.2 m relative to Tokyo Peil (TP) along the coastal areas to deal with L1-type tsunamis. In order to address L2-type tsunamis, which exceed this height, raised

Figure 21: Conceptual diagram of the Ishinomaki City Plan for Earthquake Disaster Reconstruction

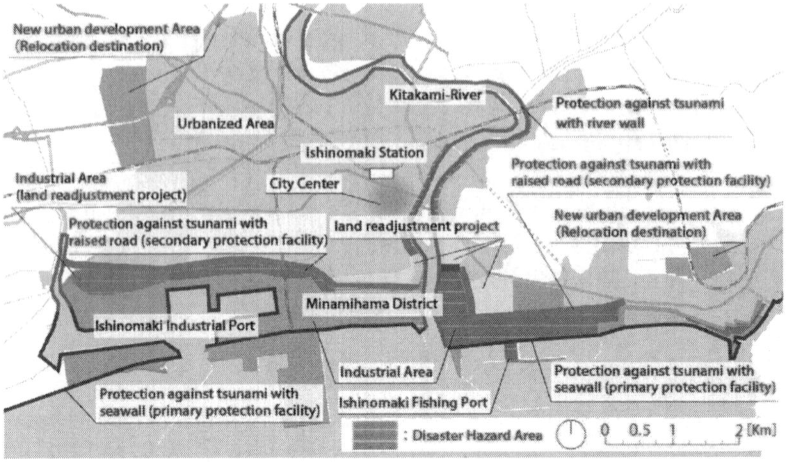

roads are being constructed approximately 500m inland at a height of generally 5m relative to TP to act as a secondary embankment. As well as designating the intervening strip of land as a disaster hazard zone in which the use of land for residential purposes is prohibited, the city intends to resettle residents under the auspices of the Disaster Prevention Group Relocation Promotion Project. Five resettlement housing developments have been constructed inland by utilising land readjustment projects. Following the relocation, the original sites will be put to industrial use or developed as parklands through land readjustment projects or other projects.

In addition, despite having suffered devastating damage in the recent tsunami, some of the zones inside the secondary embankment that had previously lacked sufficient infrastructural development and that will now also be safe from L2-type tsunamis will now be developed for mainly residential use with sufficient parklands or roadways by carrying out land readjustment projects. Moreover, with respect to the city centre, which suffered variable damages, in addition to individual instances of reconstruction, large-scale development is also being planned through urban redevelopment projects that span multiple sites. In other areas, it is basically assumed that reconstruction will take place on an individual basis.

Onagawa Town

The town of Onagawa is a fishing and fish-processing community that had a pre-disaster population of 10,000 people. The town was struck by a tsunami that reached a maximum height of 14.8m and inundated approximately 320 hectares, resulting in over 800 dead or missing casualties and causing tremendous damage that destroyed approximately 3,000 buildings (almost two thirds of the total). Although a reconstruction plan for Onagawa Town was formulated in September 2011, this remained at the level of setting basic policy. Specific land-use plans were formulated afterwards, in January 2012.

In order for Central Onagawa not to be subjected to similar damage a second time, the town is building a breakwater at the entrance to the bay. However, a seawall along the coast will not be constructed to preserve the view of the sea from the town. Instead, most of the lower area will be raised up to protect from tsunamis, as mentioned later.

Hillsides will be quarried out at four locations in the vicinity of the old town, which suffered devastating damages, with the earth used to fill in 200 hectares of low-lying areas, raising them by an average height of 10m (and maximum height of 17m) (see Figures 22 and 23). The high ground and raised zones created by this construction, which will not be flooded by L2-type tsunamis, will be used primarily as sites for residential facilities. Lower zones raised by an average of 4m will be used as sites for tourism and commercial facilities. In addition, low-lying areas that will not be subject to infill construction will generally be used as a location to site fisheries facilities; moreover, they are also planned to be maintained, in part, as a memorial park

Figure 22: Reconstruction planning for the town of Onagawa (ground plan)

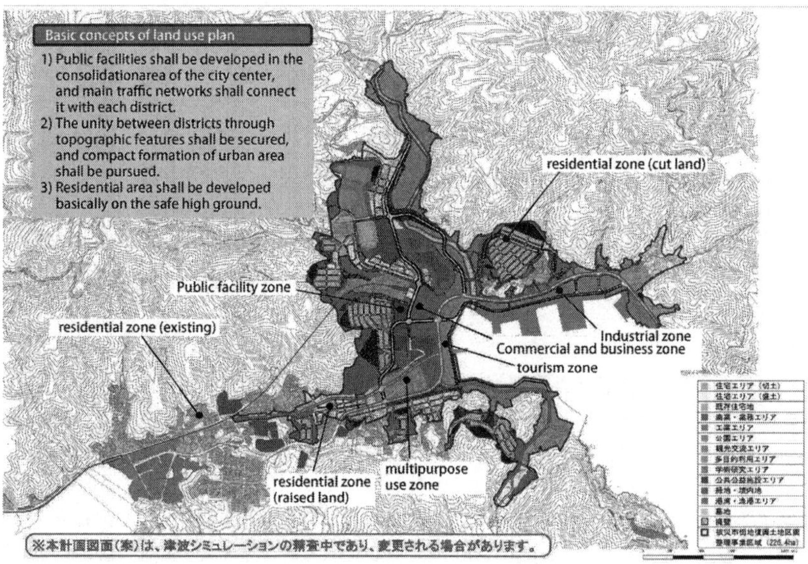

Figure 23: Reconstruction planning for the town of Onagawa (cross-sectional plan)

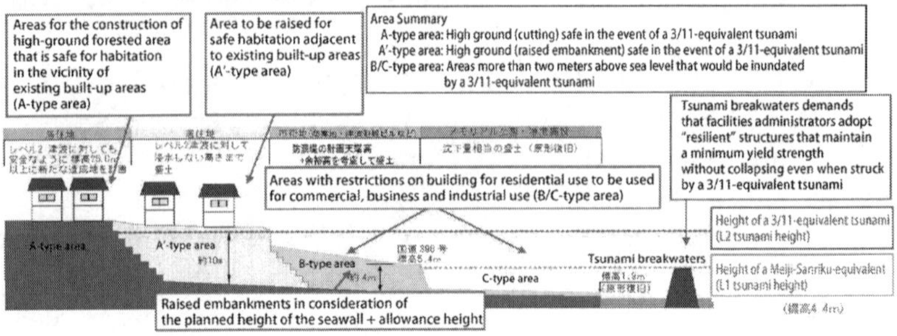

provided with coastal and circuit routes in order to pass on a record of the disaster.

At the same time, Onagawa is aiming for a compact and liveable model of community development, with various public and commercial facilities being consolidated in the town centre under the three basic policy pillars of 'maximising the use of the sea', 'maximising the use of the original topography' and 'maximising the use of historical assets and other resources such as public facilities'.

Challenges in planning process of land use

Now, let us examine the areas that must be considered when formulating this sort of plan. In doing so, we will also examine how these are actually being addressed within the current reconstruction process. The following discusses items that must especially be considered when formulating this sort of Reconstruction Land Use Plan from the two perspectives of *identity* and *comprehensiveness* at the local level, as well as the realities of their treatment.

District-level identity: respect for social value criteria

The first point is that planning should be determined in accordance with the characteristics of the district level. The nature of material components in urban spaces, which is to say things like the height of seawalls or the details of land use (or else the details of regulations for its implementation), should be determined by considering a variety of elements in a comprehensive manner. This is because seawalls, for example, not only have the effect of preventing inundation by tsunamis and storm surges, but may also have an adverse effect on the local landscape or else on the fishery and seafood-processing industries by cutting off access to the sea. The same applies to land use as control does not simply mediate the risk potential of damage such as from flooding, but may also impact on people's livelihoods by making residential and industrial activities impossible. Accordingly, it is necessary to give comprehensive consideration to the advantages and disadvantages of prospective material components, and to determine their details based on the criteria of their respective social values.

On such occasions, the scope of decision-making parties becomes the scope of interested stakeholders. With the construction of seawalls, for example, this is the scope of those who will be affected. Whereas this becomes extremely broad in the case of low-lying areas, in the case of Ria coastal areas where small fishing villages are physically cut off from one another, this will be confined to the unit of individual settlements.

Let us examine the current situation with regard to the ongoing reconstruction process. One of the largest problems facing the reconstruction effort following the Great East Japan Earthquake has been that of making decisions with respect to the problem of seawall implementation. An illustration of contrasting decision-making processes can be seen in the prefectures of Miyagi and Iwate, among the largest of the affected areas.

Whereas officials in Iwate Prefecture have, as a general rule, shown a tendency to implement seawalls that correspond to the L1 height, they have not necessarily stuck to this precept; in the event that a given district requests implementation at a lower height, the prefecture has respected their decision provided that they have satisfied certain criteria. For example, in the Akamae district of the town of Otsuchi, it was revealed that the previous seawalls, which had been 6.4m in height, were to be newly replaced with L1-height seawalls of 14.5m. This led local residents to raise a dissenting view, including from the perspective that this would ruin the local landscape. Whereas the town had secured L2 safety in terms of residential land use by relocating to higher ground, it had been decided that industrial land use would accept L1 risk, and the prefecture decided to implement seawalls at the same height as previously in accordance with the town's decision.

Miyagi Prefecture, on the other hand, with the exception of abandoned farmland and other hinterlands that are more or less completely fallow, has basically pursued a policy of implementing L1 seawalls across the board. This has led to the occurrence of opposition movements in places like fishing villages, where the policy will have an industrial as well as a scenic impact through the isolation of the fishing ports from the hinterlands, and other communities, where tourism will suffer when hinterlands are cut off from sandy beaches used for ocean swimming.

Comprehensiveness at the district level: spatial coordination between projects

The second point is the necessity of avoiding a siloed implementation of government-led projects, and, instead, for various departments to work in tandem. At present, the affected areas are faced with the task of rebuilding new towns from the ground up in areas devastated by the tsunami. As a result, a variety of projects being carried out by a variety of authorities, including road construction, dike construction, land readjustment and disaster public housing construction, have become congested within a confined area. For this reason, project coordination at the district level may be said to be necessary from the following three perspectives.

First, project coordination is necessary to ensure the progress of the individual projects. When project components, such as the configuration of project zones and the details of plans, time schedules, and explanatory meetings and other measures to deal with local residents, are carried out individually, this gives rise to a variety of problems and the individual projects are unable to move forward as a result. Specifically, for example, project zones may overlap or road networks fail to interconnect. Unless mutual adjustments are made and brought into alignment when pushing various projects forward, what originally seemed just an isolated project will be unable to advance smoothly.

A second imperative is to eliminate waste. Given that reconstruction projects are carried out with public funds, it is necessary that they be carried out efficiently. By carrying out project coordination, for example, it will be possible to build multipurpose facilities that can also function as tsunami shelters and make decisions on seawall implementation standards in conjunction with land use in the hinterlands.

Third, coordinated projects may and can make cities and towns more appealing. For example, there are plans for the implementation of river embankments along the old Kitakami River to cope with L1-height storm surges. Ishinomaki City has a long history of more than 400 years as a river port dating from the Edo period, and prior to the earthquake disaster, there had not been any embankments in the city centre, the river and the town coming together to shape both the landscape and the lives of the people. In response to the recent disaster, redevelopment projects have been planned centring on the replacement of bridges and the construction of embankments, as well as commercial facilities and public housing in the hinterlands. These projects have been separately led by prefectural, municipal and private entities, and if they are carried out separately, as is so often the case in urban areas, the river and the town will be separated, and Ishinomaki's unique character will be lost. Therefore, by coordinating projects relevant to particular areas, and by implementing the design and planning of these in a unified manner, it will be possible to create a river space with a new allure while still preserving Ishinomaki's character, such as by, for example, allowing direct access from the upper level of the town to the top of the embankment, and then to the riverside, or by integrating the utilisation of commercial facilities and public squares (see Figure 24). Therefore, coordination to this end has been taking place between project leaders.

Figure 24: A proposed plan for the areas in Ishinomaki City's downtown core and along the Kitakami River

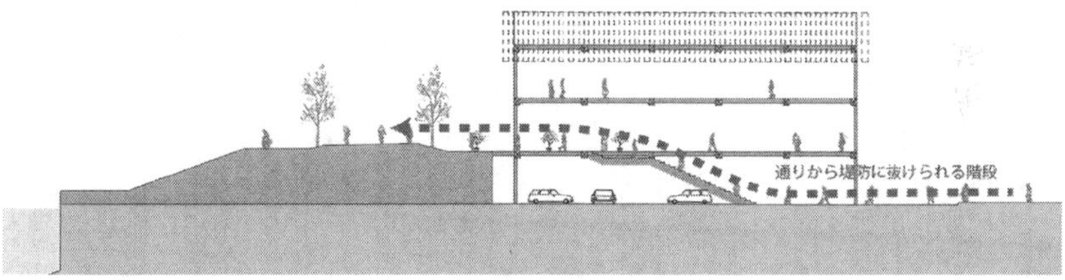

Tangible space formation

Now, let us examine the kinds of spaces that are taking shape in reality as these plans are formulated and these projects are brought forward in the face of such challenges. Here, after categorising zones as 'built-up or settled areas', 'group relocation sites' and 'low-lying areas', I will clarify the spatial images now taking shape after the passage of the four years since the disaster, as well as discuss their related challenges, based on land-use planning relating to reconstruction, regulations and the realities of projects' adoption. Here, 'group relocation sites' is meant to indicate housing complexes, especially those on higher ground, intended for the Disaster Prevention Group Relocation Project. In addition, 'low-lying areas' indicates those areas that were inundated by the recent tsunami. Specifically, in addition to disaster hazard zones, these also include land readjustment project zones for reconstruction in the original location, as well as areas that were affected but that were not slated for special urban planning area development projects (so-called 'white areas').

Macro-consolidation

Among the affected areas are many fishing villages along the Ria coastline, including some small settlements of only several dozen inhabitants. Many of these spaces are structured in such a way that residential spaces are integrated with the fishing port, docks and work houses. However, on the other hand, these smaller fishing villages are more likely to experience severe depopulation and demographic ageing, and even if there had not been a disaster, the sustainable management of the settlements and the difficulty of providing them with social services had given rise to the challenge of inefficiency.

In addition, the low-lying plains were scattered with farming villages that were facing similar challenges. Many of these rural villages have been forced to relocate in the wake of the recent disaster.

On such occasions, one option is to use the relocation as an opportunity to attempt a consolidation of numerous settlements. By doing so, it becomes possible to ensure a certain scale, thereby enabling the efficient provision of social services, in other words, to enjoy what from residents' perspectives are better social services. This has the effect of increasing the sustainability of such communities.

On the other hand, in view of the fact that the industries that support such communities are agriculture and fishing, this is detrimental in the sense that the fields, fishing ports and fishing grounds that underlie such industries will now be farther away. Furthermore, fishing villages, in particular, are characterised by a long history of competition for fishing grounds with neighbouring fishing villages, and do not therefore necessarily enjoy friendly relations with nearby villages.

Given this background, there are few areas that favour consolidation, particularly in the case of fishing villages, with reconstruction on higher ground in the vicinity of the original settlement being favoured as a general rule, no matter how small the village. In these communities, it seems likely that the question of how to maintain the settlements will become a challenge as demographic ageing advances to an even further stage. On the other hand, with regard to the farming villages on low-lying plains, it is rather the convenience of urban living on the outskirts of the metropolitan city of Sendai, like Natori and Iwanuma, that has been prioritised, and more than a few cases are also evident that have favoured relocation and consolidation to the vicinity of built-up areas away from the coast.

The formation of urban areas by individual reconstruction in existing urban settlements: coexistence of infill-style and sprawl development

Instead of the framework of the Disaster Prevention Group Relocation Promotion Project, there are also a number of cases in which disaster victims have taken advantage of a scheme known as 'Projects to Relocate Residences Away from Cliffs and Other Hazardous Areas' to purchase residential land on their own and rebuild independently. The impact of such initiatives on the urban spatial structure can be divided into two categories.

The first of these is infill-style development, namely, a case in which vacant sites or farmland inside built-up areas or existing settlements are converted to residential use. Places where such situations can clearly be seen, that is, examples that highlight only infill-style development rather than the sprawl development discussed later, include the cities of Sendai and Ishinomaki, as well as settlements along the Ria coastline. Metropolitan areas have designated urbanisation promotion areas and urbanisation restricted areas, and urban developments are, as a rule, prohibited in the latter areas. For this reason, development is necessarily restricted to urbanisation promotion areas, where it takes place on the vacant land therein. In the settlements along the Ria coastline, the fact that the natural topography means that usable land is restricted to within the existing settlements means that development is necessarily carried out inside the existing settlements, where it is not accompanied by large-scale construction under the auspices of the Disaster Prevention Group Relocation Promotion Project. Such developments may be described as desirable in terms of land-use planning since they bring a higher density of land use where such had previously become less dense as a result of demographic ageing.

On the other hand, a contrasting case can be seen in areas where sprawl development is progressing. This takes place primarily in small and mid-sized peri-urban areas where urbanisation restricted areas have not been designated. Here, because of the lack of development pressure as a general rule, there are very few regulations relating to land reclamation in ordinary situations, the use of buildings or the forms that they take. Therefore, while sprawl progressed historically against the background of motorisation, these conditions may now be said to be accelerating. This has led to an increase in infrastructural maintenance costs, the deterioration of the landscape and land-use problems such as inefficiency and mutual interference caused by the mixture of farmland and residential land.

Land-use challenges in group relocation areas

In order to obtain ministerial consent for a Disaster Prevention Group Relocation Promotion Project, there needs to be some amount of evidence that the demand for relocation actually exists. Thus, the polling of local citizens represents a basis for the calculation of such demand.

However, the inclination of residents changes significantly with the passage of time. In particular, even though they might have initially considered a move to higher ground as a form of 'hope', more and more people are abandoning such ideas in the face of the realities of factors such as the size of the financial burden and the rigours of acquiring financing from the banks. In addition, owing to the sharp rise in construction costs due to increases in the materials and labour costs associated with the reconstruction, there have also been some cases of people opting to delay such a move, even among households that have indicated an intention to rebuild on their own. These people choose to live in public houses instead of rebuilding their own houses. Furthermore, in some cases, people have relocated to other locations, including to other municipalities, and to rebuild their own houses by themselves since the development of relocation sites sometimes takes too long to wait, being more than six years.

Such changes in demand are extremely difficult to anticipate. Moreover, although some reductions in the scale of housing developments have been made in response to such changes in intention, responding to every such change is difficult and could lead to planning delays. For this reason, the initial scale planned for housing developments has become excessive, and vacant lots emerged sporadically from the beginning of construction. For example, despite the planned construction of 274 lots in Disaster Prevention Group Relocation Residential Housing Developments across three districts in the town of Yamamoto in Miyagi Prefecture on the basis of prior polling, as of March 2015, 32% of these remained vacant (Anonymous, 2015).

Notably, the sale of relocation housing developments, previously limited to disaster victims, has been opened to the public at large. While this is expected to lead to some degree of improvement in the state of vacancies in the relocation housing developments, the extent of such improvement remains unclear.

In addition, a not insignificant proportion of households that were relocated in the first place were those of elderly residents. As a result, the demographic ageing of the relocation housing complexes is proceeding extremely rapidly, and this is soon expected to be accompanied by the appearance of vacant lots and vacant houses. On the other hand, at least in view of recent trends, it hardly seems an exaggeration that there are barely any communities in which there is expected to be new demand to fill these vacancies. In other words, many of these vacant lots and houses represent properties for which there is no outlook for reuse on any permanent basis.

It is necessary, therefore, for the local community to manage the use of land. Land-use management is indispensable to avoid the external diseconomies caused by the abandoned vacant lots and to keep the local environment optimal. In such cases, the separation of the right to own and the right to use becomes a critical issue, although the details of this issue will not be discussed here.

Land use in low-lying areas

There are a variety of projects planned for low-lying areas hit by the recent tsunami. However, these must confront land-use inefficiencies resulting from the uneven distribution of public and private land and the original lack of demand for its uses, which have resulted in the issue of low-density land use. In the following, let us look at these issues in turn.

First of all, there are the so-called 'white areas'. These are areas that were inundated by the recent tsunami, but which have been determined to be safe for construction due to the construction of seawalls or so-called secondary embankments (ie the anticipated inundation depth is generally less than 2m in L2 simulations following the construction of the seawalls), and that have not been slated for special urban planning area development projects as the result of having been developed to some degree with basic urban infrastructure. Individual and voluntary reconstruction has been possible since immediately after the disaster, and repairs and reconstruction proceeded quickly in regions where the degree of damage was relatively small, such as in cases where the inundation remained below or only slightly above floor level, or where the flow of the tsunami current was slowed. However, on the other hand, depending on the area, there are also some communities that despite having been devastated by the recent tsunami, are now safe even in L2 tsunami simulations as the result of seawall construction and have been allowed to remain as they are. Ishinomaki's Watanoha district is a typical example of such an area (see Figure 25).

Next, there are land readjustment project implementation zones, where reconstruction is carried out in the original location. Here,

Figure 25: The actual situation of Watanoha District in Ishinomaki City

Source: Photo by author

despite having suffered catastrophic damage in the recent tsunami, the development of basic urban infrastructure is carried out under the auspices of land readjustment projects to improve the residential environment over what had existed previously, with the aim of providing an incentive to disaster victims so that they will have the intention to continue residing in the area. In addition, since the development of public facilities such as parks and roads is carried out without a significant increase in land prices, these become, in effect, depreciation compensation zones. Furthermore, because it is normal to build disaster public housing in these zones, the purchase of land for this purpose is also carried out using public funds. This also has the added effect of being in keeping with the intentions of disaster victims who wish to sell their land in the area.

Nevertheless, depending on the location, in some cases, the total area of land owned by victims who hope to relocate exceeds the total area of lands that are purchased for public use based on these arguments. In such cases, the public is obviously unable to purchase all of the available land. Despite the fact that, at this stage, there can be no predicting whether the lands that residents had hoped to sell but were not bought will be put to effective use or else abandoned as vacant lots once the project has run its course, the latter possibility cannot be discarded. Moreover, although an individual may continue ownership for the time being owing to a 'Japanese obsession' with property, in some cases, the decisions as to how such properties will be used will not have been made because the owner in question has already rebuilt elsewhere. Those who wish to sell their property or relocate outside their location account for approximately 43% of property owners for in-land readjustment projects across the entire affected area, while those who wish to continue living where they are, or who intend to retain their property, account for approximately 50% (Anonymous, 2014). Thus, despite ongoing land readjustment projects, low-density urban areas are gradually taking shape without land being effectively put to use.

Third, despite the designation of disaster hazard zones, in some cases, rather than prohibiting the construction of housing across the board, the contents of building regulations permit construction – housing included – on conditions such as the height of ground floor or ground level. In Sendai, it has been decided that residential victims in L2-inundation prediction zones outside of disaster hazard zones are entitled to various subsidies to pay for the independent reconstruction of housing in the city's development zones, and the framework is the same.

In such cases, on the one hand, if residents hope to relocate, they will be able to sell their property and move ahead with relocation while receiving assistance under the framework of the Disaster Prevention Group Relocation Promotion Project. On the other hand, if they wish to rebuild in their current location, it is possible to repair or rebuild their residence. In other words, individual landowners can select options in line with their own intentions. Conversely, however, situations have occurred in which lots on which houses have been rebuilt have become mixed up with lots that have been turned into vacant properties and sold to municipalities by dismantling their land-use restrictions. In particular, it has been found that extremely low-density land use is taking place in Nobiru district, for example (see Figure 26). This is conceivably because those who wished to rebuild in their original properties on their own have already rebuilt, and it seems unlikely that further homes will be rebuilt in future. Quite to the contrary, there is a high possibility that a situation like this could become permanent. The distortion of adjustments to regulatory and project frameworks affecting land use to be able to fulfil victims' desires as much as possible may thus be said to be impacting subsequent land use.

The problem of making use of the original areas targeted by the Disaster Prevention Group Relocation Promotion Project

In addition to the disaster hazard zones in which the reconstruction of residences in their original location is permitted, mentioned at

Figure 26: The actual situation of Watanoha district in Higashimatsushima City

Source: Photo by author

the end of the previous section, the problem of land ownership also remains even in disaster zones, where residential usage is prohibited. In conjunction with the designation of disaster hazard zones, municipalities purchase those sites that are the objective of Disaster Prevention Group Relocation Promotion Projects. This signifies a considerable increase in municipally owned lands, whose maintenance, and if left unused (such as dealing with weeds), will come to represent a large financial burden.

A further problem is that lands that are available for purchase are strictly limited to lands that had previously been used residentially. Lands that had previously been used for commercial and industrial purposes, or were otherwise parking lots or farmland, are not the objective of the Disaster Prevention Group Relocation Promotion Project and are not subject to public purchase. This means that they will remain the property of their original owners. For this reason, lands that are private property and lands that are purchased by municipalities have become intermingled. Accordingly, this has led to difficulties in the wholesale utilisation of lands acquired by municipalities.

Therefore, first of all, the challenge is how to proceed with the consolidation of land. In that case, while it is necessary to carry out land readjustment projects, the biggest problem is the question of whether there is enough demand for land use that would produce benefits commensurate with the costs of carrying out said projects. In addition, on top of the demand for land use, other data points that should be considered, for instance, include the reduction of maintenance costs associated with consolidation and the cost savings associated with the abolition of roads and other infrastructure. That having been said, there are not many places where such conditions are met. For example, in the areas around the harbour in metropolitan areas like Sendai, where there is any expectation of land use that is to some degree industrial, commercialisation is already under way (see Figure 27). In small- and medium-sized cities, however, such commercialisation is a difficult prospect, there being few places where there is any concrete expectation of sufficient demand.

In such cases, it is necessary to consider land use in temporally and spatially small units. That is to say, while the previous solution consolidated land on a large scale and assumed that it would be used over the relatively long term, in cases where this proves difficult, it is possible to conceive of utilising land in small spatial units in the relatively short term.

For example, in the Ushihama and Hamaichi districts of Higashimatsushima City, the city office is replacing the topsoil on the site of its former residential buildings to be converted for use as farmland. This will be leased free of charge for 10 years to a local agricultural corporation, which will use these 5 hectares of land to

Figure 27: Land Readjustment Project Blueprint for the Gamo Hokubu district, Sendai City

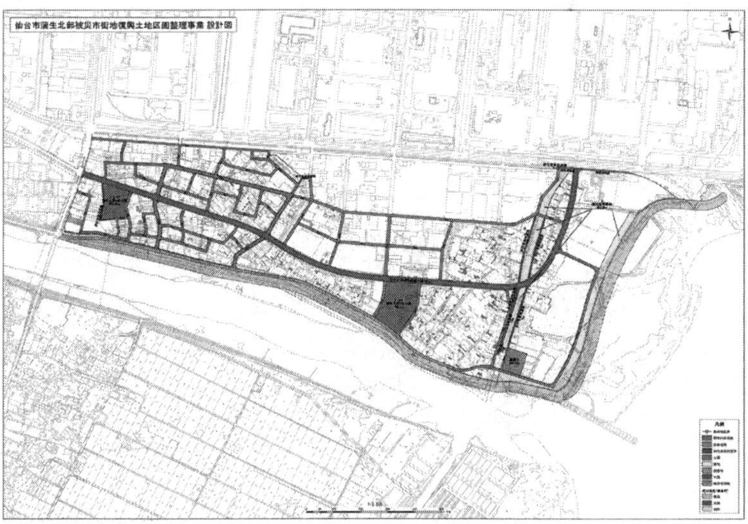

Source: Sendai City Materials

cultivate soybeans using existing agricultural machinery, without any new capital investment. Although such residential sites are scattered throughout the district, since the scale of residential land in the original agricultural settlement was around 1,000 m^2, this hardly represents a major operational problem. That said, converting land for agricultural use in this way can face challenges in terms of poor profitability and a scarcity of labour. Absent expected demand for land as populations decline, it could be said that there is a need to give an active role to the proliferation of such small-scale short-term utilisations.

Conclusion

The Great East Japan Earthquake was the largest natural disaster to strike Japan since the end of the Second World War. The reconstruction effort pursuits, in addition to the safety and secureness, sustainability through a paradigm shift in urban planning. Regarding the former aspect, resilient urban and rural spaces that correspond to the tsunami risk of once in several hundred years are now being developed. Regarding the latter aspect, the density of some urban settlements, in which vacant lots and houses were prevalent because of population decline, is getting higher. Other cases, however, face difficulties, especially from the latter aspect, for example, the difficulty of respect for district-level decision-making and project coordination, as well as the expansion of sprawl development and the formation of low-density urban areas in the affected and relocation areas. However, even faced with such problems, and despite criticism that it has been delayed, the reconstruction effort has made steady progress.

It is precisely by overcoming such challenges that a contribution may be made to ordinary urban planning, namely, the formation of urban spaces that will be sustainable in an era of a declining population.

References

Anonymous (2015), 'Transferring the lands in Yamamoto town', *Kahoku Shimpo*, 20 April, p 16

Anonymous (2014), '40% will not "live" after the land readjustment projects', *Yomiuri Shimbun*, 9 March, p 1c

Central Disaster Management Council (2011), *Report of the Committee for Technical Investigation on Countermeasures for Earthquakes and Tsunamis Based on the Lessons Learned from the '2011 off the Pacific coast of Tohoku Earthquake'*, Cabinet Office, Government of Japan

CHAPTER A2a

Politics in spatial planning in Aceh recovery post-tsunami 2004

Togu Santoso Pardede and Gita Chandrika Munandar

Indonesia	
Population: 255,461,700 (2015 estimate)[1]	**Government:** Unitary presidential constitutional republic
Area: 1.9 million sq km (742,308 sq miles)[2]	**Administrative structure:** Indonesia consists of 34 provinces, five of which have special status. Each province has its own legislature and governor. The provinces are subdivided into regencies (*kabupaten*) and cities (*kota*), which are further subdivided into districts (*kecamatan*), and again into administrative villages. Village is the lowest level of government administration in Indonesia.
Hazard profile: Landslides (Nation wide), Volcanic eruptions (along the ring of fire: Sumatra, Java, Nusa Tenggara, Sulawesi), Tsunamis (coastal areas face Indian Ocean: Sumatra Jawa, Bali-Nusa Tenggara, Mollucas) Earthquakes (nationwide) River floods (nationwide) Drought (Kalimantan ,Sulawesi, Papua) Forest fires (Sumatra, Kalimantan, Jawa, NTT)	**Authorities in charge of risk assessment and management:** National Disaster Management Authority (BNPB) on national level and Local Disaster Management Authority (BPBD) on provincial and regencies/ cities level.
Analysed event: Earthquake in the Indian Ocean (8.9 on the Richter scale) followed by tsunami in Aceh (26 December 2004. Second earthquake (9.1 on the Richter scale) followed by tsunami in Simeulue Island (Aceh) and Nias (North Sumatra) (26 March 2005) Estimated return period: Estimated damages: $4.45 billion (equal to 97% of Aceh's GDP) Casualties: 170,000 people were killed and about 500,000 were left homeless (1st earthquake & tsunami) 905 people were killed and displaced tens of thousands people (2nd earthquake & tsunami)	**Role of spatial planning:** Responsible for risk management of some hazard that are spatially relevant (volcanic eruptions, earthquakes, tsunamis, floods, drought, and landslides). Responsible for post-disaster recovery. Although now the focus shift to the prevention phase (mitigation).

Notes

1 https://en.wikipedia.org/wiki/Indonesia#Administrative_divisions
2 http://www.bbc.com/news/world-asia-pacific-14921238

Introduction

When Aceh was hit by an earthquake followed by a devastating tsunami in 2004, the government of Indonesia introduced a master plan as the basis for the rehabilitation and reconstruction efforts in Aceh to create a post-disaster resilient community. Spatial planning policies and participatory planning processes were adopted during the reconstruction. Buffer zone and infrastructure development, as well as other mitigation efforts, were designed. Community-based village plans were adopted and implemented in all of the affected areas.

However, almost a decade after the tsunami, the cities that were destroyed by the tsunami in 2004 were revived at the same sites as before the tsunami. Plans to relocate the communities to a safer place almost did not happen. Plans to rebuild spaces that are more resilient to disaster were not achieved either. An earthquake of 7.9 on the Richter scale in April 2012 proved the failure of the spatial-based disaster mitigation in Aceh. People were running around frantically, not knowing where to go. The Early Warning System did not work; evacuation routes and shelters were not used properly.

On the other hand, the speed of the reconstruction process in Aceh received a lot of praise. It was considered as having been done in a much faster way than the reconstruction process in other disaster-affected areas in the world, such as New Orleans, USA (Taufiqurrachman, 2006), and Sendai, Japan. In Aceh, three years after the tsunami, most of the victims of the tsunami had been resettled. The economy of the city of Banda Aceh was recovered quickly, even exceeding the conditions prior to the tsunami. The population of Banda Aceh was also increased rapidly. After being hit by the tsunami, the population of Banda Aceh decreased from 265,098 in 2004 to 177,881 in 2005, but began to increase again in 2006 and finally achieved 228,562 people in 2011 (see Figure 28).

While the reconstruction process in Sendai-Tohoku was criticised for its slowness and had to deal with a population decline in the affected areas, three years after the tsunami, almost all damaged municipalities have just completed examining and designing their recovery plan according to their circumstances. They are taking the time to plan carefully to make sure that the cities hit by the tsunami will be resilient against future tsunamis.

This chapter argues that the socio-political conditions in Aceh prior to the 2004 tsunami had a significant influence on the decisions determining the rehabilitation and reconstruction efforts to be done. After being hit by the tsunami in 2004, the socio-political situation

Figure 28: Population in Banda Aceh, 2001–2011

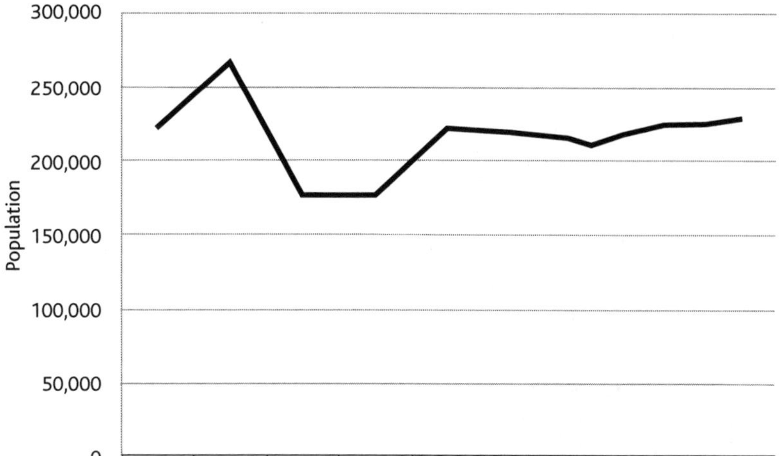

	2001	2004	2005	2006	2007	2008	2009	2010	2011
Total BA	223,223	265,098	177,881	178,380	219,659	217,918	212,221	234,448	228,562

Source: Author, BPS Aceh (2011)

in Aceh became more complicated since this region had been ravaged by conflict for 32 years. Given the socio-political background, the Rehabilitation and Reconstruction Agency (BRR) tends to avoid conflict with society. Although the government of Indonesia had initially formulated a master plan for the rehabilitation and reconstruction of Aceh, the BRR ultimately decided to use a village plan – a community-based spatial plan at village level – as a basis to rebuild settlements that were destroyed by the tsunami. Village plans were implemented on a massive scale throughout the entire affected areas. The authors argue that, despite the participation of the community, mitigation efforts to rebuild spatial resilience after the disaster in Aceh have not met the target of building resilient communities. Politicisation affected the implementation of spatial plans. The community and government (BRR) tended to look at short-term needs and interests, while overlooking the long-term ones. A technocratic approach and plan must be tailored to the socio-political conditions of the time in order to be implemented without a hitch. For a resilient Aceh in the future, collaboration and coordination between multiple stakeholders, especially the government (central, provincial, local) and community, is a crucial factor.

Literature on and practice of land use for post-disaster recovery

During the first decade of the millennium, there has been an increasing number of disasters, many of which were high-impact disasters (intensive risk events), namely, Gujarat Earthquake, India (2001), Indian Ocean Tsunami (2004), Hurricane Katrina, USA (2005), Kashmir Earthquake, Pakistan (2005), Yogyakarta Earthquake, Indonesia (2006), Shenchuan Earthquake, China (2008), Cyclone Nargis, Myanmar (2008), Haiti Earthquake (2010) and, most recently, the Great East Japan Earthquake and Tsunami (2011) and Typhoon Yolanda, Philippines (2013).

The planning for rebuilding a city/community after disaster is generally a dilemmatic and complex process. There is a trade-off between speed and quality, quick action and broad participation, human interest and political interest. It is a complex process, compressing all aspects of urban development into a short time period. The planning process in a post-disaster situation is somewhat different than normal development planning. Based on academic studies and planning practice, Olshansky and Chang (2009) provided a summary of the indicators of successful recoveries as follows: substantial external funding, provided quickly and with few restrictions; strong local leadership; cooperation between city, state and federal officials; local, citizen-based processes for making and reviewing reconstruction decisions; previous planning documents that describe consensus policies for future development; and pre-existing planning institutions. Recovery processes are complex and unique to location, time and context. It is not possible to measure the length of the process or to identify the endpoint of recovery (Olshansky and Johnson, 2010). People are not in 'ideal' condition, either physically or mentally. It is difficult to respond to the real needs of those people in such a condition. Thus, post-disaster planning is different from normal planning because in normal planning, people are not under the extreme stress that they are in disaster planning.

During the reconstruction process, there is a good opportunity to engage in activity that will increase the level of development and reduce vulnerability to future disasters (Berke et al, 1993; Milleti, 1999). 'Building Back Better' during post-disaster recovery and reconstruction should address the underlying vulnerabilities and should avoid 'emergency reconstruction' efforts. Missing such opportunities will expose communities to future risks and trap them into a disaster cycle. As the primary tool for hazard mitigation at the community level, effective spatial planning is critically important to build community resilience. The regulation of the spatial plan can reduce the exposure of residents to natural hazards (Olshansky and Kartez, 1998; World Bank, 2010). The *Sendai framework for disaster risk reduction 2015–2030* promotes the incorporation of disaster risk management into the post-disaster recovery and rehabilitation process through land-use planning (UNISDR, 2015). Spatial planning is an essential component of a community's long-term resilience.

However, in practice, there is a challenge in implementing disaster risk reduction during recovery, especially the implementation of the spatial plan in Aceh, to convince residents to relocate to less hazardous areas in the neighbourhood or within the city as planned. The majority of residents in Banda Aceh resisted being relocated to higher ground or safer places prepared by the government. They preferred to stay on their original land due to their emotional ties to the land and their livelihood, while refusing to see the ongoing dangers of their situation.

Aceh reconstruction efforts

Massively damaged conditions in the tsunami-affected areas required the development of cities and regions with comprehensive planning that considered aspects of sustainable development and disaster mitigation in order to build spatial resilience after disasters in the future. The government of Indonesia divided the rehabilitation and reconstruction of Aceh into four stages (see Figure 29):

1. *The Emergency Stage* (26 December 2004–28 March 2005). The National Coordination Agency for Disaster Management (Bakornas PB) conducted the emergency response, and the Coordinating Minister of People Welfare coordinated all stakeholders. As the emergency stage ended, the President appointed the Minister of the National Development Planning Agency (BAPPENAS) to prepare the rehabilitation and reconstruction plan (Master Plan) and the institution to implement the plan.
2. *Rehabilitation and Reconstruction Stage* (April 2005–2008). At this stage, a mandate was given to the BRR (Aceh-Nias) to coordinate and execute the recovery efforts in Aceh and Nias (North Sumatra) based on the Master Plan. In 2007, the plan was reviewed and enacted as Action Plan (2008–09) to be followed until April 2009.
3. *Transition and Continuation of Rehabilitation and Reconstruction Stage* (April 2009–2012). The transition stage started from April 2009 until the end of December 2009, to be continued until the end of December 2012. Post-BRR in April 2009, when coordination

Figure 29: Planning and Policy Framework for Aceh Reconstruction

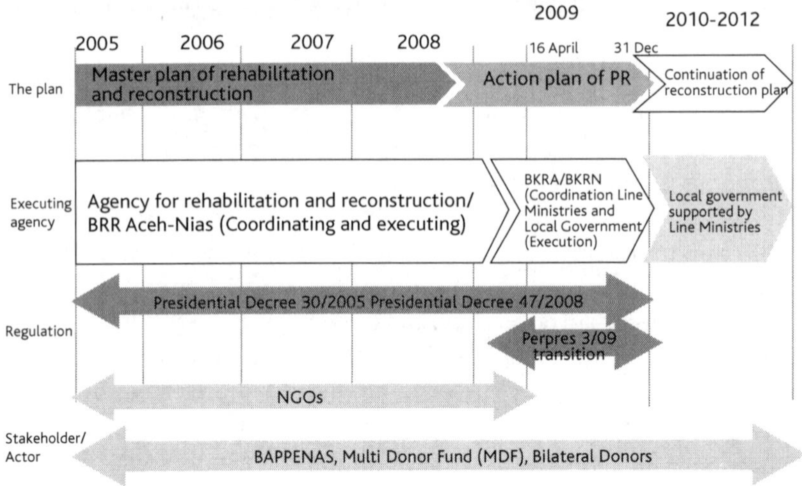

Source: Author, Republic of Indonesia (2009)

returned to normal, the coordination of the government/ line ministries' works during the stage of continuation of rehabilitation and reconstruction was conducted by BAPPENAS at the national level and by ad hoc agencies of reconstruction (BKRA and BKRN) at the provincial level. The Completion and Continuation of Rehabilitation and Reconstruction Action Plan (2010–12) was created as a guideline to complete the works until 2012.

4. *Beyond 2012.* Back to regular development by local government.

The development of the Master Plan of Rehabilitation and Reconstruction

Soon after the tsunami, the national government took actions directly to handle Aceh. The local government was disabled as most of its officers had been victims. Within two weeks, the national government, together with the international community, prepared damage and loss assessments with findings that still remain the best overall evaluation of the disaster's impact. Based on this assessment, under the leadership of BAPPENAS, a Master Plan of Rehabilitation and Reconstruction was formulated on 26 March 2005 using a multi-stakeholder approach. It involved a wide range of stakeholders, including international donors, line ministries, local government representatives, civil society groups, religious leaders of Aceh and local and national universities, organised around 10 different thematic groups and working simultaneously in Jakarta and Banda Aceh. The Master Plan was then legalised on 15 April 2005 as Presidential Decree No 30 Year 2005.

The Master Plan consisted of 12 books: the Main Book of Master Plan and 11 detailed sectoral books.[1] One of the 11 sectoral books consisted of a spatial plan of Aceh Province, Banda Aceh City and eight other districts, as well as three main cities, which were heavily damaged. All activities of the sectoral ministries concerned were accommodated in the 11 books, which would be implemented within four years. The spatial plans that were developed were used as guidelines from the national government during the reconstruction process, which would then be used as a reference to revise the existing provincial and local spatial plans.

The establishment of the Rehabilitation and Reconstruction Agency

On 30 April 2005, a single agency with full authority to coordinate and implement the rehabilitation and reconstruction of Aceh was established, the BRR, led by Kuntoro Mangkusubroto. The BRR was given a four-year mandate (ending in April 2009) to coordinate and implement the rehabilitation and reconstruction using the Master Plan as a basic framework. The establishment of this agency was aimed to minimise bureaucracy, delegate authority, increase efficiency and speed up the process of rehabilitation and reconstruction without compromising the government's safeguards and integrity.

There was an evident tension between the need to show quick results and the need for careful and comprehensive planning. The scale and scope of the disaster and the involvement of so many agencies, such as all line ministries, the private sector, donor agencies and non-governmental organisation (NGOs), compounded this complexity. The establishment of the BRR as a single agency with full authority to coordinate and implement the rehabilitation and reconstruction of Aceh was expected to address the complexity of the coordination issue. Due to its unique characteristics (experimentation, flexibility, limited duration, smaller size and full authority), the BRR could be an ideal vehicle for experimentation within the restrictions that regulate government action.

Spatial planning policies in the Aceh reconstruction

According to 'Book 2: spatial plan and land affairs', the spatial planning policies in the reconstruction efforts included (Bappenas, 2005):

1. Creating a safer and better life. Disaster-prone areas should be equipped with mitigation infrastructure such as escape routes, escape building and an early warning system.
2. Giving citizens the freedom of choice in settling down (at relocation areas or original places).
3. Involving local community and social institutions in disaster management and recovery efforts.
4. Highlighting cultural and religious characteristics.
5. The spatial planning process being a combination of top-down and bottom-up approach (participative).
6. Restoring local governments' role in spatial planning. The concept of spatial planning in the Master Plan was prepared by the national government because the local government (at that time) was not fully functional. Furthermore, the local government can use the concept of spatial planning in the Master Plan to finalise the planning process and continue the legal process of the definitive local spatial plan.
7. Protecting citizens' civil and land rights.
8. Accelerating the land administration process.
9. Providing fair and affordable compensation.
10. Revitalising the economic activities of the community.
11. Restoring environmental-supporting capacity.
12. Rehabilitating the spatial structure and pattern in Aceh Province that were damaged by the tsunami to rebuild the linkage between western and eastern coastal cities in order to increase equitable development.
13. Rebuilding the cities towards economy, physical and social resilience involving the local community.

The spatial plan provided a spatial guideline for general policies on rehabilitation and reconstruction in Aceh, which included: (1) rebuilding communities; (2) rebuilding the economy; (3) rebuilding infrastructure and housing development; and (4) rebuilding the governance. The general policies were translated into action plans in the emergency response phase and rehabilitation and reconstruction phase by considering the direction of spatial plan/spatial integration in the Aceh and Nias (North Sumatra) regions based on the principle of sustainable development. Furthermore, the integration of sectoral and regional policies was translated into work plans based on location, activities undertaken, person/institution in charge, implementation time and source of funding. Every phase – policy and strategy formulation, regional development, the formulation of work plans, implementation, and monitoring and evaluation – was done through the involvement of the community's aspirations, hope and participation (see Figure 30).

Figure 30: Planning framework for the rehabilitation and reconstruction of Aceh and Nias (North Sumatra)

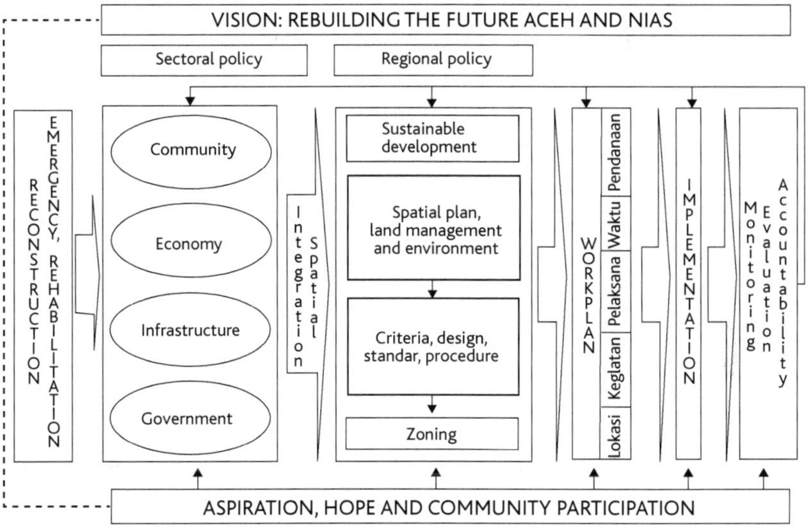

Source: Main Book Master Plan (Bappenas, 2005)

Implementation of the spatial plan

There was resistance from the community to the spatial plan formulated in the Master Plan. The plan was considered too macro and difficult to implement in the field. The resistance occurred particularly because of the buffer zone policy that prohibited construction within two kilometres of the coast. The Indonesian government planned to divide the two-kilometre area into three zones. The first zone, comprising of mangroves, palm trees and pine trees, would be separated from the sea by break walls and were extended to 300 metres inland. The second zone, within which only fishermen would be permitted to live, would extend a further 1.6 kilometres inland and included the construction of some power generators and infrastructure. Trees would be planted in the third zone, a 100-metre zone on the edge of Aceh's coastal towns and cities. The land-clearing process could present problems because many landowners did not agree with the buffer zone policy. Some had tried to rebuild their homes or had stuck flags in the ground to mark their property within the buffer zone, while other groups had sent petitions opposing the relocation plan.

After the Master Plan was legalised as Presidential Decree No 30 Year 2005 and was going to be implemented, conflicts occurred, for example:

1. Conflict between an ideal spatial plan and the community's wish to be quickly resettled. The Master Plan mandated the need to reformulate the local spatial plan (city/regency) to include disaster mitigation in order to reduce the risk of future disaster. However, the revision and preparation of the local spatial plan (Banda Aceh City and Aceh Province) until it was legalised into local regulation took quite a long time because it had to go through the process of ratification by the local parliament. The people could not wait until the spatial plan was reformulated and legalised to rebuild their home towns. In this case, Kuntoro Mangkusubroto –the head of the BRR – supported the community to rebuild their settlements without the existence of a local spatial plan: "Nothing is the role of [spatial plan] province or district plan. It doesn't work. I forget them. It's a political process that means it must go to local parliament process. We can't control that political process" (Interview, 2010).
2. Conflict between the Master Plan and the BRR strategy. The BRR reconstruction strategy on spatial planning did not entirely follow the Master Plan. The village planning concept implemented by the BRR was a concept that was developed outside the defined spatial planning within the Master Plan.

According to the Law on Spatial Planning in Indonesia, spatial planning was the authority of local government. This was the biggest obstacle to the BRR. Local government initiative was not as fast as needed to follow the pace of reconstruction. The BRR had helped the local government in developing a spatial plan through the deliverance of several studies related to the spatial plan, but the process of the legalisation and implementation of the plan had not begun. Village spatial planning was a breakthrough for the BRR as it was considered a middle ground that could resolve the need for rapid spatial implementation. The Village Spatial Plan or Village Plan (VP) was formulated in a short amount of time by the community accompanied by facilitators, legalised only by approval from the *Keuchik* (village leader) and operationalised as a reference for reconstruction and development in the village. This was also in line with the mandate of the Master Plan to accommodate community participation in the reconstruction process.

Village Spatial Plan

The VP is a community-based spatial plan and community recovery plan. The Head of the BRR issued guidelines for VP through BRR Decree No 1/1.02/01.01/2005 in June 2005 to all parties involved in the process of village planning throughout the Aceh region. These guidelines were published six months after the tsunami hit Aceh, whereas some housing assistance from NGOs had already been conducted. It covered around 647 villages in Aceh Province and 63 villages in Banda Aceh City in less than two years. The partners of village planning included UN Habitat– UNDP (UN Development Program), United States Aid (USAID), Asian Development Bank (ADB), Australian Aid (AUSAID), German Technical Coorporation Agency (GTZ), Mercycorps, Yasasan Inovasi Pemerintah Daerah (YIPD) (land mapping) and the BRR (see Figure 31). Each donor agency had its own standard/guideline; however, a common standard was achieved and the BRR guideline was used as the minimum standard. This might be the first large-scale community planning effort in world history in terms of the number of agencies, time and coverage involved (Pardede and Kidokoro, 2008).[2]

An example of the VP in Lamjabat Village (USAID model) is shown in Figure 32.

Figure 31: Donors' contribution to village planning in Banda Aceh City and Aceh Province

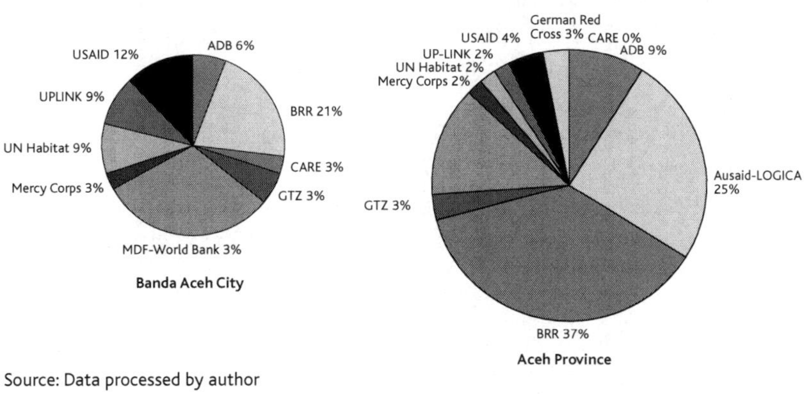

Source: Data processed by author

Figure 32: An example of the Village Plan: Lamjabat Village (USAID Model)

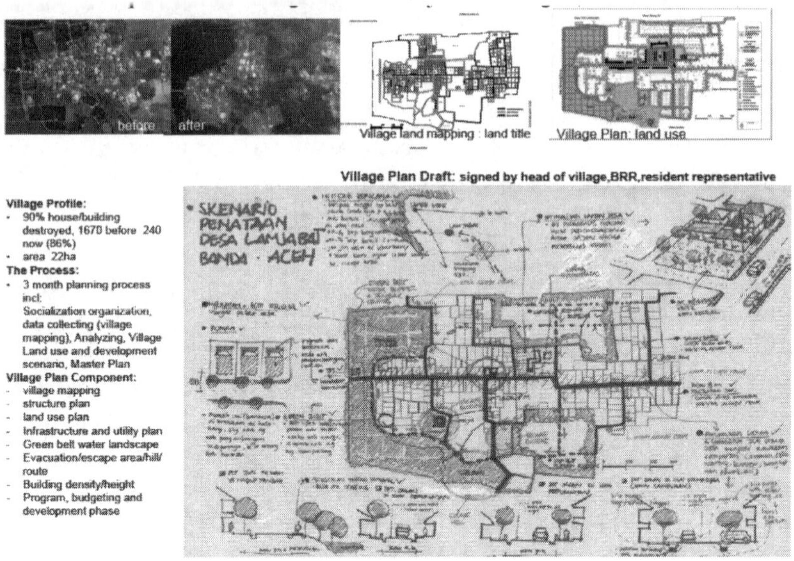

Source: Author; BRR (2005)

Discussion

Village planning: the politics of community participation planning

Village planning could provide a quick overview of the local situation from the viewpoint of the people. It also helped to transfer the communities' needs and wishes to decision-makers and planners in a direct way so that it motivated the communities and mobilised the implementation as well. Overall, it strengthened bottom-up processes. Furthermore, this participation during the planning process served as a healing process for the community since they were kept busy with making plans that could help them to forget the trauma and sadness of the past, and focus more on what they hoped for the future for their family and community. However, village planning also faced challenges, such as: social jealousy within communities or between villages; reconstruction delays; and the varied quality of houses due to land-title issues, bad contractors and institutional problems.

The usage of village planning as a basis for the reconstruction of settlements in Aceh, although considered a breakthrough and deemed appropriate for the conditions of that time, also had its drawbacks, such as:

- The quality of reconstruction of the villages was different from one village to another. This happened because the quality of the developed VPs varied between villages. The quality of the VP was determined by several factors; one of the factors was the participation of the community. In some villages, the community was eager and better prepared to participate, while in other villages, the community did not have the spirit or the patience to participate. S. Lalu (interview, March 2009), coordinator of Village Spatial Planning Ausaid, Local Governance and Infrastructure Community in Aceh (LOGICA), stated that village planning was intended to increase community participation in formulating settlement planning. However, apparently not all village planning was developed with community participation. Other factors included the quality of the facilitators and contractors. All these factors affected the quality of reconstruction in the villages.

- The lack of planning linkages between villages and with the meso/macro-planning (city) level. The VP was developed by the community itself with the help of facilitators. The VP was completed later and offered by the *Keuchik* (village leader) to NGOs/donors who were willing to do the reconstruction in accordance with the VP. Interaction and coordination only occurred between the communities with the facilitator, but there was no coordination with the surrounding villages, with the local government responsible for development in larger areas (district, city level), the contractor/NGO who built the houses, or the BRR who built the infrastructure. Therefore, after all the villages were redeveloped, we could find a few things that were not in sync (road networks that were not well connected, housing arrangements that were not suitable, etc). This was due to the non-existence of a macro-plan that could be used as a reference for the development of infrastructure networks, as the forming of city structure, and therefore the connectivity between villages or between land uses, became unclear (Dercon and Kusumawidjaya, 2007). In this context, village spatial planning cannot be used solely without a city or district spatial plan. A VP does not contain a comprehensive analysis of the linkages between villages or with a wider area. Therefore, a wider plan covering comprehensive analysis is necessary. This is especially important when many stakeholders are involved at the same time (NGOs, donors, community and government). Communication and coordination between stakeholders during the design and implementation of the plan is a must.

- The loss of golden time to develop a safer settlement from future tsunami. After the tsunami, the people actually agreed to be moved to a safer place, outside the buffer zone. M. Nurdin, Mayor of Banda Aceh, stated that at that time (from six months to one year after), the community was still traumatised by the tsunami (interview, July 2011). However, the relocation could not be realised because of several reasons. One reason was that the long-term benefits of spatial planning in the Master Plan and disaster mitigation through the implementation of the buffer zone was not well explained or campaigned for. Lack of communication led to the Master Plan being poorly understood. Only a plan that is understood by the stakeholders will be accepted, and can therefore be implemented. Another reason was the delay in the disbursement of funds for the purchase of land for the relocation of the community. Thus, many of the residents whose homes were destroyed went back to their original location in the disaster-prone areas and rebuilt their houses there since there were abundant funds offered by NGOs/donors.

The drawbacks of the VP were partly due to its formulation, which did not follow the rule as defined in Law No 24 Year 1992 on spatial planning. According to Law No 24 Year 1992 on spatial planning, a provincial spatial plan is used as a reference for city and regency spatial plans, which, in turn, become the basis for the formulation of a detailed plan. However, the BRR made the plan in the opposite way. They started with the VP in 2005–06, followed by sub-district plans in 2007, and finally the city/regency plan in 2009. Edy Purwanto (interview, January 2012), the Deputy Operational of the BRR, stated that the BRR had developed spatial plans with different approaches since the normal process would take a long time.

Political dilemma leads to failure in mitigation efforts

Disaster mitigation for the long-term future was not embodied in the reconstruction in Banda Aceh. For example, one of the policies and strategies in the Master Plan was to build residential areas away from the coast in disaster-prone areas (there is a buffer zone), but what happened was that people moved back to their original locations and built housing settlements on the coast. In another example, the Master Plan (March 2005) and the Banda Aceh City Spatial Plan formulated by Japan International Cooperation Agency (JICA) (August 2005) indicated a ring road construction plan. The ring road development plan had two functions: to reduce congestion in the city centre; and to be the boundary for settlement development. The road could also function as an embankment for future tsunamis. Shimizu, a consultant for JICA who developed the Spatial Plan of Banda Aceh (interview, 2008), stated that the plan was rejected by the BRR for political reasons. The BRR worried that the ring road would become the boundary of the buffer zone, which was a sensitive issue at that time.

Kuntoro Mangkusubroto stated that in performing its duties, the BRR tried to avoid conflict with the public as much as possible because it could hinder the process of reconstruction (Teleconference with Tohoku University–JICA, 4 April 2011). Efforts to avoid and reduce conflict were undertaken by the BRR, for example, by recruiting media workers and ex-Free Aceh Movement (GAM) members into the BRR. Kuntoro had full authority to manage the recovery process.[3] He also had full authority to manage the BRR as much as needed in accordance to the situation and conditions of that time without the need to have permission from Jakarta. The political decision to avoid conflict had an impact on the implementation of the Master Plan. The Master Plan, which included disaster mitigation such as the development of a buffer zone and the relocation of people from disaster-prone areas, was not implemented entirely. However, what happened was that people were still living in disaster-prone areas. An earthquake that hit Banda Aceh City on 11 April 2012 caused panic in the community. They did not use the evacuation route or the rescue building.[4] Public responses to the tsunami warning were varied; even the siren did not work in Banda Aceh (BNPB, 2012). It shows that community resilience to disasters has not been formed.

Science-based planning and community-based participatory planning

Science-based planning is suitable for long-term and larger-scale (macro and meso) planning, while community-based planning is more suitable for short-term and small-scale planning. When village planning was chosen as the basis for development in the reconstruction process in Aceh, it was a decision to prioritise community-based planning. This is different than the approach in Japan. The post-disaster reconstruction process in Sendai-Tohoku was done by science-based planning, in which the recovery policies, as well as land-use policy, were obtained from intensive research by the university. The community plan was only started after the macro-plan was finished. Planning was done on the national, prefecture and city level (with input from scientist). Similarly, in New Orleans after Hurricane Katrina in 2005, planning began with several plans at the macro-level (state and city) and then continued with the plan at the district and neighbourhood level. In Aceh, the planning was done at the national/provincial and village level, passing the plan at city/regency level.

Since the local government in Aceh became paralysed by the tsunami, the President commissioned the National Planning and Development Agency to formulate a multi-sector rehabilitation and reconstruction plan in a short amount of time (three months) to be used immediately after the emergency phase was completed. Consequently, the formulation of the Master Plan did not involve the participation of the people of Aceh and was a more technocratic and top-down approach. This is in contrast with Japan and the US, which enabled massive and structured community participation for both macro- and micro-plans, thus requiring a longer time to formulate the plans.

Every attempt at recovery is faced with major issues that make the recovery efforts unique. In Aceh, there was the issue of a long-term armed conflict between the GAM and the government of Indonesia, which caused the people of Aceh to distrust the Indonesian government in rebuilding Aceh after the tsunami. In Japan, a large number of elderly residents, and the declining number of residents in the areas hit by the tsunami, led to a dilemma in rebuilding tsunami-resistant regions with large investments; while the issue of racism in the US decelerated the recovery process in New Orleans. Those issues affected every decision taken in performing recovery efforts in the area (for a detailed comparison, see Table 3 overleaf).

'Trust'

Trust is very crucial to implementing a plan successfully. Given the socio-political background in Aceh, efforts to build trust with the people of Aceh were a must in order to have a smooth rehabilitation and reconstruction programme without it being rejected by the people. For that reason, the BRR sought to avoid conflicts with the communities as much as possible and tended to meet the demands of the people. In order to build trust with the community, the BRR chose to give freedom to the community in determining their choices or decisions during the reconstruction process. However, in the absence of a macro/meso-spatial plan that could serve as a reference, the implementation of the VP was not well directed.

Sociocultural beliefs

The people of Aceh believe that disasters are the will of God, and therefore cannot be avoided. Wherever we live, if it is the will of God, disasters will continue to happen. Communities with that perception seem to dominate and are spread evenly in Banda Aceh. The paradigm that sees natural hazards as a punishment or a curse given by God, due to the sins of mankind, is still firmly entrenched in the community of Banda Aceh (Ahmad, 2011). This kind of belief can hinder mitigation efforts. On the other hand, in Simeulue Island, there is a hereditary culture to save themselves from disasters when a tsunami hits the area, known as 'Smong'.

Conclusion

The BRR may succeed in the development of Aceh, but not in building community resilience and disaster mitigation. People in Aceh, particularly those in vulnerable areas, have no operational standard to face calamities, even though the 2004 tsunami should have served as a valuable experience to anticipate similar situations. Knowledge and awareness of disasters are extremely important to communities living on the western coastal zone of Aceh, especially in areas without a tsunami early warning system.

Learning from the experience in Aceh, the implementation of rehabilitation and reconstruction after disasters should not only be limited to physical reconstruction. The development of social resilience is also required. When the physical reconstruction has been completed, socialisation and training on spatial planning and disaster risk reduction efforts must be continuous so that it becomes knowledge that is well understood by the public.

The socio-political situation in Aceh led the BRR to decide to use a VP. Although the implementation of a VP was considered a success and deemed appropriate for the conditions of that time, it also had its drawbacks. This was mainly due to the absence of planning at the macro-level, particularly at the city level, which would be the basis for the establishment of the city structure and the development of integration between villages. That condition was a dilemma faced by the BRR at the time: the choice between building quickly without rejection from the community but only for the short term, or waiting for the completion of the city spatial plan for the long term, which takes a long time and can therefore lead to conflict in society. Every decision has its consequences.

Table 3 Comparison of Aceh, Tohoku and New Orleans in the formulation of recovery plan

		Aceh after Earthquake and Tsunami 2004	
No	**Aspect**	**Macro-Wide Plan**	**Community Plan**
1.	**Type of Plan**	Master Plan	Village Plan (647 Plans)
2.	**Institution in charge of the plan formulation**	BAPPENAS (Ministry of National Planning)	Villagers assisted by facilitator team
3.	**Content of the plan**	Recovery policy, strategies, program	Action Plan: site plan, infrastructure , development plan
5.	**Schedule for plan making**	A week after the disaster, starting with damage assessment. This assessment was used as basis for developing master plan	Start 3-6 month after the tsunami, some villages 1 year after
6.	**Time frame period and budget**	6 years (for 4 years and for 2 years (US$7.7 billion)	3 years
7.	**Funding for developing plan**	National	Donors, NGOs, BRR
8.	**Time making process**	3 months	3-6 months
9.	**Social issues**	Long Arm Conflict (GAM vs RI)	
10.	**Planning capacity**	Local Government was paralysed due to the tsunami. Therefore the rehabilitation and reconstruction planning was taken by National Planning Ministry	
11.	**Plan template**	Medium Term National/Province Development Plan	
12.	**Planning stage**	2 steps planning: National/province plan followed by village plan	
13.	**Participatory planning process**	• Lack of participation in Province/City Wide Plan (top-down) • Community participation in village/neighborhood level assisted by facilitators/NGOs/donors	
Implementation of the plan			
14.	**Institution for managing recovery**	Special National Agency (BRR) (after 3 months) had considerable latitude to coordinate, monitor, and implement recovery, especially to fill the gap	BRR with Satker (Working Unit), Regional Office (after 2007)
15.	**Funding for plan implementation**	National budget, International donor, NGOs	Donors, NGOs, National
16.	**Adoption of the plan**	4 month after the event	3-months to 1 year after
17.	**Evaluation of the Plan**	2 years after (2007) Mid Term Review by BRR & Bappenas	Evaluation at some villages by Village officers, community and facilitators

Source: Author; Iuchi et al (2013); Olshansky & Johnson (2010)

Great East Japan Earthquake and Tsunami 2011		**New Orleans after Katrina 2005**	
Macro-Wide/City Plan	**Community plan**	**Macro-Wide/City Plan**	**Community plan**
National policy, Prefecture plan and City plan		1. FEMA ESF-14, ULI 2. Louisiana Speaks, BNOP 3. UNOP City Wide Plan	1. Lambert Plans 2. Neighborhood Plans
1. National Recovery Agency 2. Local Govt with committee: national academics (scientist), business association, citizen organizations,	Local communities, CSO, *machizukuri* planning team, local officers	Federal, State, Local Government with city wide planning team: ULI, Concordia	Planning Consultant (Lambert) with Community
Recovery projects, industry, economic revitalization, land use, disaster mitigation	Relocation land adjustment	Recovery strategies/ direction and recovery projects	Recovery strategies/ direction and recovery projects
• National Recovery Vision (1 month after) by National Reconstruction Council • Prefecture and Local Recovery Plan (1-3 month after)	Start 3-6 month after the concept from National and Prefecture (to be integrated)	UL: a week after BNOP: 2 weeks after Lambert Plan: 1 year after UNOP: 1 year after	6 months-1 year after
7-10 years (total 21T Yen = US$262 billion)	7-10 years	10 years	10 years
National, Prefecture and City Government	City Government	Federal, State, City Govt, Foundation	Federal Govt, Foundation
3-7 months	6 months	6 month (BNOP)	6 month
Aging population		Racism	
The community-based plan (*machizukuri*) was formulated with the support of urban planning consultants		Lack of local government planning officers after the hurricane. Strong planning team/ consultant city/district/neighbourhood	
Comprehensive City Master Plan		Comprehensive City and neighbourhood Plan	
2 steps planning: province/city plan followed by district/ community plan with refer to national recovery policy		Several steps of planning: FEMA ESF14, BNOP, Lambert Plan, UNOP	
• Holding committee of expert and representative of stakeholders in city wide plan. • Very active citizen participation – *machizukuri* in neighborhood level		• Consultants led the planning activities. There was limited administrative officer oversight, participation • Citizen participated actively in a series of both citywide, district and neighbourhood meetings.	
National Recovery Agency (after 9 months) sets guidelines for local planning, approves local recovery plans, and coordinates work of national ministries as they implement reconstruction	Local Government, Community	Louisiana Recovery Authority (State).set planning policy for recovery, to the governor and state legislature, and provided oversight of state agency recovery activities. NORA	ORM City Office Recovery Management
National, Prefecture, City Govt, Tax	City, Community	FEMA, State, City, Foundation	Federal. City Govt, NGO, Foundation
7 months after the tsunami	1 year after	6 month to 1 year after Katrina	1 year after

Acknowledgements

The author is extremely grateful to Professor Robert Olshansky, University of Illinois at Urbana-Champaign, who gave insight on post-disaster planning theory, provided time to visit the disaster in the city of New Orleans and arranged a meeting with Steven Bingler-Concordia, who led the making of Unified New Oleans Plan (UNOP), and several recovery stakeholders on July 2010. Acknowledgement is also given to Professor Kidokoro, the University of Tokyo, who supervised the author's dissertation and facilitated research between March and December 2013 so that the author had a chance to see the Tohoku recovery.

Notes

[1] The 11 detailed themes books were: (1) Spatial Planning and Land Affairs; (2) Natural Resources and Environment; (3) Infrastructure and Settlement; (4) Economics and Employment; (5) Local Institutions; (6) Education and Health; (7) Religion and Social Culture; (8) Regulations; (9) Safety, Order and Defence; (10) Governance and Control; and (11) Budgeting.

[2] Three multilateral agencies, three bilateral donors, three international NGOs and more than 2,000 planners and facilitators were despatched to villages.

[3] Before Kuntoro agreed to be the head of the BRR, he negotiated with the President and asked for special authority, which was enacted by law for speedy reconstruction.

[4] JICA had provided a tsunami evacuation building with a 4th floor and designed for an earthquake of 9 on the Richter scale with a helipad on the top. It was designed for 2,000 people standing for three to four hours (teleconference Kuntoro with Tohoku University).

References

Ahmad, A. 2011, 'Survey Conducted on 20 June–20 July, 2011', *Kompas*, 20 July.

BPS (Badan Pusat Statistik), 2011, *Banda Aceh in Number 2011*, BPS Kota Banda Aceh.

BAPPENAS 2005, *Master plan for The Rehabilitation and Reconstruction of the Regions and Communities of the Province of Nanggroe Aceh Darussalam and the Islands of Nias, Province of North Sumatera*, Government of Indonesia, Jakarta.

Berke, P.R., Kartez, J., & Wenger, D. 1993, 'Recovery after Disaster – Achieving Sustainable Development, Mitigation and Equity', *Disasters, 17*(2), pp. 93-109.

BNPB 2012, *Master Plan Tsunami Disaster Risk Reduction*, Government of Indonesia, Jakarta.

BRR Aceh 2005, The Lamjabat Village Spatial Plan,

Dercon, B. & Kusumawijaya, M. 2007, 'Two Years Settlement Recovery in Aceh and Nias: What should the Planners have learned?' *Paper presented at International Seminar on Post Disaster Reconstruction*, URDI, Yogyakarta, 8-10 July 2007.

Iuchi, K, Johnson, L. & Olshansky, R.B. 2013, 'Securing Tohoku's Future Planning for Rebuilding in the First Year Following the Tohoku-Oki Earthquake and Tsunami,Earthquake', *Spectra*, Vol. 29 pp. 479-499.

Mileti, D.S. 1999, *Disasters by Design: A Reassessment of Natural Hazards in the United States*, Joseph Henry Press, Washington, D.C.

Olshansky, R.B. & Chang, S. 2009, 'Planning for Disaster Recovery: Emerging Research Needs and Challenge', *Progress in Planning,* 72, pp. 200-209.

Olshansky, R.B. & Johnson, L.A. 2010, *Clear as Mud: Planning for Rebuilding of New Orleans,* American Planning Association, Chicago, IL.

Olshansky, R.B & Kartez. 1998, *Managing Land Use to Build Resilience in Cooperating with Natural Hazards with Land-Use Planning for Sustainable Communities*, Joseph Henry Press, Washington D.C., pp. 167-201.

Pardede, T. & Kidokoro, T. 2008, 'Aceh Reconstruction Planning, Top Down or Bottom up: Overview of Planning Theory and Learning from Community Planning in Aceh', *Proceedings 7th Asian City Planning,* City Planning Institute of Japan, Tokyo.

Republic of Indonesia, President Regulation number 3, 2009, *Termination of the Task of the Agency for Rehabilitation and Reconstruction of Aceh and Nias and Continuity Rehabilitation and Reconstruction of Aceh and Nias, North Sumatra.*

Taufiqurrahman 2006, 'EU, World Bank laud Aceh Reconstruction', *The Jakarta Post,* 14 July.

United Nation Office for Disaster Risk Reduction (UNISDR) 2015, *Sendai Framework for Disaster Risk Reduction 2015-2030,* United Nations, Headquarters (UN)

World Bank 2010, *Safer Homes, Stronger Communities: A Handbook for Reconstructing after Natural Disasters.*

CHAPTER A2b

Coastal resilience in Indonesia: from planning to implementation

Gusti Ayu Ketut Surtiari, Neysa Jacqueline Setiadi, Matthias Garschagen, Joern Birkmann, Riyanti Djalante and Yekti Maunati

Introduction

Resilience has become a prominent concept in the field of disaster risk reduction (DRR). Yet, analysis of whether and how the conceptual claims of resilience are translated from the planning stage into practical implementation is often lacking or thin (Garschagen, 2013). In this chapter, 'disaster resilience' is defined as the ability of a system, community or society exposed to hazards to resist, absorb, accommodate and recover from the effects of a hazard in a timely and efficient manner (UNISDR, 2009) while maintaining their structure and functions (Holling, 2001). The debate on disaster resilience is largely driven by the observation that disaster impacts remain high at the global scale (Guha-Sapir et al, 2015). The Hyogo Framework for Action (HFA) signals a global commitment to build resilience in order to reduce vulnerability and risk to natural hazards (UNISDR, 2007). To extend and build upon the achievement of HFA, the Sendai Framework for DRR emphasises the need to strengthen the resilience of people, communities, countries and their assets (UNISDR, 2015).

Adaptive governance is an approach that can support the management of complexity in disasters (Djalante, 2012) in order to ensure effective disaster responses (Lei et al, 2015). It focuses on the need for institutional adaptation in addition to the sheer adjustment of the biophysical and built environment, which is particularly relevant for cities and planning institutions (Birkmann et al, 2010, 2014). This is particularly important where institutions for DRR have to consider not only changes in the nature and intensity of natural hazards, but also changes to the socio-economic and cultural dimensions that take place in dynamically developing countries and cities, which affect the pathways of vulnerability and adaptive capacity towards these hazards

(Garschagen and Kraas, 2010; Garschagen and Romero-Lankao, 2015). Sudmeier-Rieux et al (2014), therefore, call for risk-sensitive land-use planning, that is, the change of planning principles and regulations, as one of the major components of adaptive urban governance.

Djalante et al (2011) proposed four characteristics of adaptive governance that can support the implementation of building resilience to disasters: multilayered institutions; participation and collaboration; self-organisation; and learning. However, there are opportunities that can be strengthened in the current state of adaptive governance, namely, the involvement of multiple stakeholders in Disaster Risk Reduction (DRR) and Climate Change Adaptation (CCA), increasing the global network, and improving the availability of strategic environment impact assessments (Birkmann et al, 2014). The involvement and collaboration of multiple stakeholders across levels are important to provide advantages for sharing and learning that contribute to the achievement of disaster resilience (Djalante, 2012; Lassa and Nugraha, 2014).

There is some research on resilience related to urban planning in Indonesia that shows the lack of capacity to translate the concept of resilience into planning (Chang Seng, 2013; Friend et al, 2014; Archer and Dodman, 2015; Jarvie et al, 2015). This can allow vastly diverging interpretations and applications of resilience from institution to institution (Garschagen, 2013; Lu and Stead, 2013). We argue that the strengthening of multi-stakeholder collaboration is needed in the process of policies and programmes to fill the existing gap on knowledge capacity and to support the sustainability of resilience programmes. This chapter aims to explore and discuss the implementation of urban planning for building disaster resilience at the city level from the perspective of adaptive governance, particularly with regards to the role of multiple stakeholder engagement.

The case studies of the cities of Semarang and Padang have been selected due to: (1) their exposure to different types of natural hazards; and (2) their similarities in multi-stakeholder collaboration in the process of resilience building. In Semarang, we explore the incorporation of resilience to climate-related hazards in urban planning, and in Padang, we explore the incorporation of tsunami risk reduction in urban planning.

The study in this chapter is conducted through reviewing legal documents relating to disaster management and spatial planning at the national and local levels, building on existing research activities in the cities of Semarang and Padang. Data and information on the Semarang and Padang case studies are derived from research conducted by the

Human Ecology Team[1] of Lembaga Ilmu Pengetahuan Indonesia (LIPI) from 2011 to 2013 and PhD research conducted by Neysa Setiadi from 2007 to 2009. Primary data and information were collected through semi-structured interviews with government officers at the city level, non-governmental organisations (NGOs) at the national and city levels, civil society, and experts from academic institutions.

This chapter consists of three main sections. The first section of the chapter describes the structure of key policies regulating the interface between DRR and development and spatial planning at the national level that become references for local government to frame risk-sensitive urban planning and administration. In the second section, we present the background of the two case studies, focusing on their responses to disasters. This will be followed by a discussion on how the institutions (policy and regulations) function at the local level, how multiple stakeholders have implemented DRR plans and how coordination was achieved across the different levels.

Architecture of urban planning and disaster risk reduction in coastal cities: from the national to the local level

National level

This section briefly describes four selected laws and one regulation that influence the incorporation of DRR into planning and become references for city governments to address DRR in urban planning. These are Law 17/2007 followed by President's Regulation 05/2010, Law 24/ 2007, Law 26/ 2007 and Law 1/2014:

1. Law 17/2007 concerning the National Long Term Development Plan for 2005–2025 was established to describe the link between national and local development planning and the authorities of local governments, which control and monitor the development process under the coordination of the national government.
2. President's Regulation 05/2010 was enacted as a derivation of Law 17/2007. It focuses on the National Medium Term Development Plan for 2010–2014, which addressed DRR and CCA as one of the priority programmes[2] entitled 'Environment and disaster management'. This requires all local governments to consider DRR and CCA in their local development plans.
3. Law 24/2007 was established with a focus on national disaster management. In comparison with the previous legislation on disaster prevention, this law reflects the strong commitment of the

Indonesian government to manage disasters through integrated action across sectors and for various types of hazards. It also clearly emphasises the need for the participation of public, private and international organisation in the effort to reduce risk activities.

4. Law 26/2007 on spatial planning was enacted to account for the decentralisation system. This law emphasises the vulnerability of Indonesia to different types of hazard and highlights a desire to strengthen disaster preparedness by designating special areas for evacuation and shelter.
5. Law 1/2014 was enacted as an updated version of Law 27/2007 on coastal and small island management. This law promotes structural and non-structural approaches to protect coastal areas such as the rehabilitation of the ecosystem, increasing disaster response capacities and increasing community resilience.

Overall, the laws and regulation have to be addressed in coordination across levels, from the national to the district level (see Figure 33).

Figure 33: The linkage between spatial planning and development planning across levels

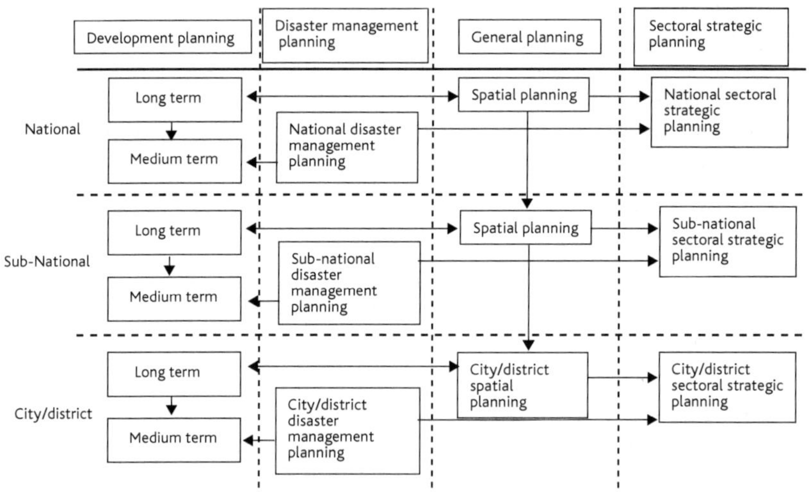

Source: Modified from Chang Sheng, 2012 that reproduced Azis, 2009

Case study at the city level

Beginning with a description of the two case studies, this section describes the extent to which DRR is integrated into urban planning processes in order to enhance resilience. The section describes the natural hazard risks that each case study faces and details the institutional process with regards to DRR and resilience. For the Semarang case study, we focus on the 'City Resilience Strategy' (CRS) document, which encourages risk-sensitive urban planning. Meanwhile, for the Padang case study, we focus on the development of evacuation plans and tsunami risk reduction. Both are analysed from the perspective of multi-stakeholder engagement and learning processes.

Two cities vulnerable to natural hazards

Semarang and Padang are two coastal cities with experience of climate-related and geophysical hazards. Both cities have enacted local regulations for spatial planning in line with national spatial planning law. As mentioned, in the spatial planning law, detailed information about the evacuation routes and shelter locations are provided at the city level. Both Semarang and Padang have not only provided local regulations to achieve resilience, but have also demonstrated the engagement of multiple stakeholders in the process, as the following sections demonstrate.

Semarang

Semarang, the capital city of central Java, is a delta city located in the northern part of Java Island. It is recognised as a dynamic metropolitan city with a large international harbour and dominated by industrial activities. Semarang City occupies an area of around 373km^2, with 1,572,105 inhabitants. As a metropolitan city and industrial area, 70% of the land area is dominated by buildings and housing, with around 10% used for agricultural activity, which is mostly concentrated in the southern part of Semarang City (BPS of Semarang City, 2014). Rapid urbanisation in Semarang has caused the rapid growth of the city. At the same time, there are 26,518 families still living below the poverty line (BPS of Semarang City, 2014). Importantly, most of these families occupy vulnerable areas, such as places near the coast, inundation zones and land prone to landslides. As a result, these families form the group most likely to be affected by tidal floods due to their exposure and limited response capacities.

As a dynamic city located in a delta area, Semarang City is exposed to several types of natural hazards. Floods (Marfai and King, 2008), landslides, droughts and typhoons are the most dominant hazards in Semarang City. From 2005 to 2010, the incidence of natural hazards increased significantly. In particular, the number of flood events

Figure 34: Map of Semarang and Padang Cities location circled (Google Earth, 2015)

increased by more than 70% and landslides increase by around 80 percent (BPBD of Semarang City, 2011). In addition to these hazards, Semarang is also heavily affected by land subsidence (Marfai and King, 2007; Lubis et al, 2011; Hakim et al, 2015). The land subsidence is triggered by groundwater extraction and the development of green areas, which are now laden with heavy construction (Marfai and King, 2007). For instance, in the northern part of Semarang, there are many industrial buildings located on unstable land that have contributed to land consolidation (Marfai and King, 2007), and this has resulted in severe inundation (Marfai and King, 2008). Such inundation is likely to increase in the future as a result of sea-level rises, which are projected to increase gradually (ACCCRN, 2010; Rahmawati et al, 2013). The places that are prone to this future inundation risk include several assets that are critical to the province, such as the international harbour and industrial facilities.

Padang

Padang, the capital city of the West Sumatera Province, is another dynamic city in Indonesia that is also located on the coast. Padang City stretches over 674.96km^2, with a population of 889,646 in 2014, distributed across 11 sub-districts (BPS of Padang City, 2015). The population density in the city is around 1,229 people per km^2, spread from the coast to the upper land. Due to its topographic conditions, in addition to the historical, political and economic activities, the city grew from the coastal areas surrounding the river mouth (Muaro - old city of Padang) towards the north and later to the east (Zaidulfar, 2002). This spatial development was planned to take the form of concentric rings but the expansion of the city was actually more sporadic. Similar to other cities in Indonesia, this phenomenon of urban sprawl has occurred in Padang City as a result of the lack of planned supporting infrastructure.

Padang is prone to flooding due to its geographic location in a coastal area, with some of the land below sea level (Lämmel et al, 2010) and other parts located on the floodplain of 23 rivers, including 10 particularly large rivers. The high number of rivers is further compounded by characteristic high rainfall and insufficient drainage infrastructure capacity (Putra et al, 2013). Padang City is also located above the Sumatran subduction zone where the Indian–Australian and Eurasian plates meet. As such and based on projections (see Borrero et al, 2006), the city has been identified as one of the most plausible places for the occurrences of major tsunamis in the future (Borrero et al, 2006; McCloskey et al, 2010). Following the tsunami in Aceh, several major earthquakes occurred in several areas of Indonesia, including Padang, which logged an earthquake of 7.6Mw on 30 September 2009 (McCloskey et al, 2010). This major earthquake caused significant damage and resulted in the deaths of more than 1,100 people, and contributed to a decrease in economic growth by around 0.3% in 2009 and 1.0% in 2010.[3]

In terms of tsunami exposure in the city, it was estimated that around 65% of the population located in the potentially affected areas are vulnerable, comprising women, children and the elderly (Setiadi, 2014). The exposure is also different depending on gender and daily activity patterns, which have important implications for evacuation planning. For instance, women are more likely to carry out activities at home for the duration of the day compared to men, who are more likely to move to other places during working hours, returning to their homes in the evening. This underlies the importance of evacuation facilities in settlement areas and the need to limit the density of settlement (with more vulnerable people) in the more hazardous zones.

Urban planning in the two cases studies: towards coastal resilience

Urban planning in Semarang City: from spatial planning to resilience to climate-related hazard

The city government of Semarang addresses the issue of disaster resilience as a central goal in many of its formal regulations and plans:

1. Local regulation 11/2011 was established with a focus on city spatial planning. As instructed in Law 26/2007, this regulation highlights the need to conserve areas that are prone to disasters and provide evacuation routes and shelters at the community level.
2. Local regulation 12/2011 concerning mid-term development planning for Semarang City (Rencana Pembangunan Jangka Menengah Daerah (RPJMD)) for the period 2010–15 has been developed by referring to the laws on national development and spatial and national disaster management plans. Normally, this regulation will also refer to city spatial planning but it was established after local regulation 12/2011 was enacted.
3. Local regulation 13/2010 provides for the establishment of a local agency for disaster management (Badan Penanggulangan Bencana Daerah (BPBD)). In addition, under regulation 13/2010, BPBD is to become the focal point for implementing DRR and coordinating with other relevant institutions. In terms of the strategic planning of BPBD (2011–15), the three laws presented earlier (Law 17/2002, Law 24/2007 and Law 26/2007) guide it to focus on establishing community resilience to disasters through the empowerment of the community and by strengthening the disaster preparedness and collaboration of local agencies and related organisations (BPBD Kota Semarang, 2011).

The effort of Semarang City to build resilience is obviously shown by regulation 12/2011, which proposed the need to mitigate disaster and to respond to climate change through mitigation and adaptation actions. The translation of the resilience concept into the medium-term development plan (RPJMD) in Semarang City is supported by the development process of the 'City Resilience Strategy' (CRS) of Semarang City. The formulation of the RPJMD of Semarang City took a similar time frame to the development of the CRS document. Therefore, Semarang City has an opportunity to gain knowledge about the resilience concept from the forum that created the CRS.

The CRS document is a dynamic plan for Semarang City to adapt to the impact of climate change, and was created by the Semarang City government in collaboration with non-state institutions as a part of the Asian Cities Climate Change Resilience Network (ACCCRN) programme (ACCCRN, 2010). One year before the RPJMD was established, Semarang City was selected to become part of ACCCRN. As part of this network, Semarang was assisted by the Rockefeller Foundation through Mercy Corps Indonesia, as a local partner, to start developing the CRS. Mercy Corps Indonesia collaborated with the city government to create a 'city team' responsible for developing the CRS document. Members of the city team comprise government officers, local NGOs and local academic institutions. Some of the government officers are also members of the RPJMD 'team' (ACCCRN, 2010). Based on the semi-structured interviews in Semarang, members of the city team were found to play an important role in influencing the integration of the resilience concept into the RPJMD process. For instance, the city team approached the city mayor about the importance of responding to climate change impacts by integrating the programme into wider development planning activities. The first phase of the integration of resilience into RPJMD was instigated through the allocation of a budget for the development of resilience-promoting programmes. While the RPJMD does not elaborate on the particular programmes that might be implemented, the allocation of a budget does signal a commitment to address and take action to enhance resilience in a manner that is complementary to the CRS.

The engagement of multiple stakeholders can be seen through the process of the development of the CRS and its integration into the RPJMD. In addition to the various organisational representatives in the city team, the process to develop the CRS consists of Shared Learning Dialogues (SLDs). The SLDs include the participation of local government representatives, NGOs and academic institutions. The aims of the SLDs are to share and discuss information on climate change and resilience building (ACCCRN, 2010). This process is thus useful, for instance, to reduce the gap between scientific data and real-life conditions at the local level in the vulnerability assessment for the CRS. Through SLDs, Mercy Corps Indonesia can collaborate with academic institutions to verify and improve the vulnerability assessments in order to ensure that they are more sensitive to local conditions (ACCCRN, 2010; Sutarto and Jarvie, 2012).

To strengthen the commitment to respond to climate change, Semarang City published the Integrated City Climate Strategy (ICCS) for the period from 2010 to 2020. This has become the legal framework for the climate change strategy of Semarang City, with instruments for synchronisation in budgeting, implementation, control and evaluation of climate change management (Government of Semarang City, 2013). Even though this document focuses more on the mitigation process, especially programmes to reduce emissions, it also includes provisions for relevant programmes linked to adaptation, such as flood mitigation through the development of an early warning system and mangrove plantation to protect coastal areas from sea-level rises.

Despite the positive support for the implementation of resilience building through the implementation of the CRS and local regulations, the involvement of multiple stakeholders in order to support the integrated assessment, particularly in terms of multi-hazard risk for new city spatial plans, remains outstanding. The multiple stakeholders' activities are still lacking in involving the impact of land utilisation in the analysis under the new regulation. For example, new city spatial plans focus on infrastructure development to the southern part of Semarang City in order to reduce the exposure of people from flood and inundation in the northern part of Semarang City (see Figure 35). Therefore, the northern part of Semarang City will be only utilised for economic and government activities and for the coastal protection area. However, while the movement of infrastructure development to the southern part of Semarang might reduce exposure to flooding, it would increase another potential risk to disaster, namely, landslide. The southern part of Semarang is expected to become conservation areas, but in order to develop resettlement, it will encourage land-use change from a green area to a settlement area (Nugroho, 2009; Cahyadi, 2014).

Figure 35: The changing land utilisation between current condition and new spatial planning in the northern part of Semarang City

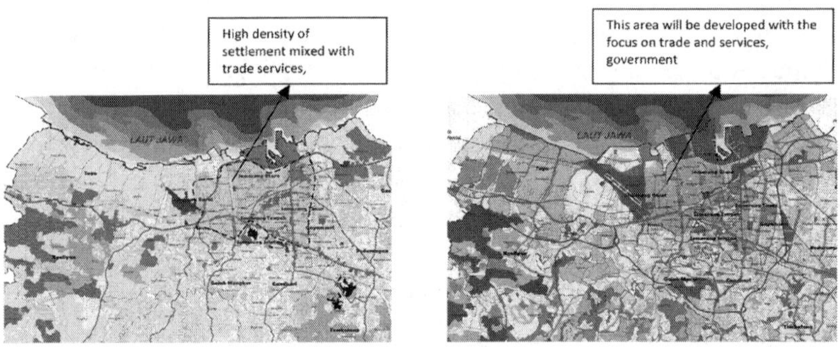

Source: Bappeda Kota Semarang (2013)

Urban planning in Padang: towards disaster resilience (tsunamis and earthquakes)

For the Padang case study, we describe local regulations on development and spatial planning in order to examine the engagement of multiple stakeholders in addressing tsunami risk reduction. Similarly to Semarang, we also discuss the change of land utilisation of the new city spatial planning, in particular, for resettlement.

The new city spatial planning of Padang City was enacted from 2010 to 2030 based on Local Regulation 4/2012. This is a revision of the previous plan, which was intended to be implemented for the period from 2004 to 2013. The regulation concerns the risk of earthquakes and tsunamis and was developed in response to the Indian Ocean Tsunami mega-disaster of 2004 and the several subsequent earthquake events. Padang City is also concerned with other hazards, such as floods, landslides and storm surges, and these have also been addressed with the aim of reducing disasters (Setiadi, 2014).

In the new city spatial plan (2010–30), the city government intends to shift the development of settlements and economic activity towards the northern and eastern parts of the city, which have less exposure to coastal disaster (see Figure 36). The coastal areas are then to be designated mainly for trade and service sector purposes; however, as instructed by the regulation on city spatial planning, evacuation routes and shelters are also planned to increase resilience.

The implementation of resettlement from the exposed areas of the city along the coast remains challenging since this area already comprises dense construction and has a long history of settlements mixed in with trade and services. However, the targeted area for the new residential areas in the east is currently used for agriculture (Setiadi, 2014). As such, the change in land use from agriculture to settlement may have implications for livelihoods and food security. The current concentration of mixed activities in the areas close to the coast is at

Figure 36: Map of existing land use and new spatial planning map of Padang City

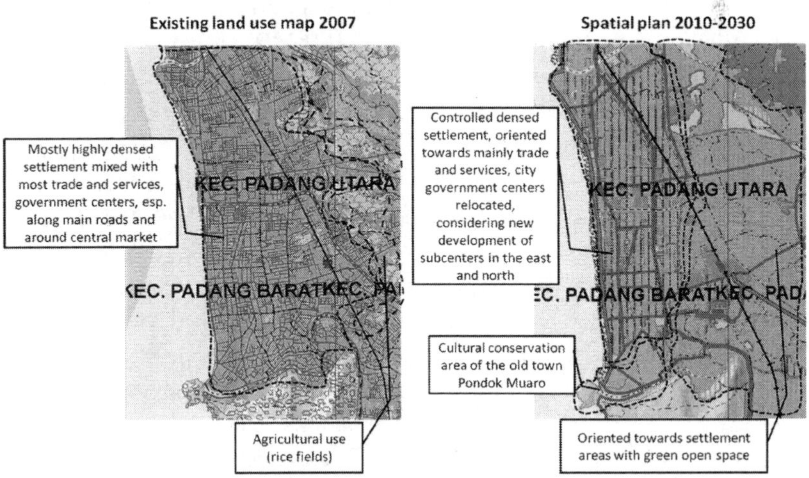

Source: Setiadi, 2014 (Basis map: Bappeda Kota Padang, 2010, own annotation)

high risk of earthquakes and tsunamis. As such, the city spatial plan can potentially be an effective tool for DRR by locating development away from high-risk areas. However, there are limits to the extent to which spatial planning can influence existing land use. In the case of Padang, the allocation of land in high-risk areas, either for resettlement purpose or evacuation facilities (evacuation routes and shelters), requires the approval of the public before it can proceed. Many people in the hazard zone would not be willing to move to safer places and give up their land for the widening of evacuation roads. However, there are examples of public cooperation in land-use changes, particularly where the process is supported by local NGOs. Where the public is involved in the planning and implementation process, it seems more likely that they will be willing to provide their land to support the development of evacuation routes.

In terms of tsunami preparedness, the local community has advanced initiatives to increase awareness among those exposed to earthquakes and tsunamis. The progressive activities of the local community are supported by local NGOs such as Komunitas Siaga Tsunami (KOGAMI) and the Red Cross, who work directly with the communities to raise awareness. They also support the local mapping of evacuation pathways and capacity-building activities within local government agencies.

In the planning process for tsunami risk reduction, many local government agencies have been included in the process, resulting in significant contributions. The local disaster management body (BPBD) was a newly established government agency at the time of this research and thus was not yet fully functional as the main agency responsible for monitoring risk and coordinating DRR. Thus, the local development planning agency's (BAPPEDA's) role in coordinating the various agencies and activities for the development of disaster mitigation-based spatial planning, including the planning of evacuation facilities and infrastructure, was more important in the early stages. In addition, the spatial planning and urban design agency (TRTB) and public works agency for regional settlements and infrastructure (PU Kimpraswil) played an important role in providing technical support for the planning and implementation of tsunami risk reduction, for example, by means of providing evacuation facilities and infrastructure, building permits, and building codes.

National agencies were also involved in reducing earthquake and tsunami risks. For instance, the Fishery and Marine Agency from the province level supported the city government by providing a vertical evacuation shelter. The National Disaster Management Authority

(BNPB) has also addressed its commitment to reduce tsunami risks by improving the early warning system and evacuation shelters (Setiadi, 2014). Such involvement from higher levels of government is very important, particularly to support the improvement of expensive infrastructure that would otherwise be beyond the capacities of local budgets (Setiadi, 2014).

Additionally, international NGOs and international scientific communities, such as from Germany, Japan and the US, supported capacity building, particularly in relation to tsunami hazards, by assisting with a vulnerability assessment and the development of an evacuation plan (Setiadi, 2014). However, the various activities that have been carried out by a diverse range of stakeholders may also potentially cause confusion among decision-makers. In the evacuation planning process, it was particularly challenging to consolidate the different information from each stakeholder into useful planning tools such as the city official evacuation plan, evacuation shelters plan, tsunami hazard and risk zoning. The need for structured coordination is important in order to manage various stakeholder engagement activities across levels and the different data and information sources. Such coordination can be achieved by developing a consensus among the different stakeholders through meetings. An example of successful consensus building can be seen in Padang, where various data and methodologies for developing tsunami hazard maps were considered and evaluated before coming to an agreement (Setiadi, 2014). Combined with the multi-stakeholder engagement process for planning in Semarang, this highlights how multiple stakeholders can play a key role in supporting city government action for DRR through city planning.

According to Setiadi (2014), there is still a gap in addressing tsunami risk reduction in urban planning observed in the case of Padang, especially in integrating vulnerability information into mapping and planning. Currently, in city spatial planning, tsunami risk zoning remains focused on the hazard exposure, and a mechanism for the integration of vulnerability assessments, such as specific consideration for vulnerable groups, in this spatial planning remains unresolved.

Conclusion and outlook

The objective of this chapter was to understand how the implementation of urban planning could be used to support the enhancement of resilience to natural hazards, especially by considering the incorporation of multi-stakeholder engagement. Based on two case studies with different types of hazards, we conclude that multiple stakeholders' engagement across scales is very important in supporting city governments to build resilience to disasters and in order to implement the laws and regulations from the national level. Both Semarang and Padang demonstrated that the engagement of multiple stakeholders at various levels plays a significant role in supporting city governments to translate the resilience concept into their plans. In the Semarang case study, a forum, namely, the city team, was effective support for knowledge sharing among the team members. In the Padang case study, multi-stakeholder engagement across levels was a success in supporting the building of resilience through early warning systems and evacuation development. However, there still remain several challenges to the implementation of existing legal documents in order to achieve resilient cities.

First, due to the various backgrounds of the stakeholders, coordination among them is important in order to optimise resources to achieve resilience to disasters. Our research found that as local agency for development planning, the main coordinator is Bappeda, but it is also important to consider local agencies for disaster management that are responsible for risk management.

Second, more formal strategies and approaches could improve the engagement of local participants further. In particular, it is important to note that during the research in the two case studies, local governments still received support for their stakeholder engagement activities from national and international organisations, but in the future, local governments will have to be able to support themselves to continue existing and develop new programmes. Therefore, the local initiation of participation still needs to be strengthened as it will be important for the sustainability of programmes.

Third, based on the case studies in Semarang and Padang, city spatial planning related to the shifting orientation of land use in vulnerable areas from dense human activity to conservation areas is still lacking for the multi-hazards risk assessment. Without a comprehensive study, for instance, the objective of reducing the risk of inundation affecting people in northern Semarang City may be counterproductive. This is because the change in land use and consequent reduction of vulnerable areas is likely to cause an increase in the risk of landslides. Likewise, in Padang, relocating people from the coast to the upper land will cause a land-use change from agriculture to housing development that will have an impact on the livelihoods of the people in the surrounding areas.

Referring back to the adaptive governance concept as an approach for government to respond to the complexity of disasters, the engagement of multiple stakeholders has to be enhanced through the whole process, from planning to implementation, with a clear coordination mechanism.

Acknowledgements

To my colleagues in the research team in Semarang City: Rusli Cahyadi, Laksmi Rachmawati, Fadjri Alihar, Toni Sutopo and Rusida Yulianti. I would like to thank PPK LIPI who financed the research under the Daftar Isian Pelaksaan Anggaran (DIPA) budget scheme. Analysis for Semarang is part of the research and particularly elaborates the focus on multiple stakeholders' engagement. Analysis for Padang could only be done through the funding of Deutsche Forschungsgemeinschaft/Das Bundesministerium für Bildung und Forschung (DFG/BMBF) 'Last-Mile – Evacuation' and the courtesy of the people of Padang, especially colleagues in the University of Andalas, KOGAMI, and in city government agencies of Padang.

Notes

[1] Research team for Semarang City for the period of 2010-2014.

[2] The National Priority Programme is a part of national development planning, in which the national government commits to the development of Indonesia. The National Priority Programme began after the reformation era, from 1999 onward, and consists of 11 priority programmes (law, education, health, poverty reduction, food security, infrastructure, business and investment, energy, environment and disaster management, borders, disadvantages and conflict, culture, and technical innovation).

[3] See: http://siteresources.worldbank.org/INTINDONESIA/Resources/Publication/280016-1235115695188/5847179-1257228519234/exsum.bh.pdf

References

ACCCRN (Asian Cities Climate Change Resilience Network) (2010) *City resilience strategy: Semarang's adaptation plan in responding to climate change*. Semarang: ACCCRN.

Archer, D., & Dodman, D. (2015). Making capacity building critical: Power and justice in building urban climate resilience in Indonesia and Thailand. *Urban Climate*.

Birkmann, J., Garschagen, M., Kraas, F. & Quang, N. (2010). Adaptive urban governance: new challenges for the second generation of urban adaptation strategies to climate change. *Sustainability Science*, 5(2), 185-206

Birkmann, J., Garschagen, M., & Setiadi, N. (2014). 'New challenges for adaptive urban governance in highly dynamic environments: Revisiting urban planning systems and tools for adaptive and strategic planning', *Urban Climate*, 7, 115-133.

Borrero, J.C., Sieh, K., Chlieh, M., & Synolakis, C.E. (2006), 'Tsunami Inundation modeling for Western Sumatra', *Proceedings of the National Academy of Sciences of the United States of America*, 103(52), pp. 19673–19677.

BPBD of Semarang City (2011). *Renstra SKPD Badan Penanggulangan Bencana Daerah Kota Semarang 2011-2015.*

BPS of Padang City (2014). *Kota Padang Dalam Angka 2014* (Padang City in Figures, 2014).

BPS of Padang City (2015). *Kota Padang Dalam Angka 2015* (Padang City in Figures, 2015).

BPS of Semarang City (2014) *Kota Semarang Dalam Angka 2014* (Semarang City in Figures, 2014).

Cahyadi, R. (2014). 'Respon Pemerintah Kota Semarang Terhadap bencana Longsor (Response of Semarang City Government on Disaster of Lanslide)', in Y. a. Supriyatna, *Adaptasi Pemerintah Daerah Terhadap Dampak Perubahan Iklim: Peluang and Tantangan. Pemberlajaran dari Kota Semarang.* Pintal, Jakarta.

Chang Seng, D. (2013). Disaster Risk Preparedness: The Role of Risk Governance, Multi-Institutional Arrangement and Polycentric Framework's for a Resilient Tsunami Early Warning System in Indonesia. Graduate Research Series. PhD Publication. *Publication Series of UNU-EHS Vol. 15.* UNU EHS. Bonn

Djalante, R. (2012). 'Adaptive governance and resilience: the role of multi-stakeholder platforms in disaster risk reduction', *Natural Hazards Earth Sci.* 12, 2923, 2012

Djalante, R., Holley, C., & Thomalla, F. (2011). 'Adaptive governance and managing resilience to natural hazards', *International Journal of Disaster Risk Science*, 2(4), 1-14.

Friend, R., Jarvie, J., Reed, S.O., Sutarto, R., Thinphanga, P., & Toan, V.C. (2014). Mainstreaming urban climate resilience into policy and planning; reflections from Asia. *Urban Climate*, 7, 6–19.

Garschagen, M. (2013). Resilience and organisational institutionalism from a cross-cultural perspective: an exploration based on urban climate change adaptation in Vietnam. *Natural Hazards*, 67(1), 24-46

Garschagen, M. & Kraas, F. (ed). (2010*). Assessing future resilience to natural hazards : the challenge of capturing dynamic changes under conditions of transformation and climate change* 2010/05/30 – 06/03 Davos. Davos: n/a, 2010.

Garschagen, M. and Romero-Lankao, P. (2015). Exploring the relationships between urbanization trends and climate change vulnerability. *Climatic Change*, 133(1), 37-52.

Government of Semarang City (2013). *The Integrated strategy of climate Change in Semarang City, 2010-2030.* Semarang.

Guha-Sapir, D., Hoyois, Ph., & Below. R. (2015). *Annual Disaster Statistical Review 2014: The Numbers and Trends.* Brussels: CRED.

Hakim, B. Al, Wibowo, M., Kongko, W., Irfani, M., Hendriyono, W., & Gumbira, G. (2015). Hydrodynamics Modeling of Giant Seawall in Semarang Bay. *Procedia Earth and Planetary Science,* 14, 200–207.

Holling, C.S. (2001), 'Understanding the complexity of economic, ecological and social system', *Ecosystem* (2001) 4: 390-405.

Jarvie, J., Sutarto, R., Syam, D., & Jeffery, P. (2015). Lessons for Africa from urban climate change resilience building in Indonesia. *Current Opinion in Environmental Sustainability*, 13, 19–24.

Lämmel, G., Grether, D., & Nagel, K (2010). The representation and implementation of time-dependent inundation in large-scale microscopic evacuation simulations. *Transportation Research Part C: Emerging Technologies*, vol. 18, no. 1, 84–98.

Lassa, J.A. & Nugraha, E. (2014). Building Urban Resilience to Climate Change in the Secondary Cities in Indonesia. *IRGSC Working Paper Series No. 9.* ISSN 2339-0638

Lei, Y., Liu, C., Zhang, L., Wan, J., Li, D., Yue, Q., & Guo, Y. (2015). Adaptive governance to typhoon disasters for coastal sustainability: A case study in Guangdong, China. *Environmental Science & Policy*, 54, 281–286.

Lu, P. and Stead, D. (2013). 'Understanding the notion of resilience in spatial planning: A case study of Rotterdam, The Netherlands', *Cities,* 35 (2013) 200-212.

Lubis, A. M., Sato, T., Tomiyama, N., Isezaki, N., & Yamanokuchi, T. (2011). Ground subsidence in Semarang-Indonesia investigated by ALOS–PALSAR satellite SAR interferometry. *Journal of Asian Earth Sciences*, 40(5), 1079–1088.

Marfai, M.A. & King, L. (2007). 'Monitoring land subsidence in Semarang, Indonesia'. *Environmental Geology*, vol. 53, pp. 651-659.

Marfai, M.A. & King, L. (2008). 'Coastal flood management in Semarang, Indonesia'. *Environment Geology*, vol. 55, pp. 1507-1518.

McCloskey, J., Lange, D., Tilmann, F., Nalbant, S.S., Bell, A.F., Natawidjaja, D.H., et al. (2010). The September 2009 Padang earthquake. *Nature Geoscience*, 3, 70–71.

Nugroho, P.S. (2009). 'Studi Kebijakan Pembangunan terhadap Perubahan Tata Ruang di Kota Semarang' (Policy analysis on spatial change in Semarang City), *Riptek*, vol. 2, pp. 41-51.

Putra, A., Triyanto & Semeidi H. (2013). Analisa Bencana Banjir di Kota Padang (Studi Kasus Intensitas Curah Hujan Kota Padang 1980 – 2009 dan Aspek Geomorfology). *Seminar Sains Atmosfer 2014*. Bandung 28 Agustus 2013. ISBN 978-979-1458-73-3

Rahmawati, N., Vuillaume, J.F., & Purnama, I.L.S. (2013). Salt intrusion in coastal and lowland areas of Semarang City. *Journal of Hydrology*, 494, 146–159.

Setiadi, N. 2014, People's early warning response capability to urban planning interventions to reduce vulnerability to tsunamis. Case study of Padang city, Indonesia. PhD Thesis. UNU EHS, Bonn.

Sudmeier-Rieux, K., Fra Paleo, U., Garschagen, M., Estrella, M., Renaud, F.G. & Jaboyedoff, M. (2014). Opportunities, incentives and challenges to risk sensitive land use planning: lessons from Nepal, Spain and Vietnam. *International Journal of Disaster Risk Reduction*, online first.

Sutarto, R., & Jarvie, J. 2012, Integrating Climate Resilience Strategy into City Planning in Semarang, Indonesia. *The Urban Climate Resilience Working Paper Series.*

UNISDR (2007). *Hyogo Framework for Action 2005–2015: Building resilience of nations and communities to disasters*, UNISDR, Geneva.

UNISDR (2009). *Terminology on Disaster Risk Reduction*. UNISDR, Geneva.

UNISDR (2015). *Sendai Framework for Disaster Risk Reduction 2015-2030*. UNISDR, Geneva.

Zaidulfar, E.A. (2002). Morfologi Kota Padang. PhD Thesis, Universitas Gadjah Mada, Yogyakarta.

CHAPTER A3

Planning for resilience in the New York metro region after Hurricane Sandy

Donovan Finn, Divya Chandrasekhar and Yu Xiao

United States	
Population: 322.1 million (2015)	**Government:** Federal constitutional republic
Area: 9,833,517 km²	**Administrative structure:** The United States is comprised of 50 states (each state has its own state constitution), one federal district (District of Columbia or D.C.), one incorporated territory and 15 unincorporated territories.
Hazard profile: Coastal flooding (Atlantic and Pacific coasts) Inland flooding (nationwide) Heat waves (nationwide) Blizzards (northern half of country) Alpine hazards (Rocky Mountains) Earthquakes (up to 9.2: multiple faults)	**Authorities in charge of risk assessment and management:** Federal Emergency Management Agency (FEMA) within the Department of Homeland Security (DHS)
Analysed event: Coastal flooding caused by Hurricane Sandy (2012) Estimated return period: 1:500 – 1:3500 Estimated damages: $65 billion Casualties: 117	**Role of spatial planning:** Responsible for post-event recovery and long-term mitigation.

Introduction

Although disaster planning, response and recovery have long been the purview of emergency management professionals, the role of spatial planning vis-à-vis disasters has been increasing in recent years in the United States (US) as we better understand the importance of planning to both help reduce risk from future disasters and to rebuild and recover afterwards. Superstorm Sandy, which struck the east coast of the US in late 2012, illustrates this dynamic clearly. The storm has been the catalyst for an unprecedented amount of spatial planning in a compressed time frame as communities have attempted to rebuild and recover from, as well as mitigate against, the effect of future events.

These efforts also clearly illustrate how planners have engaged federal disaster funding policies that, more than ever, prioritise local decision-making and long-term resilience in the recovery process.

Increasingly, disasters trigger a flurry of spatial planning efforts that emerge to guide decision-making once immediate response needs have been met. A vivid example of this growing connection between spatial planning and disaster recovery comes from Hurricane Katrina, which struck New Orleans, Louisiana, in 2005 and decimated the city. Katrina prompted a dizzying array of often overlapping and competing spatial planning processes, which are well documented by Olshansky and Johnson (2010) in their book *Clear as mud: Planning for the rebuilding of New Orleans*. There are multiple reasons for this convergence of spatial planning with disaster planning. One reason is the increasing recognition that the core concerns of spatial planners are the same issues that must be dealt with in the aftermath of a disaster, such as supplying adequate housing, protecting open space from capricious development, operating a functioning transportation network and assuring vibrant economic activity. Meanwhile, money flows into disaster sites in the form of insurance payments, government aid and humanitarian aid, thus facilitating spending, which, in turn, requires planning. Finally, disasters can also act as a trigger not just for rebuilding (Berke et al, 1993), but for, in the oft-used parlance, 'building back better' because old and often outdated or inadequate systems and assets are destroyed (Olshansky et al, 2006). All of this activity requires planning so that funding can be acquired and allotted strategically and communities can be engaged in sometimes-contentious deliberations about how it should be spent.

Recovery planning for large disasters in the US happens through the complex interaction of federal, state and local government programmes. Traditionally, the federal government has shouldered a heavy burden for short-term response and recovery actions through the Federal Emergency Management Agency (FEMA). Long-term community recovery has also increasingly become part of FEMA's mandate, particularly since the adoption of the 2004 National Response Plan (NRP) and later the National Disaster Recovery Framework (NDRF) and its Community Planning and Capacity Building recovery support function, which aims to help communities 'effectively plan and implement disaster recovery activities, engaging the whole community to achieve their objectives and increase resilience' (FEMA, 2011a, p 45). With these changes, FEMA has increasingly taken on responsibility to support and enhance community-scale recovery planning. As this chapter illustrates, the revised NDRF has been critical in shaping post-Sandy planning efforts by providing flexibility in the way federal funds are spent and prioritising community participation and conversations about future risk reduction in the recovery and reconstruction process. While FEMA represents the primary vehicle for the federal government's short-term response and recovery planning functions, the funding for long-term recovery projects has largely been administered through the Department of Housing and Urban Development (HUD) and its Community Development Block Grant (CDBG) programme. CDBG is HUD's main avenue for channelling federal funding to communities for capital funding and community and economic development in normal times, but it has also become an important tool for post-disaster funding under the label CDBG-DR ('Disaster Recovery').

Individual states typically act as coordinating agents between federal and local governments, which traditionally shoulder most of the disaster recovery burden. States often act as intermediaries between federal funding sources and local governments, and help serve a coordinating role among multiple affected municipalities since disasters ignore jurisdictional boundaries and because state resources are needed to supplement local capacity. Local governments provide immediate on-the-ground disaster response or must explicitly request state or federal help, and they retain that control in the recovery phase as well. However, given the scale of federally declared disasters like Sandy, financial and technical assistance from federal and state sources is critical in facilitating local recovery planning and project implementation. Thus, while recovery planning, hazard mitigation and resiliency building are both the right and responsibility of individual communities, these processes typically exhibit substantial interplay between local governments, their state government and the federal government. The Sandy experience illustrates this dynamic vividly.

Superstorm Sandy in context

Superstorm Sandy, beginning as a late-season Atlantic hurricane, lashed the north-eastern coast of the US after first causing significant damage across Jamaica, Cuba and the Bahamas. The storm, despite its massive size, was downgraded to a post-tropical cyclone (hence the Superstorm Sandy moniker that was quickly attached to it) before making landfall on the east coast of the US on 29 October 2012 near Atlantic City, New Jersey. Lacking the sustained winds typical of a hurricane, Sandy was nonetheless a devastating event due mostly to the confluence of storm surge with the Atlantic coast's high tide. Impacts were devastating for coastal communities along the eastern seaboard. The New York City metropolitan region, including the city, as well as much of the state of New Jersey and the suburban Long Island region of New York State, were hit particularly hard by the storm. The National Hurricane Center attributes 147 deaths worldwide directly to Sandy, with 72 direct and 87 indirect deaths in the US (Blake et al, 2013). Striking the densely developed New York City metropolitan region, Sandy's cost far outweighed its lethality. Reinsurance firm Aon Benfield estimates that the disaster was responsible for $68 billion in economic losses in the US, the vast majority in New York State ($33 billion) and New Jersey ($30 billion). Sandy caused more damage than any other Hurricane in US history excepting Katrina, which generated economic losses of $147.2 billion in adjusted 2013 dollars (Aon Benfield, no date).

Due to the aggressive and rapid preparations, such as closing New York City's transit system well before the storm's landfall, the response to and short-term recovery from the storm were relatively rapid and effective, and within days, much of the affected region was fully functioning, with some key exceptions. For recovery planners, Sandy presented two key challenges. First, while the region as a whole rebounded quickly, specific areas were devastated. Infrastructure, housing, businesses, transportation networks and economic activities were all affected. Second, across the entire region, debates began almost immediately about how best to simultaneously rebuild from Sandy, protect against future hazards and also move forward with pre-existing economic development goals. While many residents and policymakers viewed Sandy as a chance event and urged rebuilding exactly as before, others saw the storm as a wake-up call and a rare opportunity to strengthen and enhance communities' protections against future severe weather and other potential hazards.

These mitigation and resilience goals are hampered to some degree by the fragmented nature of governance in the US, with its strong tradition of local autonomy and a persistent resistance to federal and regional planning. The context for land-use planning varies widely among US states. A strong home-rule tradition has devolved most planning powers to local governments and states differ widely as to whether communities are required to prepare comprehensive plans and as to what information those plans must contain. Some states require municipalities to plan, others make planning completely optional and still others make it optional but provide incentives or require plans in certain situations. Spatial planning has remained a fiercely hyper-local undertaking in the US, perpetuating what Anthony Downs (2005, p 370) calls a 'disjointed incrementalism', making it difficult to effectively plan regional transit systems or to address issues such as the equitable or logical distribution of jobs, housing or regional services like waste transfer stations throughout a region. These issues are very salient in the New York metropolitan region. Based on commuting patterns and economic ties, the US Census Bureau considers New York City, with a population of about 8.4 million, as part of a larger region that includes seven additional counties in New York State, 12 counties in New Jersey and seven counties in Pennsylvania. This region of 20 million residents thus encompasses not only New York City, but also Newark and Jersey City in New Jersey, White Plains, New York, and hundreds of smaller municipalities. Yet, despite the presence of a regional Metropolitan Transportation Authority (MTA) and a federally mandated regional Metropolitan Planning Organization (MPO), known as the New York Metropolitan Transportation Council (NYMTC), it is a politically balkanised region in practice and regional planning is essentially non-existent.

Neither New York State nor New Jersey requires local comprehensive land-use plans, though many municipalities in both states have adopted plans. New York City has never had a comprehensive land-use plan adopted. As coastal states, both have detailed coastal planning requirements, as outlined in New York's Coastal Management Program of 1982 and New Jersey's Coastal Area Facility Review Act (CAFRA) of 1973, both of which have been substantially updated since adoption. There have, in fact, also been three regional plans created for the New York metropolitan region, including the New Jersey portion, in 1929, 1968 and 1996, by the Regional Plan Association (RPA), and a fourth regional plan is currently under development. However, the RPA plans, while thorough, are merely advisory as the RPA is a not-for-profit advocacy organisation whose plans have no formal power over local or regional policy.

The federal government's post-Sandy recovery and resilience efforts

Although the federal government provided billions of dollars in assistance to be spent by states and municipalities affected by Sandy, there was also a clear mandate for direct federal action, as evidenced by many of the recommendations of President Obama's Hurricane Sandy Rebuilding Task Force found in the *Hurricane Sandy Rebuilding Strategy* released in August 2013. Given the relatively moribund state of regional planning in the New York region, Sandy's widely distributed impacts across such a large, dense and politically factionalised region demanded a federal response that could address the regional issues – environmental, economic and social – inherent in recovery and resilience planning. That response was, in part, the US government's Rebuild by Design (RBD) programme, conceived by Henk Ovink, former Director General of Spatial Planning and Water Affairs and Director of National Spatial Planning for the Ministry of Infrastructure and the Environment in the Netherlands, in his role as a senior advisor to then HUD Secretary Shaun Donovan (Shorto, 2014). The RBD process was funded and coordinated by a set of philanthropic, non-profit and academic partners with oversight from HUD, with implementation paid for via Community Development Block Grant Disaster Recovery (CDBG-DR) funds. RBD began in June 2013 with a worldwide call for proposals to address resilience issues in the Sandy-impacted region, including Connecticut, Maryland, New Jersey, New York and Rhode Island. From the 148 applicants, 10 interdisciplinary teams, including designers, planners, engineers, artists and other specialties, were chosen in August 2013 to proceed to the design stage.

With assistance from non-profit partners such as the Municipal Art Society, the Regional Plan Association and the Van Alen Institute, the 10 teams spent the autumn of 2013 engaged in an intensive research and analysis phase, culminating in 44 project proposals. A jury of disaster experts and RBD sponsors selected one project per team to advance to the design phase, beginning in November 2013; each team received a second instalment of US$100,000 in funding and an additional three months to turn their selected proposal into a refined design. Each team also undertook an intensive community participation process coordinated by the non-profit partners; by the end, RBD teams had participated in more than 350 stakeholder meetings and 64 public workshops, engaging thousands of residents. The 10 final proposals included a series of artificial barrier islands to protect the New York and New Jersey harbours, a comprehensive resiliency strategy for the Hunts Point section of New York City, and an integrated storm-water management programme for parts of the New Jersey coast.

In June 2014, almost one year from the RBD's launch, HUD Secretary Donovan announced six winning proposals and awarded them a total of US$920 million in implementation funding (see Table 4). As of the time of writing, detailed designs for each proposal are being prepared and additional public outreach is being conducted. In May 2015, for instance, New York City, which will administer RBD funds awarded to projects located in the city, held the first of three public workshops about the East Side Coastal Resiliency (ESCR) Project, the city's label for the first section of the BIG U proposal (sometimes also called The Dryline), stretching from Montgomery Street to 23rd Street along the city's lower east side. The city hopes to have a completed design for the ESCR by the end of 2015 and to begin construction, after the necessary environmental reviews, in late 2017.

Table 4: Rebuild by Design proposed and final projects

Team name	Design opportunities proposed	Final projects
Interboro Team	[1] Living with the Bay: Resiliency-Building Options for Nassau County's South Shore [2] Living with the Coast: A Better Day at the Beach [3] Living with the Creek: Options for Monmouth County Watersheds [4] Living with the Marsh: Options for Staten Island's Shore	Living with the Bay: A Comprehensive Regional Resiliency Plan for Nassau County's South Shore Awarded $125M for implementation of the Slow Streams portion of team's comprehensive resiliency plan for Nassau County, NY's south shore (Long Island).
PennDesign/OLIN	[1] Hunts Point / Lifelines [2] Flood-Adaptive Design on the Hudson Peninsula: Jersey City/ Hoboken [3] Folding the Coastal Plain: Staten Island East Shore [4] Reorienting Living on a Shifting Estuary: Toms River, NJ	Hunts Point Lifelines Awarded $20M for analysis, planning, and implementation of a pilot project; part of comprehensive resiliency strategy for the Hunts Point section of New York City's South Bronx
WXY/WEST 8	[1] Designing with Nature for the Future of the Mid-Atlantic Coast [2] Atlantic Ocean: Eco-Government Strategies [3] East River: Double Agenda [4] Hudson River: Communication of Risk [5] Jamaica Bay: Tidal Society [6] Long Island Sound: Equitable Risk Assessment	Blue Dunes – The Future of Coastal Protection Unfunded proposal to create a chain of barrier islands to protect the New York and New Jersey harbors.
OMA	[1] Resist, Delay, Store, Discharge: a comprehensive strategy for Hoboken [2] Communication Systems [3] Infrastructure Catalyst [4] Planning Principles	Resist, Delay, Store, Discharge: A Comprehensive Strategy for Hoboken Awarded $230M for implementation of an integrated stormwater management strategy for Hoboken, NJ and parts of Weehawken and Jersey City, NJ
HR&A Advisors, Inc. with Cooper, Robertson & Partners	[1] Coastal Commercial Resiliency Financing [2] Barrier Island: Beach 116th Street, Rockaways [3] Dense Urban Edge: Red Hook, Brooklyn [4] Mainland Coastal: Asbury Park, New Jersey Shore	The Commercial Corridor Resiliency Project Unfunded proposal to enhance resiliency of at-risk commercial areas in Asbury Park, NJ and The Rockaways and Red Hook neighborhoods in New York City through infrastructure, policy and capacity-building.

(continued)

Team name	Design opportunities proposed	Final projects
SCAPE / Landscape Architecture	[1] Living, Growing Breakwaters: Staten Island and Raritan Bay [2] Barnegat Bay Remade: Barnegat Bay, NJ [3] Gardening the Bay: Jamaica Bay, NYC [4] Hudson Habitat: Piermont, NY [5] More Wet Meadow, Less Lands: Hackensack River, NJ	Living Breakwaters Awarded $60M for implementation of the Tottenville Reach section of a larger proposed offshore breakwater project in the Staten Island borough of New York City
MIT CAU + ZUS + URBANISTEN	[1] New Meadowlands: Productive City + Regional Park [2] Hoboken/Jersey City [3] Lower East Side [4] Making Resilient Districts [5] Newtown Creek	New Meadowlands: Productive City + Regional Park Awarded $150M for pilot wetland restoration and creation project in the Little Ferry/Moonachie area of New Jersey, including a flood protection berm; part of larger wetlands restoration and enhancement proposal
Sasaki/Rutgers/ Arup	[1] Resilience + The Beach [2] Barrier Island [3] Headlands [4] Inland Bay	Resilience + The Beach Unfunded proposal for comprehensive and localized resiliency strategies for the New Jersey coastal communities of Union Beach, Keansburg, Asbury Park, Berkeley Township and Tom's River that account for its unique tourism economy.
BIG TEAM	[1] Long Term Perspective – Harbor District: Red Hook [2] The BIG 'U' [3] Long Term Perspective – South Bronx	The BIG 'U' Awarded $335M for phase 1 of a proposed 10 mile berm around lower Manhattan in New York City
unabridged Coastal Collective	[1] Bridgeport [2] Far Rockaway [3] Long Branch [4] Rockville Centre [5] Toms River	Resilient Bridgeport Unfunded integrated coastal management plan for Bridgeport, CT

Post-Sandy recovery and resilience planning in New York City

In New York City, as the short-term emergency and humanitarian response to Hurricane Sandy evolved into a long-range recovery effort in the days and weeks after the storm, the city's Office of Emergency Management (OEM) began to cede responsibility to the numerous other city agencies involved in recovery and resilience building, including the Department of City Planning (DCP), the Department of Housing Preservation and Development (HPD), the New York

City Housing Authority (NYCHA), the Department of Small Business Services (SBS), the Department of Buildings (DOB), the Department of Transportation (DOT) and the Economic Development Corporation (EDC), among others. New units were also created, such as the Housing Recovery Office (HRO), designed to work with affected residents and to implement the city's Rapid Repair and subsequent Build It Back housing repair programmes.

The spatial planning component of the city's recovery was primarily the Special Initiative for Rebuilding and Resiliency (SIRR), established less than a month after Sandy and tasked with developing a long-term strategy for rebuilding that simultaneously accounted for future climate change risks and promoted long-term resilience. Led by the New York City EDC, with staff members on loan from various city agencies, the six-month SIRR process included detailed analysis of climate science and projections to develop an up-to-date snapshot of the city's vulnerability to sea-level rises and other climate risks, as well as stakeholder and public outreach to storm-affected communities. This process occurred in early 2013 in the midst of the active recovery process, with the goal of engaging city agencies and residents in long-term strategic thinking about resiliency while the experiences of Sandy were still fresh.

The SIRR's 438-page final report, *A stronger, more resilient New York* (City of New York, 2013), was released on 11 June 2013, less than eight months after Sandy. Including a thorough Sandy post-mortem and a detailed risk analysis for the city, the plan focuses on a 257-item agenda aimed at 'taking decisive and comprehensive steps to prepare and adapt' (City of New York, 2013, p 2) to the city's increasing risk from climate-induced hazards. Strategies focus primarily on infrastructure, the built environment, critical systems and public awareness, including coastal protections, building retrofits, resilient utility systems and others. Five hard-hit Sandy neighbourhoods (the Brooklyn–Queens waterfront, east and south shores of Staten Island, southern Queens, southern Brooklyn and southern Manhattan), containing altogether 8% of the city's population, received special attention with their own detailed recovery and resilience sub-plans.

Implementation of the SIRR recommendations will be complex and expensive. Developed as a special inter-agency project within the EDC, the city's Office of Long-Term Planning and Sustainability (now the Mayor's Office of Sustainability) was initially responsible for the implementation of SIRR. In March 2014, the city's new mayor, Bill de Blasio, announced the creation of a new Office of Recovery and Resiliency (ORR), which took over responsibility for SIRR's implementation and will coordinate all of the city's resilience efforts in partnership with other agencies. In the context of spatial planning, this includes working with dozens of existing agencies, as well as newly created units such as the DCP's Resilient Neighborhoods initiative and the DOT's Recovery & Resiliency Unit. The cost for full implementation of the SIRR is approximately US$20 billion, for which 75% of the funding has been identified.

New York State's recovery and resilience programme

The federal RBD process was an attempt to address cross-jurisdictional resiliency needs in the aftermath of Superstorm Sandy, but it focused mostly on large infrastructure projects. New York City undertook its own SIRR resiliency planning process, but as the largest city in the US, it possesses a degree of local capacity and expertise that is unmatched in the nation, let alone the region. Indeed, New York City's US$4,213,876,000 CDBG-DR allocation was almost as much as the state itself received (US$4,416,882,000). Other municipalities around the state, however, had also been heavily impacted by Sandy but were not a focus of RBD or capable of mounting their own recovery and resilience planning processes on the order of the SIRR. The state's US$650 million New York Rising Community Reconstruction Program (NYRCR)[1] was designed to fill this void.

NYRCR was a community-based planning process for communities affected by Sandy, as well as Hurricane Irene and Tropical Storm Lee, which struck the state in 2011. NYRCR builds heavily on the FEMA Long-Term Community Recovery (LTCR) process that helped create recovery plans for five upstate New York communities after Irene and Lee, which focused heavily on rapid recovery planning, community engagement and coalition building (FEMA, 2011b). Taking cues from the LTCR model, NYRCR was intended to leverage federal disaster funding and to help communities conduct detailed risk assessments and garner extensive community input to develop locally appropriate plans addressing both Sandy recovery and long-term resiliency. It was funded by approximately one third of the state's US$1.7 billion received in the first tranche of HUD CDBG-DR allocations.

First unveiled in April 2013, the programme was launched in July 2013 when New York Governor Andrew Cuomo officially announced the first 102 'localities' chosen to undertake state-funded planning processes, including municipalities, counties and 18 neighbourhoods within New York City; 22 new localities were added to the programme in January 2014. Each was also allocated an implementation funding grant eligibility amount, ranging from US$3 million to US$25 million based on FEMA damage assessments, totalling US$612,226,846 for all 124 localities. Localities were required to complete Community Reconstruction Plans following a participatory process laid out by the state, with the idea that the grants would provide seed money to begin plan implementation, while further implementation funding would come, it was hoped, from other federal and state programmes (such as the FEMA Hazard Mitigation Grant Program), debt financing, and other sources. The NYRCR was overseen by the newly created Governor's Office of Storm Recovery (GOSR), which also operates the state's housing, small business and infrastructure recovery units. Experienced planners from the New York Department of State and the state Department of Transportation coordinated on-the-ground planning efforts in coordination with the GOSR.

Local steering committees of 10 to 20 members for each locality were appointed by the Governor and chaired by local civic leaders and advocates. The state hired teams of planning consultants at an estimated total cost of US$25,000,000, and assigned them to work directly with communities and the GOSR to facilitate planning processes and write the plans. Some localities partnered with neighbouring jurisdictions to maximise the effectiveness of funding allotments. For instance, the neighbouring suburban Long Island localities of Cedarhurst, Hewlett, Lawrence, Woodmere, Hewlett Neck, Hewlett Harbor, Meadowmere and Inwood merged into one NYRCR 'community', sharing their combined US$27,609,814 in potential grant funding. In total, 66 community plans were ultimately created, with every locality successfully completing its planning process. Planning committees and their consultant teams were given wide discretion regarding the design of the process and the content of the plans as long as they followed general parameters outlined by the GOSR, including a public engagement process consisting of at least three public planning workshops, detailed asset inventories and risk, needs and opportunities assessments. The state also created the New York Rising to the Top (NYRTTT) competition and eight winning plans were awarded an additional US$3 million each for implementation funding for excellence in categories such as regional collaboration, use of technology in the planning process and inclusion of vulnerable populations.

The first of the 66 plans were released on the state's website in the spring of 2014 and the rest later that year. Action items in the plans are categorised as Proposed (expected to be funded by CDBG-DR money), Featured (no funding currently identified) and Additional

Resiliency Recommendations (aspirational). While all of the plans include direct rebuilding projects and some 'hard' infrastructure mitigation projects, such as seawalls and dune restoration, as well as adaptation tactics, such as road raising, most also contain numerous economic development, sustainability and social resilience projects. This breadth of projects illustrates some of the flexibility that has increasingly become part of the broader conversation about resilience, as well as the flexibility that HUD has increasingly championed in the spending of disaster recovery funds.

The GOSR is now working to implement the plans. Although local planning committees are no longer officially part of the process, some have remained as ad hoc working groups to address ongoing recovery and resilience issues in their communities. Private consulting firm Hunt, Guillot & Associates (HGA) of Ruston, Los Angeles, was retained by the state to help implement the plans. HGA worked extensively on post-disaster implementation and programme administration after Hurricane Katrina and will work with local governments and the GOSR to spend each community's implementation funding based on recommendations in their plans. The expectation is that, barring another large-scale disaster, the GOSR will cease operations in or around the year 2020, by which point Sandy CDBG-DR funding will have been exhausted.

New Jersey's recovery and resilience programmes

While both New York State and New Jersey were heavily damaged by the storm and each received just over US$4 billion in CDBG-DR funding (though New York City received an additional US$4.2 billion of its own), the way the two neighbouring states have addressed community-level recovery has been markedly different. One challenge to recovery planning in New Jersey is the sheer number of local governments in the state and the density of population. New Jersey is the fourth smallest state in the US by land area (8,722.58 square miles) but the 11th largest by population (8,938,175 in 2014), making it the densest state in the nation. New Jersey also contains a total of 565 municipalities, of which 57% have fewer than 10,000 residents. Thus, hundreds of New Jersey municipalities have no local capacity to plan for resilient rebuilding. Similar to New York State, this hyper-local decision-making structure also results in only minimal regional planning. Another challenge in New Jersey is the political environment. Unlike other coastal states in the region such as New York, Maryland, North Carolina, Delaware and Connecticut, the state of New Jersey has not integrated climate change as a driver of public policy. For instance, the state has mostly pushed for rapid reconstruction along the New Jersey coast, with little consideration for increased risk or for the integration of resilience measures, with the exception of dune enhancements, and, indeed, has actually made moves since Sandy to make coastal development easier and less regulated. This is a stark contrast to neighbouring New York State, which has adopted a strong climate change policymaking orientation under Governor Cuomo.

New Jersey Governor Christie created a Governor's Office of Recovery and Rebuilding (GORR) after Sandy, but while New York's state recovery office has primary responsibility for designing, operating and coordinating all of New York's disparate recovery programmes, New Jersey's GORR has mainly an oversight role, with responsibility for individual recovery programmes assigned to existing state agencies. While New York created an entire new infrastructure with the NYRCR to assist communities with local planning, New Jersey instead chose to place its community recovery programme in the Office of Local Planning Services (LPS), housed within the Department of Community Affairs (DCA), which administers the state's housing, business and community recovery programmes under the umbrella 'reNew Jersey Stronger'.

Post-Sandy spatial planning in New Jersey is funded through the New Jersey Post Sandy Planning Assistance Grant Program (PSPAG), available to nine New Jersey counties identified by HUD as heavily damaged, as well as municipalities within those counties with a rateable loss of 1% or US$1,000,000 from Sandy. In the first phase, localities apply for US$30,000 grants to complete Strategic Recovery Planning Reports outlining local government recovery and mitigation objectives. Phase 2 grants of up to US$50,000 are available in each of sixteen different categories for the preparation of plans to 'address conditions created or exacerbated by the storm, identify approaches to rebuilding that will be more resistant to damage from future storm events, and encourage sustainable economic growth' (State of New Jersey, no date). The PSPAG contains no implementation funds, and grant stipulations mandate that the grants cannot be used for staff salaries; grantees must hire consultants with American Institute of Certified Planner (AICP) and New Jersey Board of Professional Planners (PP) credentials to develop plans, reports and ordinances consistent with grant guidelines. From New Jersey's US$4,174,429,000 in CDBG-DR funds, US$13.77 million has been allocated to the PSPAG programme. Critics of the state's recovery efforts like to compare this amount to the state's 'Stronger than the Storm' advertising campaign, which has so far cost the state US$25 million (State of New Jersey Department of Community Affairs, 2013).

One challenge under the PSPAG programme is the planning and administrative capacity of small municipalities in Sandy-affected areas. As of May 2015, the state's Sandy Federal Funds Tracker website reports that US$5,780,000 in planning grants have been allocated to 36 communities to fund 110 plans. Berkeley Township, for instance, has received nine grants totalling US$303,000 for a master plan re-examination report, a floodplain management plan, a debris management plan and others. Meanwhile, some communities have only received a grant for the initial Strategic Recovery Planning Report. This is due, at least in part, to a lack of technical capacity and administrative resources in affected communities, many of which are seasonal beach communities with small permanent populations. Recognising this capacity deficit, the non-profit advocacy group New Jersey Future has been supplying planning assistance to six small communities to assist them in applying for state and other grants, and creating community-based planning processes.

Discussion: spatial planning after Sandy

In the New York region, Superstorm Sandy has prompted an array of spatial planning programmes to help communities recover from the storm and protect against future hazards. The planning processes for New York State outlined earlier illustrate the flexibility that states and localities have in spending federal funding and designing recovery processes, as well as how ideas about recovery and resilience are increasingly becoming intertwined in post-disaster spatial planning. Likewise, the inclusion of the public in recovery decision-making processes, which was especially robust in the New York State and federal programmes, is also permeating recovery planning, which has long relied on a more centralised and bureaucratic command-and-control decision-making structure. Increased focus on resilience and community participation also illustrate the increasing linkage between recovery planning and traditional spatial planning, which has long prioritised these concepts more generally.

All of these shifts are consistent with US federal policy, which has been heavily overhauled since Hurricane Katrina. The NYRCR programme is perhaps the most vivid example of that shift. In the two years since NYRCR's inception, 66 recovery plans have been produced and thousands of residents have been engaged in a community-based process that has filled an existing need in communities where participatory planning of any kind has been largely absent; many of the NYRCR communities have never had a comprehensive plan (New York City, for instance) or have not created a new plan for decades. The plans address rebuilding and resilience in a holistic way, with storm mitigation infrastructure, enhanced emergency services and other disaster-specific projects alongside more general community development, economic development and sustainability projects. While funding such projects may prove difficult because of the constraints that still limit CDBG-DR funding to a relatively narrow range of rebuilding and recovery projects, local communities – many with very limited internal technical capacity – now have a thorough and publicly vetted framework around which to focus future planning and advocacy. By contrast, New Jersey's community recovery programmes largely appear to address recovery planning as a traditionally administrative and bureaucratic function; community participation is not mentioned once in the PSPAG application. Furthermore, while 'resiliency plans' are one of the eligible PSPAG grant categories, the available grant amounts are modest and not especially focused on innovation or holistic ideas about resilience, particularly compared to the NYRCR plans. Instead, PSPAG grants are oriented mostly to the fairly routine planning documents that communities might normally prepare as a matter of course.

The state of New York chose to fund and operate its own community-based planning effort – a massive undertaking – as opposed to simply providing funding for local communities to manage their own planning efforts, as New Jersey did. This decision allowed New York State to develop a set of 66 plans that are relatively consistent in terms of the process used to create them, as well as their content and structure, with locally appropriate variations. It may also help facilitate much-needed regional planning and infrastructure spending, which would increase efficiency and stretch limited recovery resources. Additionally, the NYRCR system provided a built-in planning infrastructure regardless of existing local capacity. This proved especially useful in the New York City neighbourhoods that were part of the NYRCR programme since they lack most of the necessary administrative capacity to do their own planning, as do many of the state's smaller municipalities. This allowed many communities to develop relatively sophisticated plans that would have been beyond the means of many of them under any other circumstance and especially in the wake of a major disaster.

It is worth noting that New Jersey's more modest PSPAG programme, nonetheless, provides funding directly to municipalities. While this requires some level of existing administrative capacity in order to apply for and administer the grants, which many communities lack, the resulting plans are officially sanctioned and adopted municipal documents. By contrast, NYRCR plans, sophisticated as they may be, are not formally adopted municipal policy. It will be important to track what happens to the NYRCR plans once CDBG-DR project funding has been exhausted with significant portions of the plans still left to be implemented. In some communities, such as the City of Long Beach just outside New York City, the NYRCR planning process has been a catalyst for more planning. The city of 33,553 is currently updating its 2007 comprehensive plan to reflect some post-Sandy realities but also to take advantage of the ideas and momentum for community change created by the NYRCR process.

Finally, the planning processes utilised by the various jurisdictions will be immaterial if the projects within them cannot be implemented. The NYRCR plans, New York City's SIRR report and the federal RBD programme all include ambitious projects that have been publicly vetted and illustrate holistic and creative solutions to long-term community recovery and resilience. However, many challenges are still ahead. New York City's SIRR has a price tag of US$20 billion and might go higher, and only US$15 billion in funding has been identified. The federal RBD programme includes US$930 million in implementation funding, but that will only get most of the seven winning projects started. For instance, one winning project, The BIG U, was a proposed 10-mile multi-use storm surge barrier around lower Manhattan. The project won US$335 million in the RBD competition, but that will only fund what is known as Compartment 1, a 2.2-mile section on Manhattan's east side. The initial budget for the state's NYRCR programme was US$650 million, or approximately one-third of New York State's first tranche of CDBG-DR funding. Distributed among 66 planning communities, local allocations for project implementation range from US$3 million to US$25 million, while most of the 66 plans contain projects totalling US$30 million or more, leaving an enormous gap to be filled from other sources, few of which have yet been identified. Additionally, among the hundreds of projects, many will not easily be paid for with CDBG-DR funds, which come with very specific limitations. The money will also take a long time to spend. The GOSR is spending each community's allotted funding one project at a time to avoid cost overruns. Detailed planning and design, competitive bidding, environmental reviews, and other aspects of implementation add additional time to this process.

Conclusion

The role of spatial planning in the context of disaster mitigation and recovery in particular is relatively new in the US, where disaster management grew out of Cold War-era civil defence programmes designed to protect and assist communities in the event of nuclear war. This tactical and reactive mindset has proved pervasive, and disaster management has continued to focus heavily on short-term emergency response, with hazard mitigation and long-term recovery receiving substantially less attention (Smith and Wenger, 2006). This is changing, albeit slowly. Beginning with the National Flood Insurance Program (NFIP) of 1968 and continuing with other important legislation, such as the Disaster Relief Act 1974, the Robert T. Stafford Disaster Relief and Emergency Assistance Act 1998 and the Disaster Mitigation Act 2000, the importance of mitigation through spatial planning, as well as the scope of federal disaster preparedness and recovery programmes, has continued to grow. The Sandy recovery process represents the most recent example of recovery from a large-scale disaster in the US and shows clearly how spatial planning and risk reduction continue to be integrated ever-more fully into disaster management.

There are two central challenges in effective recovery planning: one is balancing the public's desire for speedy action with the need to deliberate carefully about how to spend scarce resources (Olshansky and Johnson, 2010); the other is the tension between a desire to simply replace what was lost and a conflicting desire to rebuild better and more resiliently (Kates et al, 2006). Since Sandy, there have been concerted efforts to address these tensions in the New York region, with detailed scientific analysis, public engagement, local capacity building and robust deliberation promoted as cornerstones of the RBD, NYRCR and SIRR programmes. For the most part, these appear to have gone beyond being merely a 'talking shop' (Gaffikin and Sterrett, 2006, p 174) of the type that characterises many participatory planning efforts, but instead were sophisticated participatory processes with robust technical assistance and substantial pre-identified implementation funding. Deliberation has also been rapid and efficient and the spatial planning efforts since Sandy have shown what is possible when governments engage communities in recovery planning from the outset and move quickly and decisively, establishing a new baseline upon which future disaster-affected communities can build. Resilience and smart rebuilding has been front and centre in these conversations and these concepts have rapidly permeated planning and policymaking in New York City, as well as the state and federal government. The central question is whether the rapid process has produced the kind of projects that can be achieved. While the sophisticated plans and beautiful renderings produced for these planning processes foretell an exciting, resilient future for the region, timely and effective implementation of these visions will require ongoing diligence.

Acknowledgements

Research for this chapter was supported, in part, by funding from the National Science Foundation (NSF Award #1335109).

Note

[1] The NYRCR was originally called the Community Reconstruction Zone Program (CRZ) but later rebranded as the New York Rising Community Reconstruction Program. It was often referred to simply as 'New York Rising' but that was also the name of the state's private housing recovery programme, causing occasional confusion.

References

Aon Benfield (no date) *Hurricane Sandy Event Recap Report*, http://thoughtleadership.aonbenfield.com/documents/20130514_if_hurricane_sandy_event_recap.pdf

Berke, P, Karetz, J & Wenger, D (1993) 'Recovery after Disaster: Achieving Sustainable Development, Mitigation and Equity', *Disasters*, vol. 17, no. 2, pp. 93–109.

Blake, E, Kimberlain, T, Berg, R, Cangialosi, J & Beven, J (2013) Tropical Cyclone Report: Hurricane Sandy (AL182012), National Hurricane Center, Miami. Online: http://www.nhc.noaa.gov/data/tcr/AL182012_Sandy.pdf

City of New York (2013) *A stronger, more resilient New York*, City of New York, New York.

Downs, A (2005) 'Smart Growth: Why We Discuss It More than We Do It', *Journal of the American Planning Association*, vol. 71, no. 4, pp. 367 – 380.

FEMA (Federal Emergency Management Agency) (2011a) *National disaster recovery framework: Strengthening disaster recovery for the nation*, Federal Emergency Management Agency Washington, DC.

FEMA (2011b) *Lessons in Community Recovery: Seven Years of Emergency Support Function #14, Long-Term Community Recovery from 2004 to 2011*, Federal Emergency Management Agency Washington, DC.

Gaffiken, F & Sterret, K (2006) 'New Visions for Old Cities: The Role of Visioning in Planning', *Planning Theory & Practice*, vol. 7, no. 2, pp. 159–178.

Kates, R, Colten, C, Laska, S & Leatherman, S. (2006) 'Reconstruction of New Orleans after Hurricane Katrina: A research perspective', *Proceedings of the National Academy of Sciences*, vol. 103, no. 40, pp. 14653–14660.

Olshansky R. & Johnson, L (2010) *Clear as mud: Planning for the rebuilding of New Orleans,* American Planning Association, Chicago.

Olshansky, R, Johnson, L & Topping, K ((2006)) 'Rebuilding Communities Following Disaster: Lessons from Kobe and Los Angeles', *Built Environment*, vol. 32, no. 4, pp. 354 – 374.

Smith, G & Wenger, D (2006) 'Sustainable disaster recovery: operationalizing an existing Framework'. In Rodriguez, H, Quarantelli, E & Dynes, R (Eds.), *Handbook of disaster research*, Springer, New York, pp. 234-257.

Shorto, R (2014) 'How to Think Like the Dutch in a Post-Sandy World', *New York Times Magazine*, online: http://www.nytimes.com/2014/04/13/magazine/how-to-think-like-the-dutch-in-a-post-sandy-world.html?_r=0

State of New Jersey (no date) Post Sandy Planning Assistance Grant, Program Description and Guidelines, State of New Jersey, Trenton. Online: http://www.nj.gov/dca/services/lps/pdf/Post%20Sandy%20Planning%20Assistance%20Grant%20Program%20Guidelines.pdf

State of New Jersey Department of Community Affairs (2013) *Superstorm Sandy Community Development Block Grant – Disaster Recovery, Action Plan Amendment Number 7,* State of New Jersey Department of Community Affairs, Trenton. Online: http://www.renewjerseystronger.org/wp-content/uploads/2014/11/NJ-Action-Plan-Substantial-Amendment-7-R-FINAL-formatted-5-23_CLEAN-ve-.pdf

CHAPTER A4a

Spatial planning focusing on risk management in Slovakia

Alena Kučeravcová and Ján Dzurdženík

Slovakia	
Population: 5.4 million (2014), 109 inhabitants per km^2	**Government:** Parliamentary republic
Area: 49 035 km^2	**Administrative structure:** Slovakia is divided into eight self-governing regions, which are further subdivided into 79 districts and just below 3000 municipalities
Hazard profile: Floods (nationwide), especially flash floods Landslides hazard (nationwide) Drought hazard	**Authorities in charge of risk assessment and management:** Different ministries and their managed organisations (Water Management Enterprise, Slovak Hydro Metrological Institute, State Geological Institute etc.)
Analysed event: Floods in Eastern Slovakia (2010) Landslides in Eastern Slovakia (2010) Estimated return period: Estimated damages: 695,1 million € (floods 2010) Casualties: -	**Role of spatial planning:** Permanently and complexly solves the spatial configuration and functional utilisation of the territory. Most frequently are taken into account areas with increased flood risk and landslide territory.

Introduction to Slovak landscape and governmental structure

Landscape

The Slovak Republic is a landlocked country in Central Europe. The Slovakian surface is very irregular, characterised by lowlands, valleys, hills, highlands and mountain chains. The Slovak topography is dominated by the Carpathian Mountains. The highest point is at the summit of Gerlachovský štít in the High Tatras, at 2,655m, while the lowest point is the surface of the Bodrog River on the Hungarian border, at 94m. A great part of the territory is covered by protected landscape areas. From a landscape structure point of view, 41% of the surface is covered by forests, 31% is arable land, 17% is pastures, 3% is cultivated land and the remaining 8% is mostly covered with human structures and infrastructure, and partly with rocky mountain ridges and others. The Slovak territory is crossed by the main European watershed. From Austria to Slovakia flows the largest Central European river, the Danube, connecting Slovakia with the Black Sea. The largest reserves of groundwater in Slovakia are located in Danube river sediments in the area of Žitný ostrov (Rye Island). The longest Slovak river is the Vah, at 390km.

Governmental structure and competences

There is a so-called parallel model in Slovakia, in which the state administration is a completely separate, independent component of the public administration. For administrative purposes, Slovakia is divided into eight regions (see the key facts box at the beginning of the chapter), which are legal entities. The regions have self-administration (original) competences, but they also perform duties transferred from the state administration. The original competences of the regions are: roads of class II and III, area planning, regional development, their own investment ventures, secondary schools, hospitals, social facilities, cultural facilities, participation in civil defence, and so on. The competences of the regions include the issuing of generally binding regulations. The local level of self-government is represented by the 2,931 municipalities, which are the most important bodies of self-government in Slovakia. The most important tasks performed by local municipalities are: local roads, public transport, environment, water supply, sewerage and communal waste, local development, housing, pre-school and school facilities, social and health-care facilities, and participation in regional plans. The municipality is self-governing and fully autonomous within its own territory (PLUREL, 2010).

With regard to the question of spatial planning, the most important body is the local municipality. As competences of territorial governance are distributed among the territorial governance levels in a 'complementary' way, there is no overlap of powers. Hence, the municipality is self-governing and fully autonomous within its own territory, as the regional bodies and districts are self-governing on regional issues.

Introduction to the spatial/territorial planning system

The planning system is decentralised and based on national legal hierarchical levels. At the national level, the relevant ministries are vested with planning competences. At the regional level, there are self-governing regional bodies (a total of eight regions), and at the local level, there are individual towns and villages, vested with competences and responsibilities for the planning and development of the respective territories.

Spatial planning is defined in Act No 50/1976 Coll on Spatial Planning and Building Regulation (Building Code) and its latest amendments. This Act regulates all three responsible planning levels (national, regional, local). According to the Act, spatial planning permanently and complexly solves the spatial configuration and functional utilisation of the territory. It determines principles and proposes and coordinates the timing of substantive activities influencing the environment, ecological stability, the cultural and historical values of the territory, territorial development and landscape creation in line with sustainable development principles.

Spatial planning documents are the main instruments of spatial planning, and shall be elaborated for the national level, regional level, municipalities and their parts (see Table 5).

Table 5: Spatial/territorial planning system in Slovakia

Level	Name of the planning documentation	Scale	Legal impact	Production/ obligatory
National	The Slovak Spatial Development Perspective 2001/2011	1:500 000	Binding - approved by the government	Obligatory
Regional	A regional territorial plan	1:100 000/ 1:50 000	Binding – approved by self-governing region	Obligatory
Local	A municipal territorial plan	1:10 000/ 1:5 000	Binding – approved by municipality	Obligatory - for every municipality with more than 2000 inhabitants is obliged
Sub-local	A territorial plan of Zone	1:1 000/ 1:500	Binding – approved by municipality	Obligatory under circumstances specified by law

The Slovak Spatial Development Perspective

The basic document on territorial planning, which covers the whole country, is the Slovak Spatial Development Perspective 2001, as amended by the Slovak Spatial Development Perspective 2011 –

Amendment No 1. The characteristics of the defined area shall be processed for the whole territory of the Slovak Republic. It shall design the spatial arrangements and functional use of the territory of the Slovak Republic and establish the framework for social, economic, environmental and cultural requirements for spatial development, care for the environment and landscape creation for the Slovak Republic and its regions. The spatial planning material elaborating on this is the Spatial Development Strategy of Slovakia. Processes shall be visualised at a scale of 1:500,000 (Slovakia) and a scale of 1:3,000,000 (international context). The Slovak Spatial Development Perspective shall establish, in particular:

- the arrangement and hierarchy of the settlement structure and nodes of residential and commercial agglomerations in international and national contexts; and
- the development of main urbanisation axes, ensuring environmental stability and territorial development aimed at the creation of equivalent living conditions in the whole territory of the Slovak Republic (see Figure 37).

Binding parts shall establish the concept of principles and regulations governing the requirements, particularly of sector conceptions, regarding the territorial arrangement and functional use of the territory in compliance with principles of sustainable development and the protection of the environment. The binding parts of the planning documentation approved by the government shall be declared by government regulation.

Figure 37: The Slovak Spatial Development Perspective (example of the complex proposal)

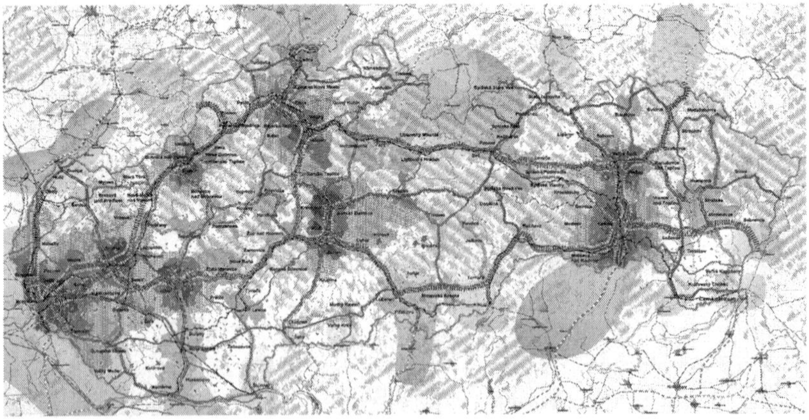

Source: Slovak Spatial Development Perspective (2011)

Regional spatial/territorial plan

A regional territorial plan is elaborated for a part of the landscape with several municipalities, in which it is necessary to design specific development plans or activities significantly affecting the spatial arrangement and functional use of the territory. The regional plan shall take into account the binding part of the Slovak Spatial Development Perspective. Graphic parts shall be processed at a scale of 1:100,000, 1:50,000, respectively, at a scale of 1:25,000 (main drawing) and 1:500,000, 1:200,000 (wider relations).

A regional spatial plan shall establish, in particular:

- principles and regulations for: the settlement structure, spatial arrangement and functional use of the territory; the arrangement of transport and technical facilities; environmental protection; the territorial system of ecological stability; and cultural heritage and the use of natural resources in terms of sustainable development; and
- mutual links between regional spatial development and its municipalities, and relations with neighbouring regions.

Binding parts shall establish a concept of the principles and regulations of the spatial arrangement and functional use of the territory as the basis for the approval document. According to the Building Code, a self-governing region is responsible for the elaboration and approval of the spatial plan of the region and its obligatory part is issued by a generally binding regulation.

Municipal spatial/territorial plan

At the municipal level, all villages and towns of more than 2,000 inhabitants should formulate their own territorial plans. Local plans have to comply with the framework given by the regional plans.

The characteristics of the defined area shall be elaborated for the territory of one municipality or for a territory of two or more municipalities. Cities and municipalities with a population of more than 2,000 are required to have a municipal spatial plan. If two or more municipalities make an agreement they may have one common municipal spatial plan. Graphic parts shall be processed at a scale of 1:10,000, 1:5,000, respectively, 1:2,000 or 1:2,880 and at a scale of 1:50,000, 1:25,000 (wider relations).

A municipal territorial plan shall establish, in particular:

- principles and regulations for the spatial arrangement and functional use of the land of the municipal territory;
- principles and regulations for the arrangement of transport and technical facilities, environmental protection, a territorial system of ecological stability, cultural heritage and the use of natural resources in terms of sustainable development;
- the acceptable, restricted and prohibited functional use of areas; and
- areas for public constructions, areas intended for sanitation and areas for protected parts of the landscape.

The binding parts shall establish a concept of regulations, with the detailed principles of the spatial arrangement and functional use of the territory expressed in the form of regulations containing binding rules. The municipality approves the spatial planning documents and declares its binding parts by a generally binding regulation.

Spatial/territorial plan of zones

The characteristics of the defined area shall be elaborated for a part of the municipality. If approved, or if the municipality has less than 2,000 inhabitants, the municipal spatial plan establishes that a spatial plan of zones for a specific part of the municipality is to be provided, and the land or construction for public purposes is to be specified. Graphic parts shall be processed at a scale of 1:500, respectively, 1:1,000 and at a scale of 1:5,000, 1:2,000 (wider relations).

A spatial plan of zones shall establish, in particular:

- principles and regulations for: the more detailed spatial arrangement and functional use of the territory, buildings and public transport and technical facilities of the mutual connection; construction locations in particular lands and urban areas; the building-up conditions of individual building lands; protected parts of the landscape; the location of greenery; and elements of a territorial system of ecological stability;
- the coordination of material and time for new construction and the sanitation of existing buildings; and
- lands for public constructions, building prohibitions and carrying out sanitation.

The binding parts shall establish a precisely formulated concept of the regulations for functionally homogeneous units. The municipality approves spatial planning documents and declares its binding parts by generally binding regulations.

Spatial planning in the Slovak Republic is a relatively complex set of instruments and methods at the national, regional and local levels, with the emphasis on applying the decision-making power of the self-government and executive authorities at these levels. The system's character and the relationships of the instruments have not yet been sufficiently reflected in planning practice due to the uncompleted state of the self-governing institutions at the regional level of decision-making. Therefore, informal instruments come to the forefront in the process of decision-making as they are applicable in vertical, as well as in horizontal, integration and in coordination with various interests in the territory.

Natural hazards and spatial planning

Natural hazards are usually defined as extreme natural events that pose a threat to people, their property and their possessions. Europe is experiencing an increasing number and impact of disasters due to natural hazards and technological accidents caused by a combination of changes in its physical, technological and human/social systems. A new report by the European Environment Agency (EEA) concludes that the number and impact of disasters have increased in Europe during the period 1998–2009. Flooding and storms were the most costly hazards. The overall losses recorded in the study period added up to about €52 billion for floods and €44 billion for storms. Flash floods are particularly dangerous in mountainous areas, where the steep slopes may increase their destructive potential.

There are growing concerns with the impacts of natural hazards related to the loss of life as well as their major costs, both financial and non-financial. The degree of natural hazard risks associated with climate change, for example, floods, forest fires and heatwaves, is actually not fully understood. However, these events occur with apparent increasing frequency and magnitude. Natural hazards and the impacts of climate change affect and will affect European regions and cities in an asymmetric way, with differences in magnitude and timescale. These incidents and consequences hamper contributions to the overall objectives of the Europe 2020 strategy of smart, sustainable and inclusive growth. European Union (EU) cohesion policy supports adaptation and mitigation measures (ESPON, 2013).

Risks due to natural hazards create a certain challenge for many institutions and stakeholders involved in dealing with risks. Spatial planning is an important part of these actions and can help to reduce the vulnerability of societies to natural hazards. From the potential natural risks point of view, in the Slovak Republic, the risks most frequently taken into account are increased flood risk (flash floods) and landslides.

Floods

Floods are natural phenomena that cannot be prevented. Floods have the potential to cause fatalities, the displacement of people and damage to the environment. Some human activities (such as increasing human settlements and economic assets on flood plains and the reduction of the natural water retention by land use) and climate change contribute to an increase in the likelihood and adverse impacts of flood events (EC, 2007).

In the last decade, Slovakia was increasingly affected by floods. These floods have resulted in substantial material damage and there are exceptional cases of loss of human life. Floods constantly point to the fact that society is very vulnerable to flooding. In the last 13 years, floods have caused the loss of at least 56 lives, and affected more than 2,500 villages and towns in Slovakia; in 2010 alone, flood damage was calculated to have cost €695.1 million. The scope and extremity of flood episodes point to the need to build a comprehensive proposal, eventually leading to the completion of flood protection measures in potential flood areas (see Figure 38) (Zeliňaková et al, 2011).

Flash floods in particular have occurred more frequently during the last decades in Slovakia. Flash flooding (Perry, 2000) is a type of flood caused by excessive rainfall in a short period of time, generally less than six hours. Flash floods are usually characterised by violent torrents after heavy rains that rip through river beds, urban streets or mountain valleys, sweeping everything before them. Flash floods are very dangerous because they can occur within several seconds to several hours, with little warning. The main characteristic of flash floods is their extremely sudden onset. The factors that contribute to this type of flood are rainfall intensity, rainfall duration, surface conditions and the topography and slope of the receiving basin.

'Flood risk' means the combination of the probability of a flood event and of the potential adverse consequences for human health, the environment, cultural heritage and economic activity associated with a flood event (EC, 2007). In recent years, extreme flood events have had huge economic, social and environmental impacts. Besides this, enhanced climate variability and climate change are expected to increase the frequency and intensity of floods. The most endangered areas to floods in Slovakia are its eastern parts, particularly the Bodrog and Hornád river basins.

Figure 38: Eastern Slovakia affected by flood in May 2010 (Slovak Water Management Enterprise, branch Trebišov, 2010)

Flood risk assessment and management

Specific flood prevention policies exist in many European countries and the European Commission's Flood Directive adopted in 2007 (EC, 2007) is a good example of concerted action at the EU level.

The directive aims at reducing the risks and adverse consequences of floods and was implemented in the member states in three stages, starting with a preliminary flood risk assessment (in 2011), followed by the development of flood hazard and risk maps for flood-prone zones (in 2013) and flood risk management plans (in 2015).

In Slovakia, protection against floods is a nationwide task. The basic rules for the prevention of and protection against floods are provided in Directive 2007/60/EC, which was transported into the legal system of the Slovak Republic by Act No 7/2010 Coll on Protection Against Floods, and generally binding regulations that establish the details of its implementation. These regulations contain a comprehensive planning system for flood risk management, which must be compulsorily implemented in all the member states of the EU:

- According to Directive 2007/60/EC, EU member states had to complete a preliminary flood risk assessment by 22 December 2011. The assessment includes maps of the river basin district showing topography and land use, a description of the floods that have occurred in the past and that had significant adverse impacts, the likelihood of similar future events, and the potential adverse consequences of future floods, taking into account hydrological and geo-morphological characteristics (see Figure 39). Based on available or readily derivable information, such as records and studies on long-term developments, in particular, the impact of climate change on the occurrence of floods, a preliminary flood risk assessment was undertaken to provide an assessment of potential risks. The Ministry

Figure 39: Geographical areas of potentially significant flood risk in the Slovak Republic

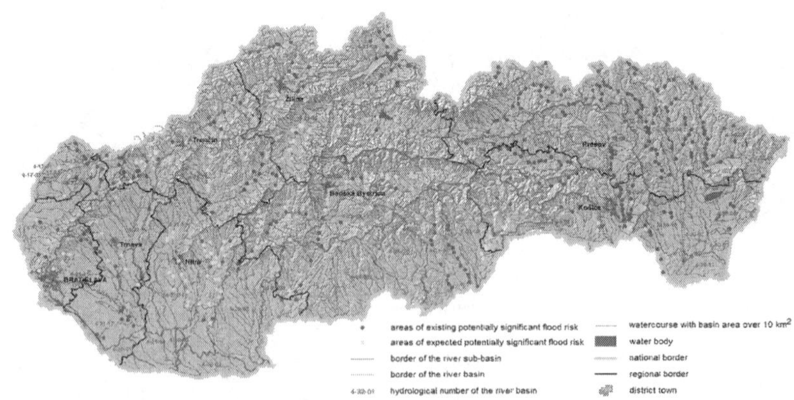

Source: Ministry of Environment SR (2011)

of Environment of the Slovak Republic (MoE SR) was in charge of the preparation of the preliminary flood risk assessment. The preliminary flood risk assessment shall be reviewed, and if necessary updated, by 22 December 2018 and every six years thereafter.

- In order to assess flood risk, it is necessary to identify both the probability and consequences of flooding. *Flood hazard maps* and *flood risk maps* were prepared by EU member states by 22 December 2013 (see Figure 40). The flood hazard maps cover the geographical area that could be flooded according to different scenarios, with a low, medium and high probability (from every 1,000 years [Qmax 1000] to every five years [Qmax 5]). The flood risk maps show the potential adverse consequences relating to human health, economic activity, the environment and cultural heritage associated with floods under these scenarios. Flood hazard maps and flood risk maps shall be reviewed, and if necessary updated, by 22 December 2020 and every six years thereafter.
- The preliminary flood risk assessment was used to identify areas that need to be considered in more detail through mapping and potentially the preparation of flood risk management plans (Gaňová et al, 2014). Based on the maps, member states established flood risk management plans by 22 December 2015. Flood risk management plans shall take into account relevant aspects, such as costs and benefits, flood extent, flood conveyance routes and areas that have the potential to retain flood water such as natural flood plains, the environmental objectives of Directive 2000/60/EC, soil and water management, spatial planning, land use, nature conservation,

Figure 40: Flood hazard maps and flood risk maps are available for public on Ministry of Environment website (http://mpomprsr.svp.sk/Default.aspx)

navigation, and port infrastructure. The purpose of the flood risk management plans is to identify means of reducing the impacts of flooding. Flood risk management plans shall be reviewed, and if necessary updated, by 22 December 2022 and every six years thereafter.

Flood risk management plans shall include *preventive measures* to reduce the adverse impact of floods in order of the urgency of their implementation. The diversity of nature does not allow to apply one method of protection against floods everywhere. This fact respects Act 7/2010 Coll on Protection Against Floods, and provides five basic groups of preventive measures for protection against floods:

- measures increasing the retention capability of territory or, in appropriate areas, supporting natural water accumulation to slow down water drain from the surface to watercourses in order to protect the territory against flooding from surface drain. Examples include measures regarding forests, agriculture land and urban areas;
- measures decreasing the maximum flow during floods, for example, water reservoirs (dams), shoes (water gates) and polders (see Figure 41);
- measures protecting territory against water from rivers, for example, the modification of watercourses, levees or flood lines;
- measures protecting territory against flooding by inner waters, for example, systems of sluiceways and pumping stations; and
- measures ensuring the capacity of watercourse beds, for example, the remove of deposits.

The flood prevention measures are decided by the MoE SR, which performs its work through its managed organisations (Slovak Water Management Enterprise, Slovak Hydro Meteorological Institute and Water Management Construction).

In flood protection, there are no universal recipes that can be applied always and everywhere without distinction since nature and landscapes are too diverse. Preventive measures that are effective in one location may increase flood risks under other conditions. The simplest, most effective and, at the same time, also cheapest measure to protect against flooding is to create the necessary space for water and not to stand in its way (Bačik, 2010).

Figure 41: Examples of preventive measures – polder (Zeleňaková, 2011) and wooden check dams in Malá Lodina Municipality (Pavol Vaľa, 2011)

Spatial planning and flood risk management

Spatial planning permanently and complexly solves the spatial configuration and functional utilisation of the territory. It determines principles and proposes and coordinates the timing of substantive activities influencing the environment, ecological stability, the cultural and historical values of the territory, territorial development and landscape creation in line with sustainable development principles.

According to the Building Code, one of the main aims of spatial planning is to determine the regulations relating to the spatial configuration and functional utilisation of the territory. It logically follows that spatial planning should be an effective tool in preventing the occurrence of flood damage and other risks caused by floods, mainly by the restriction of construction and inappropriate activities in flood-prone areas. In this respect, Act No 7/2010 Coll on Protection Against Floods has a substantial role in planning the placement of buildings. This Act works with the concept of *flood-prone areas/inundation areas* and specifies the duties of the relevant state administration of the self-governing regions, cities and municipalities during floods:

- The Act imposes a duty on municipalities to ensure the indication of any flood plain lines shown on the flood hazard maps within the municipal territorial plan or territorial plan of zones at its next examination.
- The Act defines the kinds of buildings, objects, devices and activities that are forbidden in flood-prone areas. There are also defined activities that can be permitted in flood-prone areas.

Determining flood-prone areas is important for decision-makers for planning and management activities. Defined flood-prone areas are most important on the local level (larger map scale). The municipality determines the regulations of the spatial arrangement and functional use of land in the local spatial plan or in the local plan of zones, with measures to protect against floods (see Figures 42 and 43).

The extent of inundation areas and flood plain lines depends on the functional utilisation of the territory by the water flow. Changes in the functional use of the updated spatial plan are an essential reason for changing the size of inundation areas and flood plain lines (Bačik, 2010). There is a need to determine the extent of inundation areas sensitively in order to avoid unnecessarily restricting activities near watercourses in areas where it is not necessary (see Figure 44).

On the regional level, for example, in the frame of the spatial plan

Figure 42: Example of spatial plan on local level: the territorial plan of Nižna Myšľa village with exactly defined flood-prone area and measures of flood protection (levees)

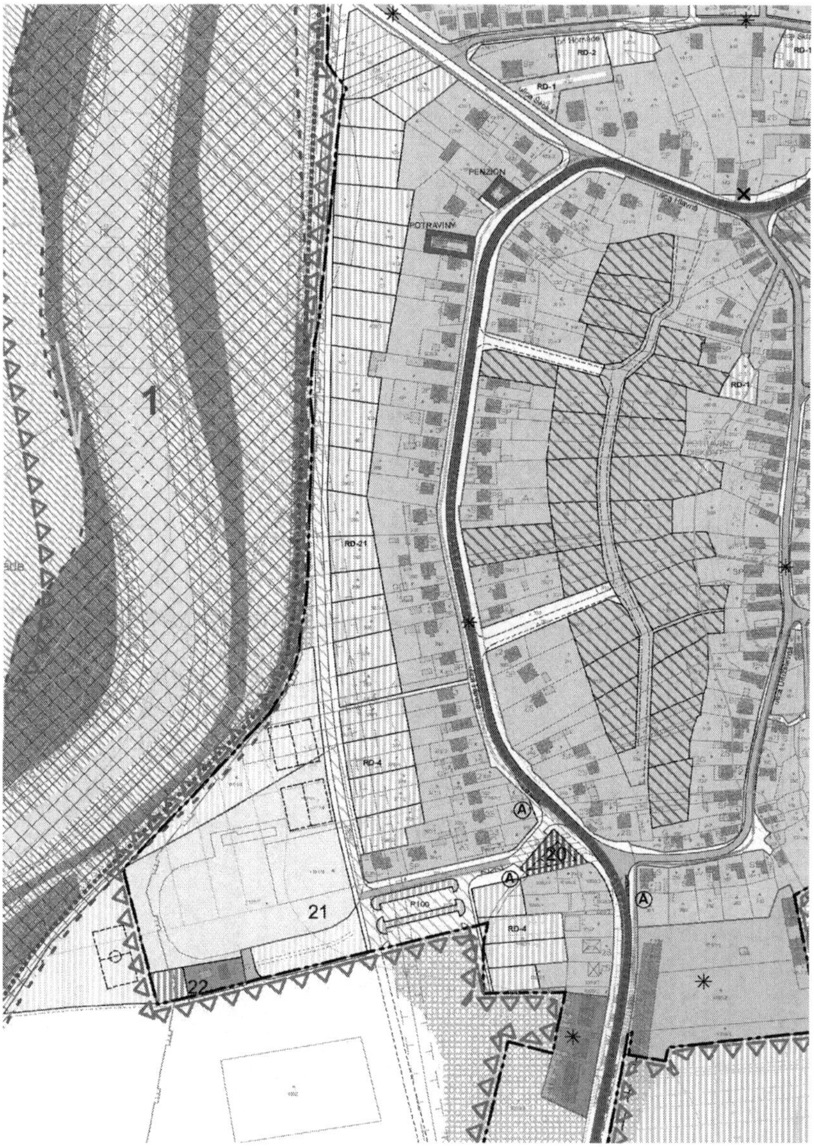

of the Košice Region, flood prevention measures are present only in textual part, without graphic localisation. Maps do not contain identified flood-prone areas, inundation areas or flood plain lines due not mainly to missing information, but to the scale – defined inundation areas are most important on the local level (larger map scale) (see Figure 45).

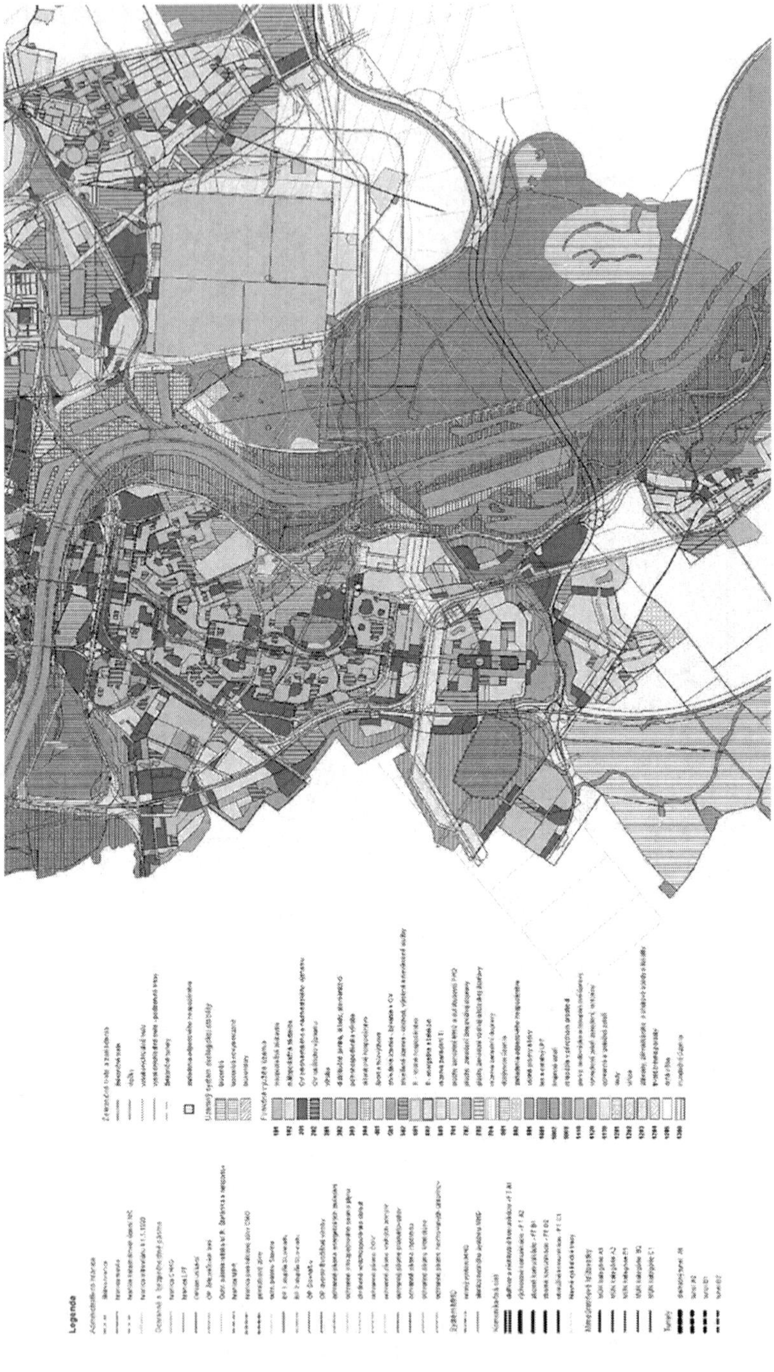

Figure 43: Territorial plan of the city Bratislava with exactly defined flood-prone area

Figure 44: Defining functional utilisation of area near river – mostly as arable land, gardens and sport facilities (Local spatial plan of Družstevná pri Hornáde village)

Figure 45: Regional spatial plan containing mitigation measurement (Košice-region, near UA-SK border)

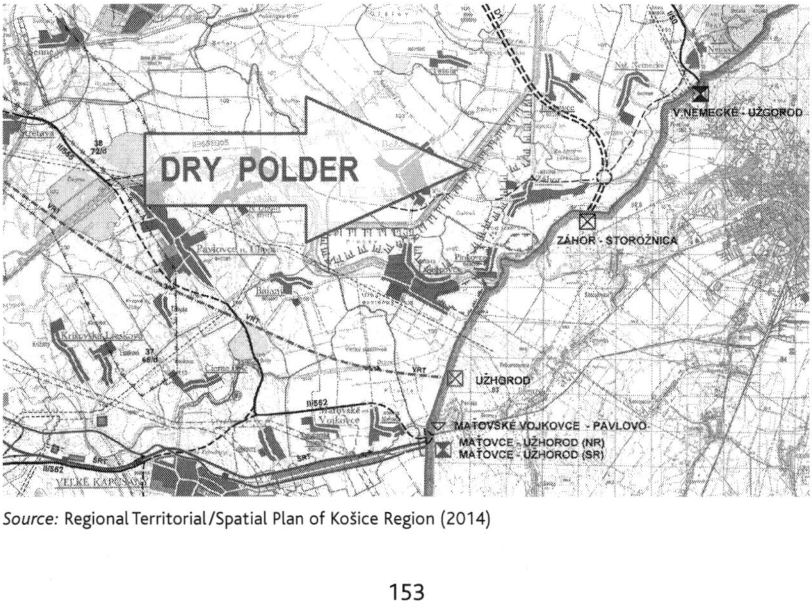

Source: Regional Territorial/Spatial Plan of Košice Region (2014)

The process of spatial planning, in coordination with the rational utilisation of the territory and basin, has an irreplaceable role. Much greater emphasis should be put on preventive flood protection measures. Flood prevention consists in taking into account the flood risk in spatial planning and the construction of buildings in a safe place, appropriate land use, and the rational management of forests and agricultural land in line with sustainable development principles.

Landslides

Slope deformations represent one of the most significant geodynamic phenomena not only in the Slovak Republic, but also in the whole of Central Europe. According to the Atlas of Slope Stability Maps, the 21,192 slope deformations registered up to now cover 2.575.912km^2, representing 5.25% of the total area of the Slovakian territory. The largest representation of slope deformations are landslides (see Figure 46), with 19,104 registered in 2006, representing 90.2% of all registered slope deformations (Ministry of Environment SR, 2013).

The main natural causes of slope deformations are climate factors in combination with erosion caused by watercourses, seepages and buoyancy forces of the groundwater. Among the anthropogenic causes we have to mention in particular undercutting of the slope, inappropriate deforestation of the slope and uncontrolled drainage of surface water. The ongoing climate change has significantly negative impact on landslide occurrence, especially in areas not affected by landslides in the past.

Figure 46: Landslide hazard (Liščák 2002, modified by Minár et al, 2006)

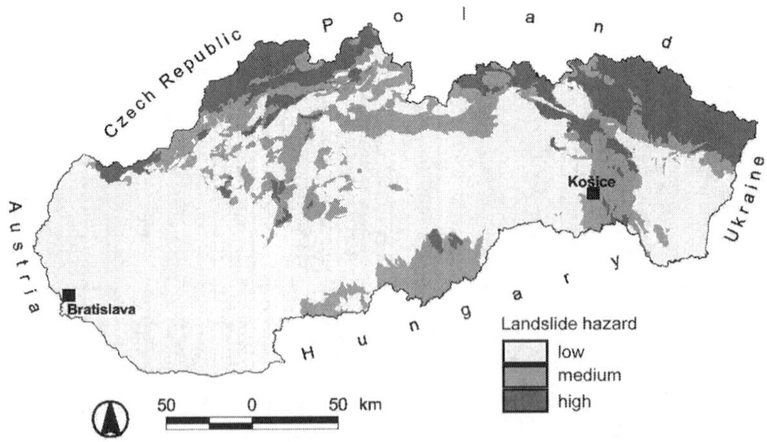

The landslides are especially concentrated in the Flysch and Neovolcanic regions. Many new slope failures occurred after extremely heavy rainfall in the year 2010; there were 577 in total in Eastern Slovakia. Up to 127 of them were marked with high (3) or very high (4) socio-economic relevance. This situation led to a number of emergencies in the municipalities in the region. The most dangerous landslides occurred in the Nižná Myšľa village, destroying dozens of family houses, local communication and engineering networks (see Figure 47). As a result of serious static damage, 29 family houses had to be demolished and several others became uninhabitable (Petro et al, 2014).

Landslide risk assessment and management

Results of a systematic inventory of slope failures, as well as results of other studies and investigations focused on landslides in Slovakia, are concentrated in the Atlas of Slope Stability Maps at the scale of 1:50,000 (see Figure 48).

The landslide prevention and landslide risk management programme for 2014–2020 ('the Programme') was developed in 2013. The aim of the Programme is to provide a comprehensive and systematic system to tackle slope movements and landslide risks in the Slovak Republic, with an emphasis on the most vulnerable areas. The Programme determines the general objectives and measures, sets targets to improve the prevention and management of landslide risks, and is intended to implement them through activities and programme measures by 2020. The Programme defines not only the progress of work in addressing landslides in order to minimise their gradual negative effects on the environment and human health, but also the financial demands and resources needed to address the issue.

Landslides and spatial planning

All geological and Engineering Geology (EG) maps in Slovakia during last 35 years have been prepared according to valid Directions and Geological Laws. The maps were compiled by different techniques (hand drawing, digitising, computing by Geographic Information System (GIS) tools) and comprise explanations, graphic supplements and cross-sections. Their accessibility for potential users is ensured by the central geological archive Geofond. In spite of the fact that EG maps represent a very good basis for the preparation of territorial/spatial planning documents, their use by planners (architects, designers)

Figure 47: Landslides occurred in the Nižná Myšľa village in 2010 (Ľ. Petro, M. Kováľčik 2010)

(a)

(b)

Figure 48: Map of slope failures of Slovakia at the scale 1:50 000 (available at the map server of the ŠGÚDŠ http://mapserver.geology.sk/zosuvy/)

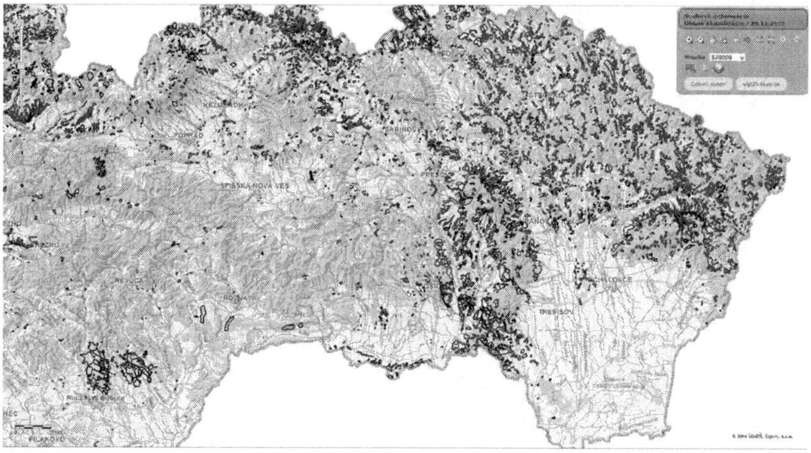

is very limited in Slovakia. The main problem consists in insufficient Slovak legislation (Petro, 2014).

According to the very old and still valid Building Law 244 No. 50/1976, there is no obligation for land-use planners to incorporate existing EG maps into territorial/land-use planning documents. Moreover, the construction of many buildings (structures), and especially domestic houses, is carried out without reviewing or investigating any EG maps. This status very often leads to the situation that structures of different types are designed or constructed in areas (regions) with high geological hazards (eg in Nižná Myšľa). As a result, a great number of structures (buildings) are damaged or destroyed and people are endangered. Material losses reach millions of euros every year. The situation is being made worse due to the reduction of suitable foundation soils, an increasing number of heavy rainfalls and the low number of insured properties (Petro, 2014).

According to Act No 569/2007 Coll on Geological Works (Geological Act) and its latest amendments (November 2013), after the experience from the landslides in 2010, spatial planning has to take into account geological data, especially those connected with possible landslides (see Figure 49). The MoE SR will therefore provide information about potential landslide areas and other geological hazards. The conditions of the utilisation of these areas also have to be defined. This should ensure so-called precautionary systems aimed at eliminating undesirable construction interventions. This can prevent future expenses for the remediation of landslides caused by human activities.

Figure 49: Spatial Plan of Zone - Vyšné Opátske, Košice – an example with identified active and potential landslide areas

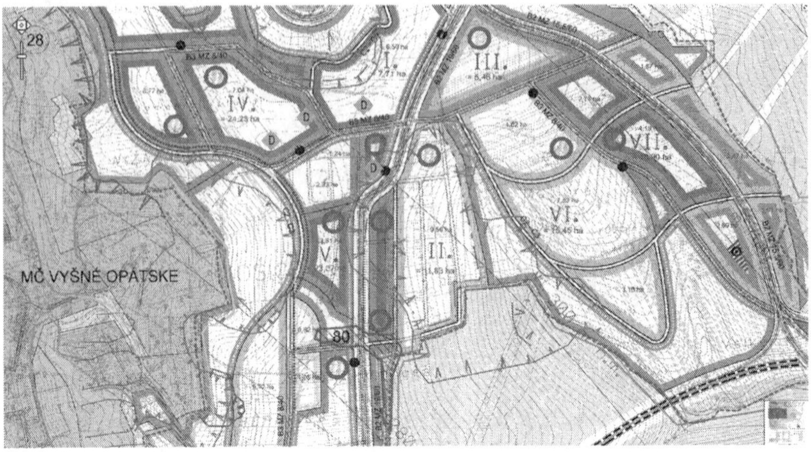

Strategic Environmental Assessment and spatial planning

Strategic Environmental Assessment (SEA) is an integral part of the spatial planning procedure after the amendment of the Spatial Planning Act according to the new Environmental Impact Assessment (EIA) Act. According to the EIA Act, the application of SEA is obligatory for:

- a substantial development policy, especially in the area of energy supply, mining, industry, transport, agriculture, forestry and water management, waste management, and tourism;
- territorial/spatial planning documentation for the regional and residential settlements of selected areas, especially the centre of a region, urban conservation areas, spas and particularly polluted localities; and
- any proposal for generally legally binding regulations that may have an adverse impact on the environment.

It is linked to a tiered process, to spatial planning levels and also to decision-making (policy, plans and programmes). According to the EIA Act, the draft of a territorial/spatial planning documentation:

- has to contain the predicted environmental impact evaluation, as well as mitigation measures, in order to prevent, eliminate, minimise and compensate for environmental impacts;

- has to have various solutions to the concept (usually three variants of the draft plan are available); and
- has to state that the spatial planning documentation designer should discuss the predicted impacts and proposed compensation with the MoE SR.

At all levels and in all categories, spatial planning documentation represents one of the fundamental instruments of environmental protection, and of sustainable development. Its role is not only to provide a comprehensive solution to the functional utilisation, coordination and organisation of investment, but also to create the preconditions to ensure the sustainable balance of all natural, civilisational and cultural values of the territory. The evaluation of spatial and cumulative effects is expected from the SEA, mainly from the SEA in the spatial planning process representing inter-sectoral planning. The SEA in spatial planning has to point out and address potential environmental risks and their originators (Belčaková, 2004).

The most important function of the SEA in spatial planning is its contribution to a stronger environmental orientation. On the one hand, the SEA directly joins the process of planning and decision-making and, on the other hand, it can meet those impacts that interfere above the framework of the actual decision-making process. Further functions of the SEA in spatial planning include (Belčáková, 2004):

- helping with the preparation of decisions, having a possibility to bring a great amount of information on predicted environmental impacts resulting from planning measures during decision-making process;
- explaining the ecological, social and economic context of the planning process and functioning as a warning in order to protect the environment against already-revealed and clearly described dangerous situations and risks, as well as those potential dangers that are not clearly, which is important at the regional level; and
- presenting information on such environmental impacts whose existence and consequences are not exactly defined.

Conclusion

The prevention of natural hazards needs comprehensive information and data about their causes and effects. Therefore, a systematic framework for the assessment and mapping of natural hazards is needed. Geological and hydrological information, such as thematic hazard maps, have a very high potential for reducing the fatality rates and losses due to disasters.

A good example of concerted action at the EU level is the Flood Directive adopted in 2007, which is being implemented in the member states in three stages, starting with a preliminary flood risk assessment (by 2011), followed by the development of flood hazard and risk maps for flood-prone zones (by 2013) and then flood risk management plans (by 2015).

In Slovakia, data (comprehensive information systems on the territory concerned, information on the designated flood-vulnerable areas, flood hazard maps and flood risk maps) that were not available or were insufficient for spatial planning documents are now published on the Internet and other appropriate places, including municipal, district and building offices. Flood hazard maps and flood risk maps are expected to increase public awareness of flood-vulnerable areas. The determination of the extent of the inundation/flood-prone areas in spatial planning documents mandatorily sets the conditions that are necessary for the implementation of various activities (defining the kinds of activities that are forbidden and that can be permitted in flood-prone areas)

From the legislative point of view, there are requirements in the modern Building Act that will also create obligations for land-use planners (architects, designers), as well as all investors, to ensure that EG investigations or reviews are carried out before starting the development of spatial planning documents. After catastrophic landslides in 2010, some improvements can also be observed in the legislation concerning landslides. According to Act No 569/2007 Coll on Geological Works (Geological Act) and its latest amendments in 2013, spatial planning documents have to take into account geological data, especially those connected with possible landslides. The effects of climate change on the future frequency and intensity of landslides are not fully understood due to a lack of information. The development of comparable guidelines for landslide risk assessment and a database on landslide occurrence and impacts should be considered.

Also, the implementation of the SEA in spatial planning is extremely important and requires adequate attention not only in the theoretical and methodological fields, but also in practice. To make the integration of the SEA and spatial planning processes more efficient, the integration has to be initiated at the very beginning of spatial/land-use planning processes, at the stage when spatial plan goals are being established, not at the stage where the draft plan is already available.

Spatial planning has to provide innovative models for regional development. The future prevention of natural hazards has to be taken into account as a priority. It is necessary to use the instruments of spatial planning to contribute to the prevention of risks and to mitigate the effects of natural hazards. Actions and measures, if well implemented, can reduce the impact of hazardous events. Spatial planning should therefore be seen as a unique tool for creating well-maintained and well-functioning landscapes.

References

Act No. 50/1976 Coll. on Spatial Planning and Building Regulation (Building Code).

Act No. 569/2007 Coll. on geological works (Geological Act) and it latest amendments.

Act No. 7/2010 Coll. on protection against floods.

Bačík, M. 2010, 'Flood maps and territorial planning', *Urbanita*, vol. 4/2010, pp. 16-20.

Belčáková, I. 2004, 'Strategic Environmental Assessment and Spatial Planning in Slovakia: Current Practices and Lessons for Practical Application of the EC SEA Directive'.

EC (European Commission) (2007) *Direcitve 2007/60/EC of the European parliament and of the Council of 23 October 2007 on the assessment and management of flood risks.*

ESPON (2013) *ESPON Territorial Observation No. 7 on Territorial Dynamics in Europe - Natural Hazards and Climate Change in European Regions*, 2013.

Gaňová L., Zeleňáková M. & Purcz P. 2014, 'Flood Hazard Assessment in Eastern Slovakia, Discovery', vol. 22, no. 72, August 2014.

Minár J., Barka I., Jakál J., Stankoviansky M., Trizna M. & Urbánek J. 2006, ´Geomorphological hazards in Slovakia', Studia geomofphologica Carpatho-Balcanica, VOL. XL, 2006: 61–78 PL ISSN 0081-6434

Ministry of Environment SR (2013) *Programme of prevention and management of landslide risks (2014–2020)*, 2013, Ministry of Environment SR.

PLUREL (2010) 'National spatial planning policies and governance typology, PLUREL Deliverable Report 2.2.1', 2010.

Perry C.A. 2000, 'Significant floods in the United States during the 20th century – USGS measures a century of floods', US Geological Survey 2000, Fact Sheet 024–00.

Petro Ľ. (2014) 'Geological hazards in Slovakia focused on spatial planning purposes (The Nižná Myšľa landslide case study)', Int. Workshop "Increasing the Resilience of our Cities against Disasters – a new Integrated Approach to Spatial Planning", Košice, Slovakia, 18 September 2014

Petro Ľ., Jánová V., Žilka A., Ondrejka P., Liščák P. & Bálik D. (2014), 'Catastrophic Landslide in Nižná Myšľa Village (Eastern Slovakia)', in K. Sassa, P. Canuti & Y. Yin (eds.), *Landslide Science for a Safer Geoenvironment*, vol. 3 Targeted Landslides, Springer, Switzerland, pp. 305-311. Springer, Switzerland.

Ministry of Environment SR (2011) *Preliminary Flood Risk Assessment in the Slovak Republic*, December 2011.

Regional Territorial/Spatial Plan of Košice Region, Košice Self-governing Region, amendments 2014.

Slovak Spatial Development Perspective 2001, as amended by the Slovak Spatial Development Perspective 2011 – amendment No. 1.

Zeliňaková M., Gaňová L., Zvijáková L., 'Flood risk assessment in the most endangered watersheds in the Slovak Republic', The 8th International Conference 2011, Vilnius.

CHAPTER A4b

Enhancement of flood management and flood protection planning in Eastern Slovakia

Jozef Šuľak and Jaroslav Tešliar

General information about the project

The project 'Enhancement of Flood Management and Flood-protection Planning of Hornád River Basin on Territory of Slovak Republic' was submitted to the Norwegian Financial Mechanism and Financial Mechanism of the European Economic Area (EEA) 2004–06 in March 2006. The grant offer letter was received in May 2007. After signing the grant contract and completion of the public procurement process, implementation started in July 2009. The project was finalised in March 2011.

The total costs were €397,718. The project was implemented by the Kosice Self-Governing Region in close cooperation with the Presov Self-Governing Region, Slovak Water Management Enterprise and Slovak Hydrometeorological Institute as project partners. It was managed by the Agency for the Support of Regional Development Kosice and the general contractor DHI, a.s.

Project background

Floods bring huge economic losses and casualties. Their causes are various, from climate change to improper water management and landscape, and indiscriminate land-use planning and settlement in areas at risk of flooding. From history, we know that floods cannot be prevented. However, it is possible to prepare for them and appropriately adapt. It is necessary to apply the principles of landscape management to eliminate the causes of excessive flood damage and to increase awareness, preparedness and the early warning of individuals and organisations.

Hornád River basin is an important area of the Slovak Republic where it is necessary to realise sustainable strategies, concepts of flood-protection and land-use planning in the areas endangered by rivers. During the years 2004–05, the most flood-endangered districts in region of Kosice were Kosice-okolie, Spisska Nova Ves and Gelnica. Overall, 60 villages were affected and damage to property amounted to more than €2 million.

The larger part of Hornád River basin is in Slovakia, and only its lower portion extends to the territory of Hungary, merging with the Slaná River and then flowing into the Tisza River. The lower part, however, is very significant for the drainage conditions on Slovak territory, especially due to its small incline. Due to backflow, this tendency is very sensitive to the rate of flow in the Tisza and Slaná rivers and, ultimately, in the river Bodrog, which may even be the cause of 'secondary' floods. For this reason, all activities were done in accordance with the principles and technology used in other parts of the Tisza River basin so that these systems are compatible with each other, allow easy exchange of data and can be interconnected into one functional unit in the future.

The spatial plan is an essential part of flood protection. It addresses the spatial arrangement and use of the territory of the region, the principles of settlement development, management, the arrangement of transport and technical facilities, nature conservation, cultural heritage, and landscaping. It must be in accordance with the Concept of Territorial Development of Slovakia 2001.

Project territory

The territory of the project is outlined in Figure 50.

Project activities

The activities of the project are outlined in Figure 51.

Project outputs

The project produced two types of output. The first consisted of text outputs such as action plans for flood protection and the description of modelled flood extensions, as well as information materials for inhabitants. Second, we produced map outputs, more specifically, maps of riverbed capacity, cross-section maps, flood extension maps, flood hazard maps and flood risk maps.

Figure 50: Target territory: River Hornád from state border with Hungary to Ružín water reservoir; River Svinka from mouth to Široké village; River Malá Svinka from mouth to Renčišov village; River Torysa from mouth to Sekčov to Tichý potok village; River Sekčov from mouth to Bartošovce village; River Ľutinka from mouth to Majdan village

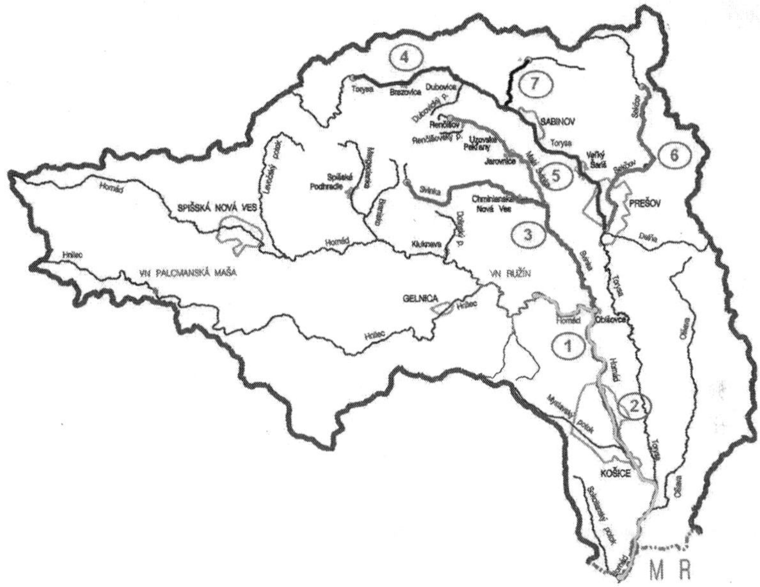

Figure 51: Project phases and activities

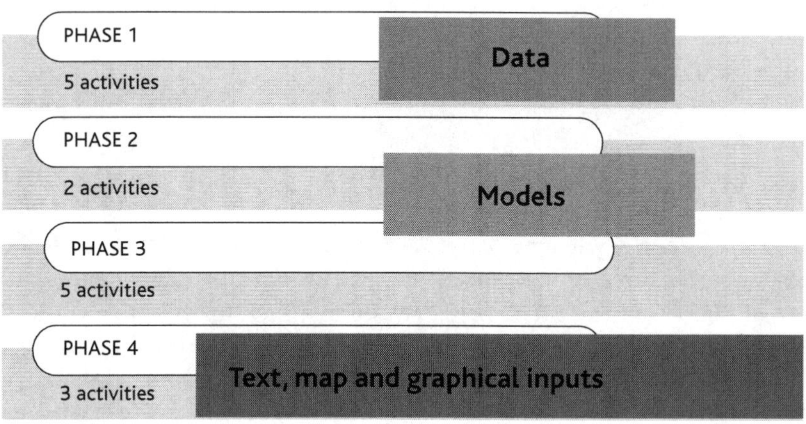

Project results

Thanks to the project and its outputs, the regional and local territorial/spatial plans can now more precisely specify risky territories. Moreover, the awareness of the inhabitants was raised, and there is also a better preparedness for floods. Finally, the project helped to improve environmental protection and the protection of biotopes, and to increase the protection of property.

Relevance for spatial/territorial planning

Most of the outputs, especially the maps, are used in spatial/territorial planning in Slovakia. All of the flood hazard maps and flood risk maps produced have been incorporated into spatial/territorial plans at the local and regional levels within the target territory.

Detailed description of project activities

Phase 1: identification and assessment of the target area

Activity 1.1: geodetic works

The survey of the area of interest was done as a first step. It consisted of the preliminary identification of new cross-sections for measuring by geodetic methods and the current state of existing river training. Preliminary resistance conditions of the area of interest were determined, as well as parts of rivers for which it was needed to provide technical documentation describing given river training, or documentation describing protective construction or eventually other objects on the river.

The DHI collected existing data from owners or operators of given parts of rivers. Available documentation was taken from Slovensky vodohospodarsky podnik (Slovak Water Management Enterprise) (SVP-OZ) Košice, from the regional office in Košice and from the regional office in Prešov. After data collection, a detailed analysis of the available documentation was done. Parts of rivers for which data about rivers could be taken directly from documentation were specified. The number, location and distance of new cross-sections that would be fixed by geodetic methods were also identified.

A geodetic company made the measurements of the missing parts of the area of interest. Based on instructions developed by the DHI, the geodetic company carried out measurements of high-water levels at predetermined points of interest in the territory. The results included current water levels, maximal (culmination) levels, the date and time of measurements, and photo-documentation of the area of interest. More than 830 cross-section profiles were measured, of which 96 were objects (bridges, foot bridges, dams, levels, slides, etc).

Figure 52: Outputs of geodetic works

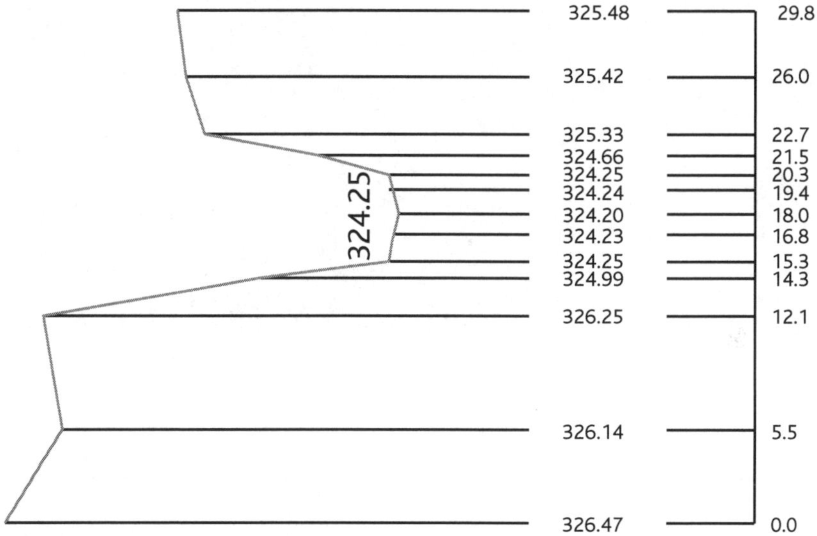

Figures 53a and 53b: Geodetic works process

Activity 1.2: orthophotomaps

The DHI, in cooperation with GEODIS Slovakia Ltd, had identified the preliminary size of the area for which it was needed to provide an orthophotomap. For this purpose, GEODIS provided a preliminary rough digital terrain model, which was created from contour maps of scale 1:10,000 and particularised by aeroplane mapping methods to an acceptable precision for preliminary analyses. Activity 1.2 was closely linked to activity 1.3, and the determination of the preliminary size of the orthophotomap areas was also connected with the preliminary size of the areas for the digital terrain model. The process of creating the orthophotomaps and digital terrain models was very close-knit.

Orthophotomaps were created by air photogrammetry methods. Their resolution is 1 pixel = 0.2m, and they are proceed by the uniform trigonometric cadastral network (S-JTSK) system. The Slovak Water Management Enterprise supplied the orthophotomaps covering an area of 114km^2. The remaining 50km^2 was supplied within the project.

Activity 1.3: digital terrain model

GEODIS supplied the digital terrain model (DTM) of the territory, which was newly processed by airborne imagery. DTM consists of two basic elements: a point and a line. The point in DTM describes the terrain characterised by its distinct location (XY coordinates) and height

(Y coordinate). The line elements in DTM describe in particular the edges of linear elements in the terrain, such as the edge of the crest and foot of dikes, railways, roads and low walls, as well as the current water level in the riverbed during the process of airborne imagery and the contours of objects such as bridges, culverts and so on.

Figure 54: Example of the orthophotomap

Figure 55: DTM partly overlapped by orthophotomap

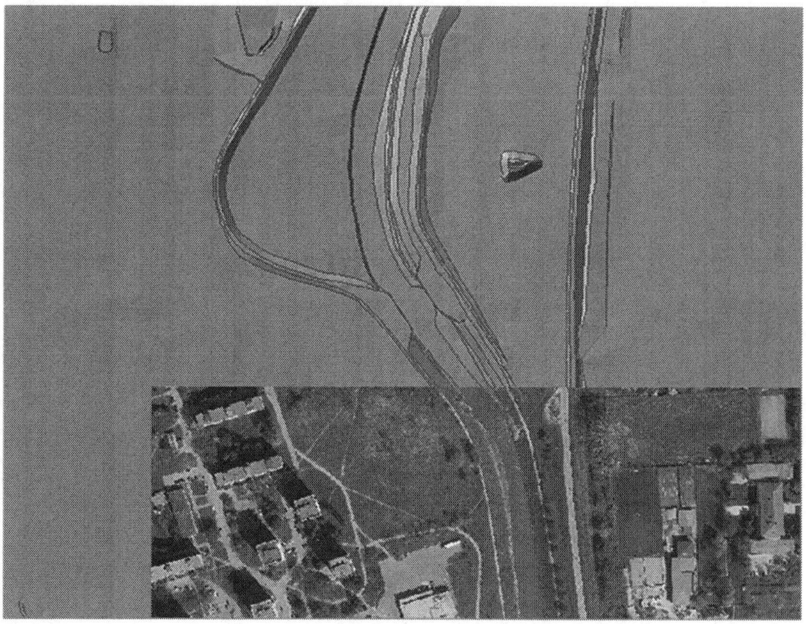

Activity 1.4: hydrological data and definition of the cross-sections of rivers

All hydrological data were provided by the project partner Slovak Hydrometeorological Institute (SHMI). They included mainly N-year discharges, shape and/or volume of particular N-year flood waves and so on.

Table 6: N-year discharges on selected profiles of river Sekčov

No.	Profile title	Hydrological no. GIS	River km	River basin area(km^2)	N-year maximum discharges ($m^3.s^{-1}$)					
					1	**5**	**10**	**20**	**50**	**100**
1.	Bartošovce – nad Pastovníkom	4-32-04-080	40.0	13.40	7	18	26	34	48	60
2.	Bartošovce – pod Pastovníkom	4-32-04-082	39.0	22.30	9	22	31	41	58	72
3.	Pod Fričkovským potokom	4-32-04-088	36.0	47.10	14	35	50	65	92	115
4.	Raslavice – pod Hrabovcom	4-32-04-096	31.0	97.60	18	46	65	84	120	150
5.	Pod Bogliarskym potokom	4-32-04-098	30.0	103.90	19	47	67	87	123	155
6.	Demjata - vodočet	4-32-04-100	26.0	123.20	20	49	69	90	127	160
7.	Pod Terniankou	4-32-04-107	21.0	193.10	24	63	83	109	153	195
8.	Kapušany - pod Ladiankou	4-32-04-115	17.0	270.60	30	80	100	130	175	215
9.	Pod *Šebastovkou*	4-32-04-117	8.0	314.00	33	87	113	147	196	242
10.	Pod *Ľubotickým* potokom	4-32-04-117	6.0	320.90	34	88	116	149	198	245
11.	Prešov – vodočet –pod Solňým potokom	4-32-04-123	0.8	352.80	35	90	118	152	202	250

There were large floods in Slovakia during May and June 2010. Thus, SHMI provided recorded water levels for the period from 15 May 2010 to 15 June 2010 from the measuring stations of Svinka in Obišovce, Hornád in Kysak and Hornád in Ždaňa needs. SHMI also supplied measurement curves for these measuring points. Such curves describe the dependence of the recorded water level and discharge and are usually used to quantify the discharge of water when the water levels are known.

DHI had requested SHMI to provide N-year water discharges for 48 specific profiles of rivers in the area of interest. Profiles were specified to reflect changes in hydrological conditions along the various rivers, particularly the increases in the discharges from major tributaries. Fifteen profiles were specified on Torysa River, four profiles on the river Lutinka, 11 profiles on the river Sekčov, five profiles for Malá Svinka and 13 profiles for Svinka. SHMI supplied a set of N-year discharges for each of these profiles. SHMI supplied maximal N-year water discharges (QN) for Hornád, Svinka, Malá Svinka and Ľutinka. These discharges were later used for the calculation of flood hazard maps for a specific QN.

Activity 1.5: digital project

The objective of this activity was focusing and organising the data collected within the project into a single form. These are primarily geodetic, mapping, morphological and hydrological data. A catalogue of objects was put together from the currently available data, as it was needed for the modelling, as well as cataloguing.

Output file formats were consulted on in April 2010 with the geographic information system (GIS) Košice Self-Governing Region Department. For the purpose of cataloguing, an HTML user interface was generated (offline web site) where all collected data related to the project are collected. Data are processed into file types that do not cause any problem for modern computers (⋆.doc, ⋆.xls, ⋆.jpg, ⋆.pdf, etc). Data are processed in GIS formats, and especially in ⋆.shp, and only subsequently in other vector formats (⋆.dwg, ⋆.dgn, ⋆.dxf, etc).

DHI developed an HTML user interface (hereinafter referred to as a 'digital project' or 'DP'), which transparently contains and offers all available data obtained within the project. The user interface is divided into three basic parts: (1) the top menu, which does not change and shows logos and phases of the project; (2) the left menu, which changes according to user-entered choices from the top menu; and (3) the right main window, which shows text and information regarding the data displayed. From the main window, the user can directly navigate to individual files. The system is configured to display the files but does not open them automatically. The users see the file/s and can subsequently decide for themselves whether they want to open it and view it on their own computer. Some files can be opened only if the computer is equipped with a special software package (such as GIS files, etc).

Phase 2: modelling, data processing and specific analyses I

Activity 2.1: 1D models

Based on the data obtained during the first phase, mainly cross-section profiles, 1D models were created for all rivers in the project territory, totalling 240km in length. We used the MIKE 11 (MIKE by DHI) modelling tool (2009 version).

At the beginning, DHI digitalised the cross-sections of the particular rivers, as well as the digital axis of the water course. As a part of processing 1D models, there was also an ongoing activity aimed at extending the measured profiles of river basins. As part of geodetic works and the provided budget, it was only possible to focus on cross-sections in rivers and it was not possible to target valley cross-sections (compound profile = riverbed + flood plain). Geodetically obtained cross-sections were afterwards extended to valley cross-sections. The heights of extended sections were taken from existing morphological specifications. 1D models were created for the rivers Hornád, Torysa, Svinka, Malá Svinka, Ľutinka and Sekčov.

Figure 56: Editing of cross-section profile in MIKE 11 software

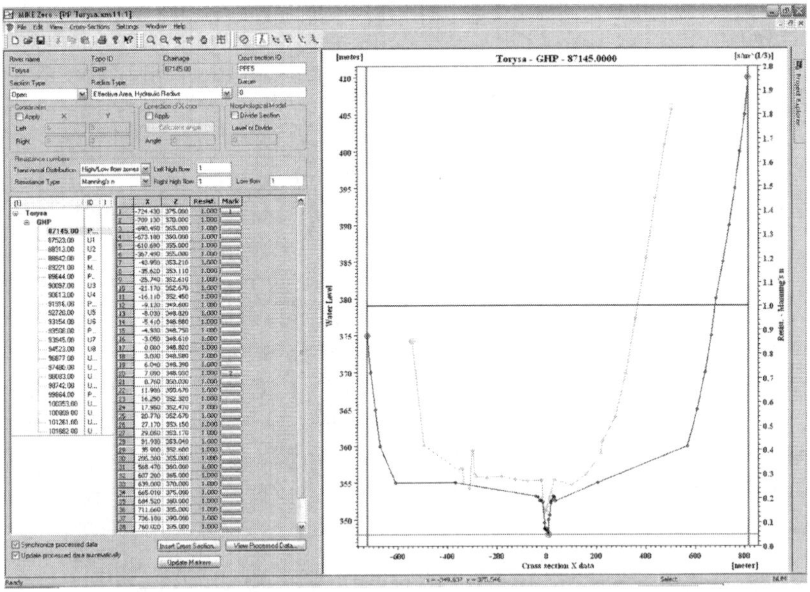

Activity 2.2: 2D models

In the part of the Hornád River from the Hungary state border to Ťahanovce (part of Košice City), 2D models were constructed. As a first step, the current orthophotomaps and DTM were taken from the Slovak Water Management Enterprise (SWME) by DHI. This data covered the area of interest of the Hornád River from the state border with Hungary to water construction ni Ružín. Later on, DHI divided this area into

Figure 57: Editing of digital axis of the water course in MIKE 11 software

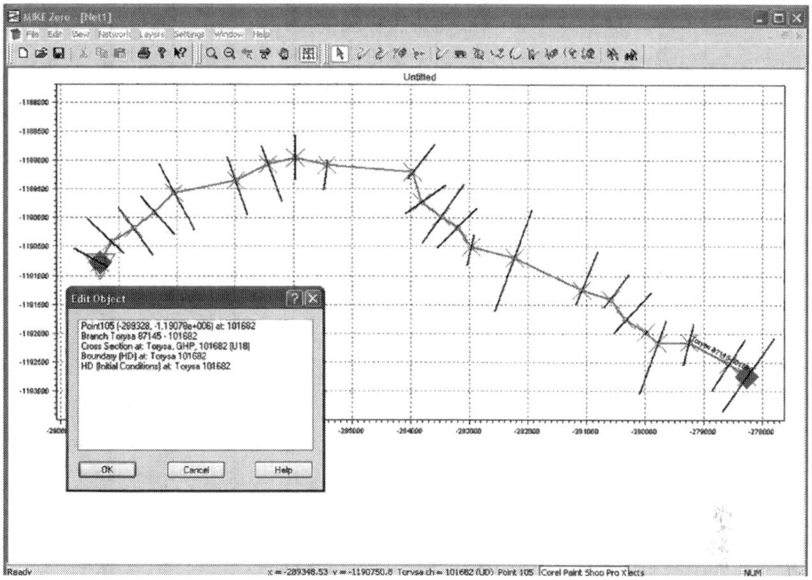

three parts: state border with Hungary–Ždaňa, Ždaňa–Vyšné Opátske and Vyšné Opátske–Ťahanovce; thereafter, 2D models were made for each part. The main reasons for such divisions were shorter computing time, easier work with the model and lower PC hardware requirements. Consequently, the managing polygons were gradually added, which describe in-line structures, rivers and the outlines of buildings, and determine the borders of the computing network.

During the creation of the model, DHI carried out field inspections. The field inspection area of interest was a potential flood area of Hornad River and the Hornad riverbed itself. Updated information on, for example, walls, dikes and so on was updated in the models. Another aim of the field inspection was to more accurately specify the boundaries between particular 2D models in order to transfer results from one model to another and vice versa.

The existing DTM contained a description of the terrain at the time the image was taken, including river water levels. For the needs of hydrodynamic modelling, the water level had to be removed from DTM and be replaced by the riverbed shape, which was under the water level. Riverbed shapes were added afterward to the DTM on the basis of geodetically measured river cross-sections. Cross-sections were measured in certain distances, and the DTM shape of the riverbed was supplied by interpolation methods.

In the case of the Hornád River, 352 geodetically measured cross-section profiles (partial result from activity 1.1) were used to create the shape of the riverbed, which was under the water level at the time the image was taken, with an overall length of 67km (from the Hungarian state border to Ružín). All four models were finally tested and adjusted based on errors that were found in the process of calculation and subsequent results of the work.

Phase 3: modelling, data processing and specific analyses II

Activity 3.1: maps of the riverbed capacity

Maps of the riverbed capacity describe the maximum riverbed capacity with regard to the ability of the riverbed to transfer concrete N-year discharges. This information is important in terms of the appraisal of the real status of the riverbed and its ability to transfer certain flows.

A simplified 2D model of Hornad was built as part of identifying risk and hazard areas. It is a network of triangular cells of a very similar size, allowing fast computing times. Another benefit of this simplification lies in neglecting line constructions, which, however, in real conditions and especially in smaller flows, can have a significant impact on flowing flood water. The resulting flood scope from this model identifies risk areas and determines areas that need more attention when mapping flood risks. The results from hydrodynamic models exactly determine flood risk and hazard areas. After calculating the whole range of N-year waters (Q1, 5, 10, 20, 50 and 100), the risk areas were identified.

Figure 58: Unstructured grid of 2D mathematical model MIKE 21FM

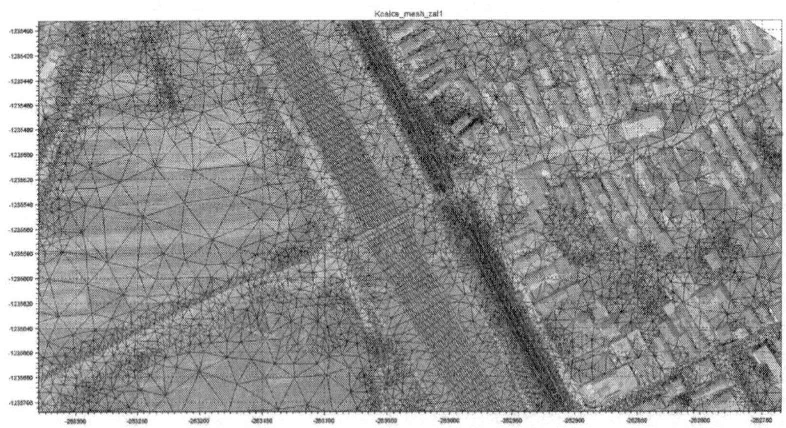

Figure 59: Morphology of 2D mathematical model MIKE 21FM (Košice city)

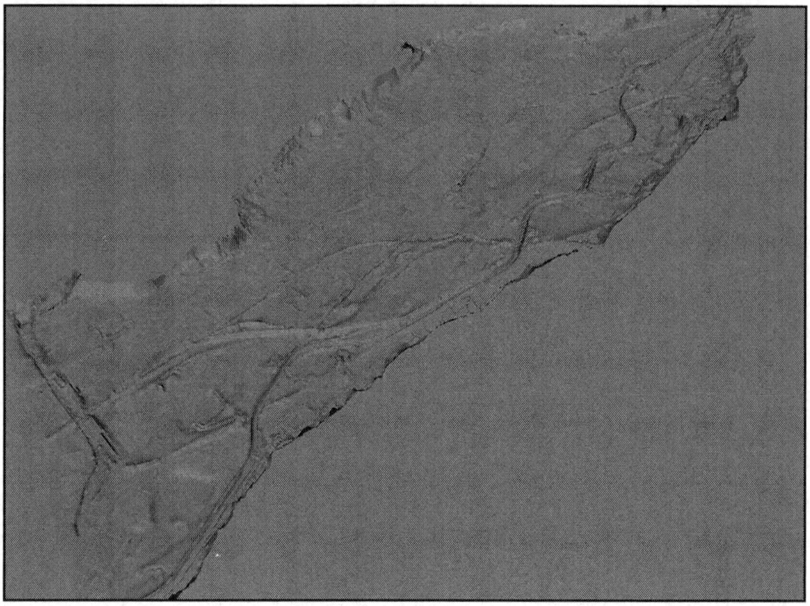

Activity 3.2: concept for flood protection

Existing flood protection in the area of interest was described in terms of modifications of the riverbed and flood protection objects. Planned investments were also described and compared to the results of modelling. The proposal and evaluation of the concept for the flood protection of risk areas consisted of two basic steps:

1. Evaluation of the concept for flood protection based on findings about the current situation and available documents. A document describing the real state and concept of flood management was prepared.
2. Proposal for flood protection. On the basis of the evaluation of the flood protection results and on the basis of consulting competent personnel, as well as those in question, a proposal for solving flood protection problem was created.

DHI, in cooperation with the SWME, created a database of available data to assess the concept for the flood protection in risk areas. All available documents including priority flood protection measures for the City of Kosice were requested from SWME for the proposal of future solutions and their confrontations with the current business development plan for investments in the years 2011 to 2016.

Figure 60: Map of the river bed capacity

Figure 61: Malá Svinka – modification of river in Uzovské Pekľany village

Based on the available materials and consultations with the SWME, a description of flood protection in the area of Hornád river (part from the state border to – WD Ruzin) was created including description of existing

flood protection within the boundaries of the city of Košice. Consequently, analyses of all the rivers in the target territory continued. All the necessary data for assessing the flood protection of the given area were completed, in particular, on the basis of communication with SWME.

Activity 3.3: flood line maps

Flood line maps represent the range of floods for particular N-year discharges (Q1, 5, 10, 20, 50 and 100). This activity followed immediately after the modelling was finished, when the final results of the calculations were available. After calculating the whole range of N-year waters, maps showing flood risk areas were created.

Activity 3.4: flood hazard maps and flood risk maps

Flood hazard maps represent water depth, altitude and water flow speed for particular N-year discharges. Flood risk maps show all the flood lines and land use, the indicative number of flood-affected residents, and important spot information such as schools, hospitals, offices, and other sources of pollution within the most extensive flooding. Flood risk maps were prepared based on polygons over the given foundation of orthophotomaps, determining acceptable risks as of a given area/utilisation of area.

Figure 62: Torysa and Sekĉov – example of flood lines map

Activity 3.5: description of the course of the flood level and rainfall–runoff model of the Svinka River basin

At first, DHI started collecting the information needed for building a rainfall–runoff model. On the basis of consultations with SHMI, already-existing gauging stations were defined in the area of interest. Basins and parts of rivers for which the rainfall–runoff models were

Figures 63a-d: Examples of flood hazard maps and flood risk maps

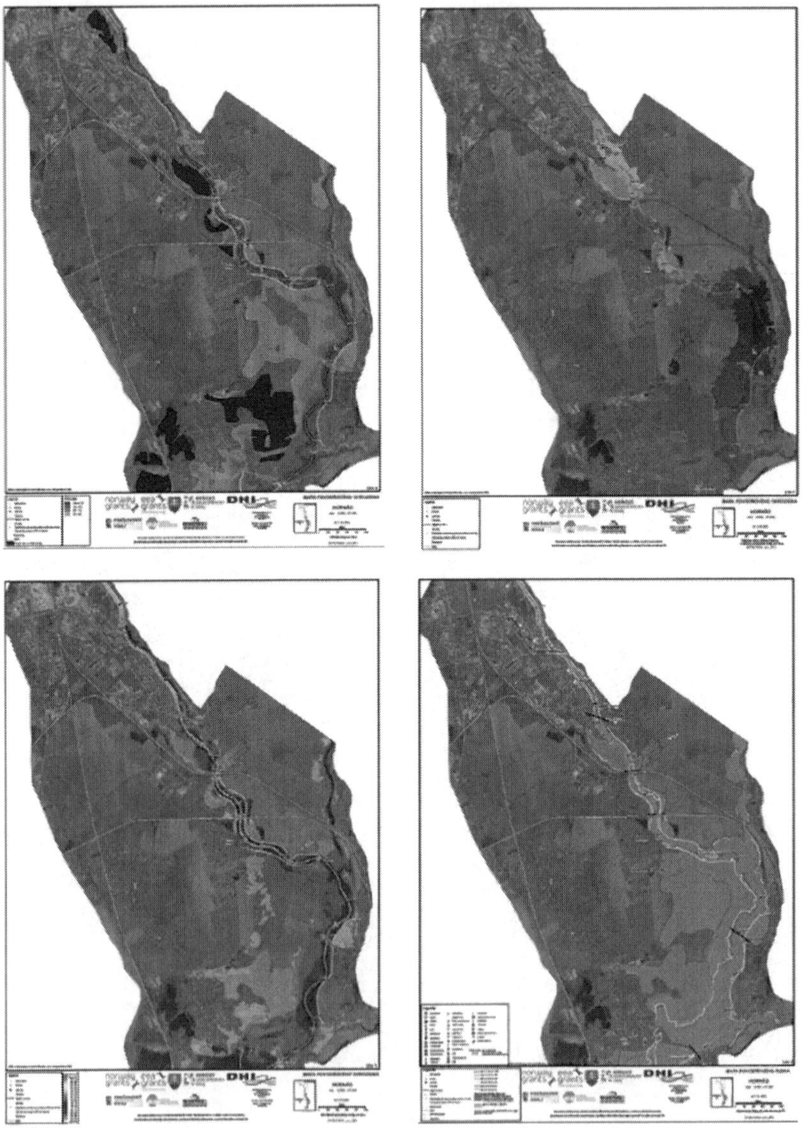

to be built were defined subsequently. The rainfall–runoff model is linked to the hydrodynamic model. In order to describe the river thoroughly by using original data on precipitation, temperature and flow for the purpose of rainfall–runoff modelling in consultation with SHMI, Svinka River was chosen up to the Bzenov profile.

Afterwards, the rainfall–runoff model of the Svinka River to the final profile of Bzenov was completed, calibrated and connected to the hydrodynamic model of the Svinka riverbed. After calculating the whole range of N-year waters (Q1, 5, 10, 20, 50 and 100), detailed descriptions of the course of the flood levels in respective parts of the modelled area were created. After completing the calculations of N-year waters, the already-existing rainfall–runoff model was linked to the 1D model of Svinka. A description of the course of the flood level for modelled areas and for each N-year discharge was created, as shown in Table 7.

Table 7: Description of the course of the flood level

Q100				
RKM	Left bank	Right bank	Location	Description
65.780-65.600	x			Flooding of the road
65.000-64.300	x		Malá Lodina	Flooding of the road
63.900-63.240	x			Flooding of the road
63.100-61.700	x		Veľká Lodina	Flooding of the road

Phase 4: improvement of regional development and spatial planning processes

Activity 4.0: maps of cross-section riverbed profiles

This activity was not originally planned within the project. Maps of cross-section riverbed profiles were created for the selected parts of the rivers where no orthophotomaps and no DTM were available. Maps show the calculated water level in particular cross-section profiles.

Activity 4.1: informative materials

DHI collected documents and data that were used for the design of the structure and content of informative materials (current legislation, the definition of flooding, the definition of flood risk area, general information on protection against flooding, information on preventive flood protection, general description of the measures, information on

Figure 64: Cross-section profiles in Google Earth

the right of protection against flooding, and so on). Eight different informative materials were produced in total, containing information on the following topics:

1. Basic project information.
2. Floods and flood protection system in Hornád River basin.
3. Flood risk areas – Hornád River, Košice City and surroundings.
4. Flood risk areas – Torysa River and Sekčov River, Prešov City, and surroundings.
5. Flood prevention.
6. Flood – what to do now?

Figure 65: Information materials

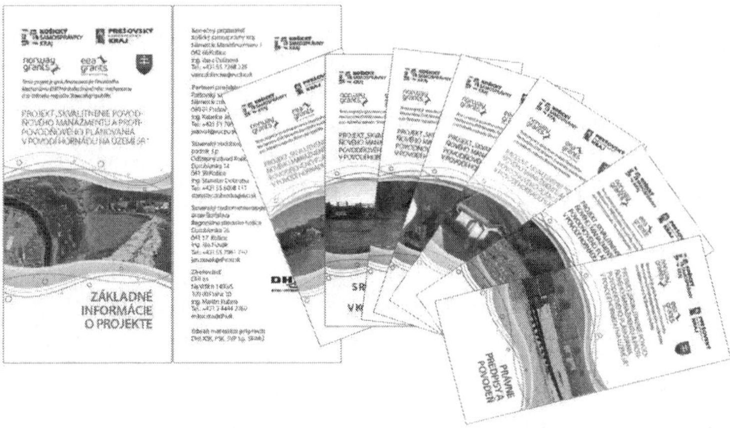

7. Threat of water pollution during floods.
8. Legislation and flood.

These leaflets are used as an awareness-raising tool – to inform inhabitants about flood risk areas within the project territory, what kind of prevention measures can be done and what to do in case of flood.

Activity 4.2: flood prevention and flood protection action plans

A basic framework of the principles and concepts that should be taken into account when developing guidelines was established as part of the work, including preparing guidelines in order to compile action plans to prevent damage to property in the event of flooding. A document named 'Flood prevention and action plans for flood protection' was created for more effective flood risk management. It describes how to create action plans, the documents that are necessary for action plans, the conflicts that can occur and an example of an action plan. These outputs are dedicated mainly to representatives of municipalities and other relevant institutions in the area responsible for flood prevention, as well as flood protection.

Activity 4.3: fixed points in the terrain

'Fixed points in the terrain' are used as practical guidance for civil protection workers. Fixed points show the amount of the calculated level of N-year discharge for a particular fixed point (usually a bridge structure), relative to the top edge of the deck. In preparing the guidelines for crisis management and evacuation plans, respective

Figure 66: Example of placement of fixed points in Svinka River

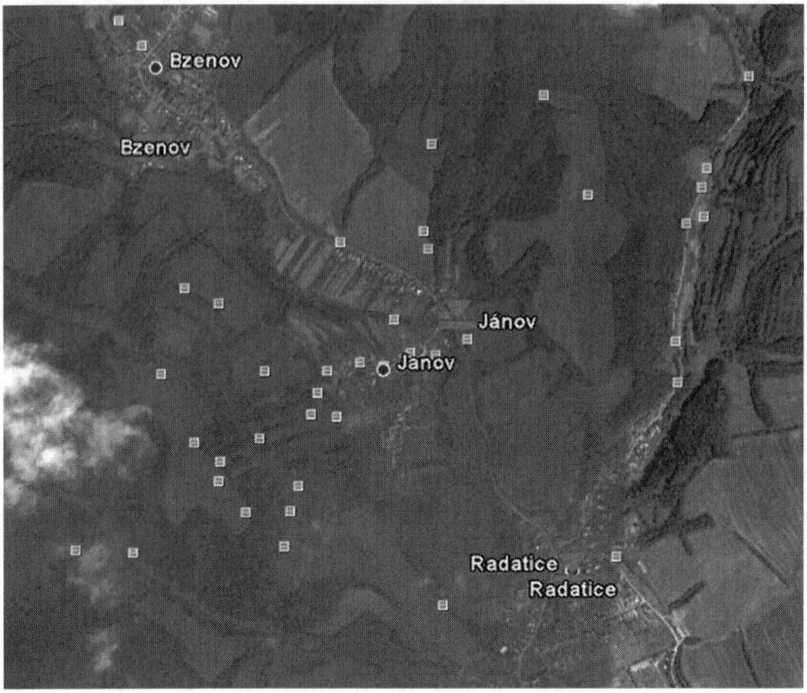

N-year floods were linked to the known fixed points in the terrain. Altitude values were recalculated into the local altitude system of water-level stations of the SHMI or the SWME; in places where water-level stations of the SHMI or the SWME could not be used, fixed points in the terrain were defined. Afterwards they were used for the orientation of an appointed person working in the terrain.

Conclusion

Project outputs provide new information, instructions and procedures to work effectively with flood risk. They will be used by the general public and professionals. The 1D and 2D hydrodynamic models can be used to solve other tasks relating to, for example, the optimisation of new flood control measures, the verification of planned investments such as flood protective liner constructions, the verification of the functioning of retention areas or the verification of imminent danger. The project is the first step in integrated river basin management and can be further extended to the higher parts of the basin and to the design and verification of the proposed measures in the basin, with the option to extend the existing models for flood forecasting.

CHAPTER A5a

Flood risk management by spatial planning

Stefan Greiving and Nadine Mägdefrau

Germany	
Population: 81.08 Mio. (2014)	**Government:** Federal parliamentary constitutional republic
Area: 357,168 km²	**Administrative structure:** Germany comprises sixteen federal states ("Bundesländer"). Each state has its own state constitution.
Hazard profile: River floods (nationwide) Coastal storm surges (North Sea) Alpine hazards (Bavaria) Earthquakes (up to 6.75 only): along the Rhine valley	**Authorities in charge of risk assessment and management:** Different federal state authorities (water management authority, geological survey etc.) No comprehensive competences on federal level
Analysed event: River flood along the river Elbe (2013) Estimated return period: <1:100 – 1:500 Estimated damages: 6.7 bio. € Casualties: -	**Role of spatial planning:** Primarily responsible for risk management of some hazard that are spatially relevant (floods, major accident hazards). Focus lies on the prevention phase.

Main characteristics of the risk management system

This chapter will give a brief overview of the main characteristics of the planning system and the emergency framework of Germany.

Brief introduction to the administrative structure

Germany is a federation (the *Bund*) and consists of the following 16 federal states (the *Länder*): Baden-Württemberg, Bavaria (*Bayern*), Berlin, Brandenburg, Bremen, Hamburg, Hesse (*Hessen*), Mecklenburg-Western Pomerania (*Mecklenburg-Vorpommern*), Lower Saxony (*Niedersachsen*), North Rhine-Westphalia (*Nordrhein-Westfalen*), Rhineland-Palatinate (*Rheinland-Pfalz*), Saarland, Saxony (*Sachsen*), Saxony-Anhalt (*Sachsen-Anhalt*), Schleswig-Holstein and Thuringia (*Thüringen*).

The federal principle provides for a dual statehood that assures the Federation and the federal states their own governmental organisation (including Parliament, organs of government and administration, and courts). This structure allows them to exercise their own governmental power (legislation, government and administration, and jurisdiction). To avoid conflicts between the Federation and the federal states, the Basic Law for the Federal Republic of Germany governs the distribution of those duties (see Maurer, 2010, p 285). According to Article 30 of the Basic Law, the responsibility for all governmental functions generally rests with the federal states, unless the Basic Law states otherwise. Besides this vertical separation of powers, the Basic Law also includes a horizontal separation of powers between the legislature, executive authority and judiciary (see Art 20 II Basic Law).

Most of the federal states divide their territory into administrative districts (*Regierungsbezirke*). These districts are further divided into county boroughs (*kreisfreie Städte*) and rural counties (*Landkreise*), which are themselves further divided into municipalities. These smaller municipalities and the county boroughs, which also act on the part of municipalities, form the basis of the German state (see Turowski, 2002, p 9).

The responsibilities of the municipalities include local transport and road construction, local land-use planning, technical and social infrastructure, and daily life protection (see Turowski, 2002, p 9).[1] This organisational structure emphasises the principle of subsidiarity, which is an important part of the German polity (see Turowksi, 2002, p 9).

Brief introduction to the spatial planning system

Spatial planning is defined as comprehensive, over-sectoral planning. In compliance with the administrative structure of the country, German spatial planning responsibilities are also divided into three levels (the Federation, federal states and municipalities) (see Turowski, 2005, p 895) and interact with other types of planning, which are partly spatially relevant, such as water management, as illustrated by Table 8.

In addition to comprehensive spatial planning, specific aspects with spatial influence are considered within sectoral planning (eg transport, water protection, nature protection). The most important one, planning approval (*Planfeststellungsverfahren*), includes the establishment of formal plans for the construction of specific facilities (eg streets, airports, waste landfills). The protection area designation formally establishes areas with a specific land use (eg water protection area, conservation area, military area). Additionally, sectoral planning authorities initiate preparatory sectoral plans that are only binding for them (eg landscape

Table 8: The planning system

Spatial level	Spatially relevant planning				Spatially non-relevant planning
		Comprehensive		Sectoral (Transport, water, geology, emergency response, etc)	Forms of non-spatial management on different spatial levels
Europe		European spatial development (no binding character)		Environmental policies, TEN, CAP	eg, Budget planning
National level		Spatial development planning		eg, National transport network plan	eg, Defence, planning, education
Federal state	Spatial planning	Regional planning (partly land-use related)	Societal planning	eg, River basin authorities in charge of management plans, partly land-use planning and management related	eg, Cultural development, education planning
Municipality (all planning on this level can be subsumed together under the term *"urban planning and management"*)		Local land-use planning		eg, Waste and sewage planning, public transport planning, municipalities are in charge of (land-use management)	eg, Lower education, municipal budget planning

Source: own figure

planning, waste management planning) (see Greiving, 2006, pp 55–6; Hendler, 2012, p 444).

Spatial planning is part of the concurrent legislative power without proof of necessity (see Art 74 I 31 Basic Law; Art 72 II Basic Law). However, in accordance with Article 72 III of the Basic Law, the federal states are able to enact regulations that vary from federal spatial law.

The Federation has used its legislative power by enacting the new Federal Spatial Planning Act (*Raumordnungsgesetz* [ROG]), which came into effect on 30 June 2009. The Act includes two substantial categories that differ in their legally binding effect. The *principles of comprehensive spatial planning* consist of statements on spatial development, regulations and protections as guidelines for subsequent weighting or discretionary decisions (see § 3 I 3 ROG). Their legally binding force is restricted to an obligatory consideration. This means that the principles have to be considered throughout the planning process but can be outweighed by other – in certain situations, more important – concerns.

The *aims of spatial planning*, on the other hand, incorporate binding standards for spatial development, regulation and protection. They have to be considered conclusively by the spatial planning authorities and be included as designations in the regional plans (see § 3 I 2 ROG). One major difference between the principles and aims of spatial planning is the criterion of conclusive consideration: while the principles only constitute input for the weighing process, the aims of spatial planning have to be included in the final decision (see Hendler, 2012, p 450).

The ROG does not set a statutory basis for a national spatial structure plan that has a binding effect on the entire territory. The ROG intends a two-tiered structure within the federal states: first, spatial planning at the federal state level; and, second, regional planning (see § 8 I ROG)

To regulate spatial planning within the federal states, section 8 of the ROG intends all states (with the exception of the federal city states of Berlin, Bremen and Hamburg) to determine their own land planning acts for their territory. Furthermore, all federal states (with the exception of Berlin, Bremen, Hamburg and the very small Saarland) have to establish spatial structure plans for subspaces of their territory (regional plans). The spatial structure plans must include arrangements about the spatial structure, especially on the structure of settlement and intended open space, as well as the location of infrastructure and paths (see § 8 ROG).

Just like on the state level, spatial planning on the local level is two-tiered. As a preparatory urban land-use plan, the zoning plan sets proposed land use based on the foreseeable requirements of the entire municipal territory. Its content is only binding for the municipality itself, not for private citizens. The binding local development plan only covers parts of the municipality's territory and includes legally binding arrangements for urban development. The binding effect in this instance comprises the administration and private individuals (see § 1 BauGB).

To conclude, the German spatial planning system's dominant regulatory function is a typical example of a so-called 'conforming planning system'. Here, a new development is legally allowed when it conforms to the regional/local plan. However, in most of the European Union (EU) member states, the development function dominates. This so-called 'performing planning system' is characterised by non-binding programmatic and/or strategic statements (Rivolin, 2008). This differentiation matters for the way in which natural hazards such as floods are handled by spatial planning (see section on "Dealing with flood risks in Germany" in this chapter).

Brief introduction to the emergency framework

In accordance with the Basic Law, the responsibility for civil protection in case of a disaster during peacetime rests with the federal states. The Federation only holds the legislative power for fields directly assigned to it by the Basic Law. Besides civil protection in the case of defence (Art 73 I 1 GG), the Federation holds exclusive legislative power for the fight against international terrorism (Art 73 I 9a GG) and protection against dangers related to the release of nuclear energy (Art 73 I 14 GG). Additionally, the competence for disease control falls under the concurrent legislative power of the Federation (Art 74 I 19 GG). All remaining competences, as well as the responsibility for administration, rest with the federal states. This is especially problematic if a disaster that exceeds the borders between federal states occurs and cooperation between the states is needed. If this is the case, Art 35 II 2 of the Basic Law enables the state to request help from the emergency response units of other federal states or the Federation (see Thiele, 2012, p 80).

If a peacetime disaster occurs, the director of administration of the municipal authority where the disaster happened will be in charge of the local response. If the situation demands it, they are supported by the staff of their administration and representatives of authorities involved in disaster management. Additionally, emergency response units (eg police, fire department, non-governmental organisations and private enterprises) help with the technical implementation of the plans. If the impact of the disaster exceeds the borders of one municipality or the local government is incapable of handling the situation on its own, the superior authority takes on the coordination role.[2]

Dealing with flood risks in Germany

Introduction and overview

Directive 2007/60/EC on the assessment and management of flood risks entered into force on 26 November 2007. This directive now requires member states of the EU (including Germany) to assess if all watercourses and coastlines are at risk from flooding. Further, it requires the mapping of the flood extent (for floods with a low probability, or extreme event scenarios; for floods with a medium probability [likely return period ≥100 years]; and for floods with a high probability) and the assets and humans at risk in these areas, and the taking of adequate and coordinated measures to reduce this flood risk. Flood hazard and risk maps have been developed until the end of 2013 and can be seen as a new, almost homogeneous, evidence basis for further actions to be taken on the part of spatial planning and other actors. Figure 67 shows such a flood hazard map for a flood with a low probability in the example of Cologne, Germany.

However, the management options of the water management authorities are limited for different reasons: these hazard maps are not binding; the legally binding flood hazard zones are restricted to those areas endangered by 1:100 flood events; and areas protected by coastal defence systems or levees along rivers are excluded from

Figure 67: Flood hazard map

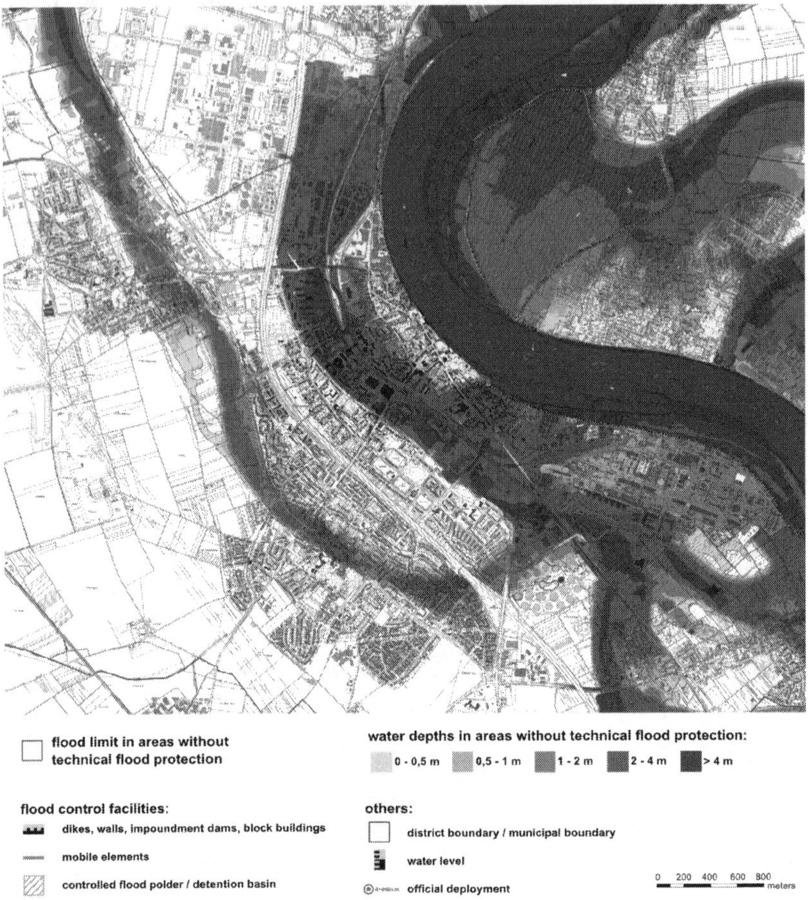

Source: Based on Bezirksregierung Köln. Own translations

these hazard zones. In consequence, water management authorities cannot prohibit further settlement or infrastructure development in protected areas.

Section 2 II 6 of the Regional Planning Act stipulates that flood risk prevention, against costal as well as river floods, has to be taken into consideration by regional planning. As opposed to water management authorities, spatial planning is entitled to take action even in protected areas. Other types of hazards are not explicitly mentioned. The principles of comprehensive spatial planning in section 2 of the ROG clarify that the protection of critical infrastructure must be taken into account (§ 2 II 3 ROG). This emphasises the importance of this topic for spatial planning. Finally, this also refers to climate change and points at mitigation, as well as adaptation needs.

At the local level, section 1 IV 1 of the Federal Building Code (FBC) demands that 'the general requirement for living and working conditions which are conducive to good health, and the safety of the population at home and at work' must be considered during the establishment of an urban land-use plan. Again, only flooding is explicitly mentioned (§ 1 VI 12 FBC). Attention is also paid to mitigation and adaptation to climate change (§ 1 V FBC).

Spatial planning and flood risk management

In the following, the different management options that spatial planning has at hand for dealing with river flooding will be discussed.

Keeping areas free of (further) development

This option is well established in the German planning system and primarily of relevance for regional planning. According to section 8 VII 1 of the ROG, priority zones can be demarked in regional plans. Here, priority is given to particular spatial functions, such as flood prevention. Any other function, such as settlement development, which may be in conflict with the prioritised function is prohibited. This means that the municipalities cannot designate the respective area within their general land-use plan as a settlement zone. However, this does not question the legality of the existing building stock due to private property rights, protected by Article 28 of the Basic Law. In consequence, the option is not effective in lowering the existing flood risk, but is only suitable for avoiding a further increase of the risk. Priority zones are also used for flood protection by means of protecting potential locations for new retention ponds, which are needed to lower the effects of a hazardous event, from other concurring land uses.

As a further option for regional planning, reserve zones come into play. Here, no strict priority, but just particular weight, is given to a specific spatial function, such as flood prevention. Further settlement development remains possible but has to be explicitly justified by the municipalities. In practice, this option is often chosen in order to mitigate residual flood risks caused by extreme events.

The option 'Keeping areas free of (further) development' is also relevant in the context of managing major accident hazards (see Chapter A5b of this book) for securing 'appropriate distances between establishments and vulnerable land-uses' (Art 13 SEVESO III Directive).

Figure 68 illustrates both priority zones (marked in diagonal hatches) and reserve zones (cross-hatches), using the example of an excerpt

Figure 68: Excerpt from regional plan of district of Cologne

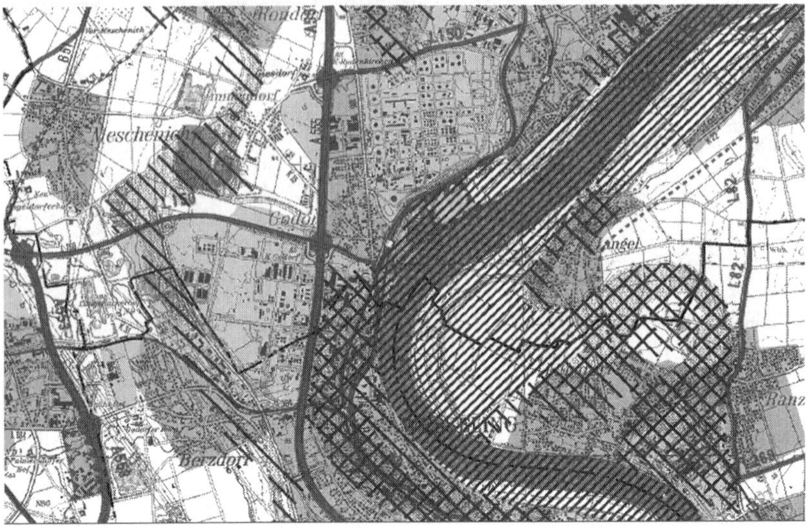

Source: Bezirksregierung Köln

from the regional plan of the district of Cologne. The river Rhine is shown in dark grey, residential zones in light grey and industrial zones in middle grey. As is clearly visible, there are some residential but also industrial zones demarked as reserve, but not priority, zones.

The necessary evidence basis for the designation of both priority and reserve zones are flood hazard maps, which provide information about flood extent and depth. In most cases, the velocity of the flood is not required, with the exception of mountain areas, where structural damages are likely to occur if the velocity exceeds 2m/s. A differentiation into hazard intensity/frequency classes is desirable but not necessary in order to weigh up the given hazard against other concerns and interests: the greater the hazard, the more a land-use restriction becomes justifiable. However, the given vulnerability of the different land uses is usually not taken into account when designating either a priority or reserve zone.

Differentiated decisions on land use

This option refers particularly to the vulnerability of different land uses to floods (or other types of hazards). Designation of acceptable land-use types is made in accordance with the given flood risk. It considers individual susceptibility to the effects of flooding (inundation, groundwater intrusion, erosion, etc) of different land-use types (such as residential zones, industrial zones and commercial zones, but also specific types of infrastructure) and even building structures.

As an indispensable prerequisite, more detailed information is needed about frequency/magnitude curves. This includes the flood depth, as more than two meters (often the case behind levees) may lead to a risk for life, and the velocity, as more than two metres per second may lead to structural damage. On this basis, spatial planning can define differentiated protection goals for several protections goods (human beings, economic assets, infrastructure, nature, etc) in accordance to both their susceptibility and their worthiness: critical infrastructure has to be protected even against extreme floods, while an inundation of industrial zones might be acceptable in seldom cases and agricultural land can be used as retention areas. However, these are partly normative questions. Thus, policy and public participation come into play (Fleischhauer et al, 2012).

Importantly, it is the definition of risk that affects risk policy. Moreover, defining a risk is an exercise of power in view of existing ambiguity. In many cases, policy discourse is not about who is correct about the assessment of danger, but whose assumptions and interests about political, social and economic conditions, as well as natural or technological forces, win in the risk assessment debate. Thus, the hazard as a potentially damaging physical event is real, but risk is socially constructed (Slovic, 1987). Due to this fact, those who manage and communicate risks to the public need to understand the emotional responses towards risk and the way in which risk is perceived by the at-risk population. To conclude, any differentiation with regard to protection goals has to be discussed with the public, and decisions have to be taken by those who are democratically legitimated to do so.

There are some examples of such kinds of differentiated decision from other European countries, like the Netherlands and Switzerland (PLANAT, 2008). In Germany, however, this option is not fully implemented in planning practices. The regional plan of the district of Cologne, however, distinguishes at least between areas that may be inundated at two metres and others, but no legal consequences are constituted on this basis.

Adaptation of building structures

This option aims at mitigating the susceptibility of those land uses that are permissible in hazard-prone areas or are already built-up. As opposed to the previous options, it is mainly relevant for the local level. In case of new developments, building protection measures can be implemented into legally binding land-use plans or included as special obligations when granting building permissions. In built-up areas, building protection is part of risk awareness-raising campaigns and addresses private households and landowners. Typical building protection measures are the definition of the minimum elevation height, a complete prohibition of basements or the avoidance of damage potential (living/sleeping rooms, electric facilities, etc) in basements by means of special obligations in building permissions.

Mitigating the hazard

The previous options refer to the reduction of the risk by means of a separation of hazard and susceptible land uses for a reduction of this susceptibility. However, although this arena is mainly up to water management, spatial planning can either support the related strategies of water authorities (primarily relevant for river flooding) or set up its own strategies for hazard mitigation – mainly in the case of urban, pluvial flooding. In both cases, the key objective is to lower the hazardous effects (frequency and or magnitude) of flooding by suitable designations in land-use plans, such as the extension of retention areas or the construction of retention ponds. These measures are up to the plan approval procedures that water management authorities are in charge of, but regional planning may define a priority zone for flood control. Alternatively, local land-use planning can designate suitable areas in accordance with section 5 X of the FBC as areas necessary for flood protection.

Local strategies are required for managing pluvial flooding. This type of flooding is not addressed by the European Floods Directive. The main causing factor for pluvial flooding is not river flooding, but surface water run-off due to torrential rainfall and the limited capacities of the local sewage system. Here, no homogeneous flood hazard and risk maps are available. This type of flooding is mainly a problem in urbanised areas due to the fact that these areas are dominated by sealed surfaces (buildings, roads, infrastructure, parking lots, etc) and also particularly susceptible to the effects of inundation due the given density of critical infrastructures (power distribution stations, hospitals, schools, etc). For proper response strategies, cooperation between urban planning, the department for heavy engineering, civil protection and the public is needed (Greiving and Lindner, 2011; Greiving et al, 2012). Typical measures are the extension of the water storage capacity of the local sewage system by constructing decentralised rainwater management ponds, a renaturalisation of urbanised areas and the creation of emergency waterways (such as public parks, parking lots, etc) for an intermediate retention of rainwater. Apart from pluvial flooding, land-use planning – together with landscape planning – may also contribute to the creation of retention ponds in the whole catchment on agricultural or forest land.

Retreat from hazard-prone areas

Most of the aforementioned options deal with new developments, but Germany's settlement structure results from long-lasting development over 2,000 years. Many historic city centres are situated on riverbanks as rivers have been important for transportation and energy production (historic water mills, but also as cooling water for modern thermal power plants). In consequence, a considerable part of Germany's economic activities still takes place along rivers. This causes fundamental conflicts with flood risk management. These conflicts may become even more relevant due to climate change as a remarkable increase in flood risks is projected for many rivers in Germany (Krahe et al, 2009; Moser et al, 2012). Moreover, coastal zones are endangered by storm surges, whose effects may be intensified by the rise of the global sea level.

Up to 2013, there was no relevant debate in Germany about a potential permanent abandonment of threatened areas, but the severe flood event that hit large parts of the Elbe and Danube catchments in June 2013 can be seen as a paradigm shift in this regard (for a more detailed discussion, see section on 'The great Elbe and Danube flood in June 2013' later in this chapter). The Federal Environment Office launched a study later in 2013 on options for the retreat from hazard zones (UBA, 2014).

This option is highly controversial due to several reasons. First, it is extremely costly as a preventive measure because full financial compensation is needed due to private property rights. Moreover, it causes (often violent) protests from landowners, which make formal, quite time-consuming, expropriation necessary because of the relatedness of the local population, who may have lived at a particular plot of land for centuries. These are lessons to be learned from one of the very few examples where retreat took place due to given flood risks: Riesa-Röderau, a newly developed residential area in Saxony, located in the floodplains of the river Elbe. After the 2002 Elbe flood event, 139 buildings – which had been built three to five years before – were deconstructed and €40 million was spent by the federal state government of Saxony in compensation (UBA, 2014, p 39). This single case was only possible due to enormous political efforts on the federal state level and cannot serve as a benchmark for further, nationwide efforts in this regard given the lack of financial capacities as well as political willingness.

However, retreat can be seen as an option in the aftermath of a disaster. The disaster itself opens a window of opportunity for retreat as the demolished building stock, as well as public infrastructure, has to be reconstructed or at least restored anyway. However, relocation may not encompass the entire building stock, but could relate to those buildings or land-use classes that are particularly susceptible to the relevant threat (eg critical social infrastructure) or dangerous for itself (eg establishments under Directive 2012/18/EU on the control of major-accident hazards involving dangerous substances). This tailor-made thinking is in line with the newly adopted Sendai Framework for Disaster Risk Reduction (UN-ISDR, 2015). Priority 4 points at 'Enhancing disaster preparedness for effective response and to "Build Back Better" in recovery, rehabilitation and reconstruction'.

Disaster risk assessment and impact assessment

The Directives 97/11/EC ('Amended Environmental Impact Assessment' [EIA]) and 2001/42/EC ('Strategic Environmental Assessment' [SEA]) offer an appropriate legal basis for risk assessment (Greiving, 2004). The corresponding SEA requirements (Annexes I and II SEA Directive) are spatially oriented (spatial extent of effect, value and vulnerability of the area, cumulative effects).

An increasing damage potential (vulnerability) or impact on the hazard potential as a consequence of the realisation of a plan/project can be understood as a significant effect on the environment. Interactions with major accident hazards and multi-hazard settings have to be considered. This statement was recently proved by an amendment of the EIA directive (2014/52/EU):

> … precautionary actions need to be taken for certain projects which, because of their vulnerability to major accidents, and/or natural disasters (such as flooding, sea level rise, or earthquakes) are likely to have significant adverse effects on the environment. For such projects, it is important to consider their vulnerability (exposure and resilience) to major accidents and/or disasters, the risk of those accidents and/or disasters occurring and the implications for the likelihood of significant adverse effects on the environment.

Moreover, guidance documents on integrating climate change and biodiversity into EIA and SEA (European Commission, 2013a, 2013b) argue that, 'In addition to climate scenarios, it is important to consider socio-economic scenarios as this will help assess future vulnerability to climate change' (European Commission, 2013a, p 39). The impact of climate change on the result of the assessment through so-called 'evolving baseline trends' has to be considered (European Commission, 2013a, p 39).

Figure 69 illustrates the match between the process of a strategic environmental assessment and a risk assessment and underlines the line of argumentation that a (flood) risk assessment has to be integrated when assessing the direct and indirect significant effects of a programme or plan on the environment.

Figure 69: Risk assessment process as integrated part of the SEA

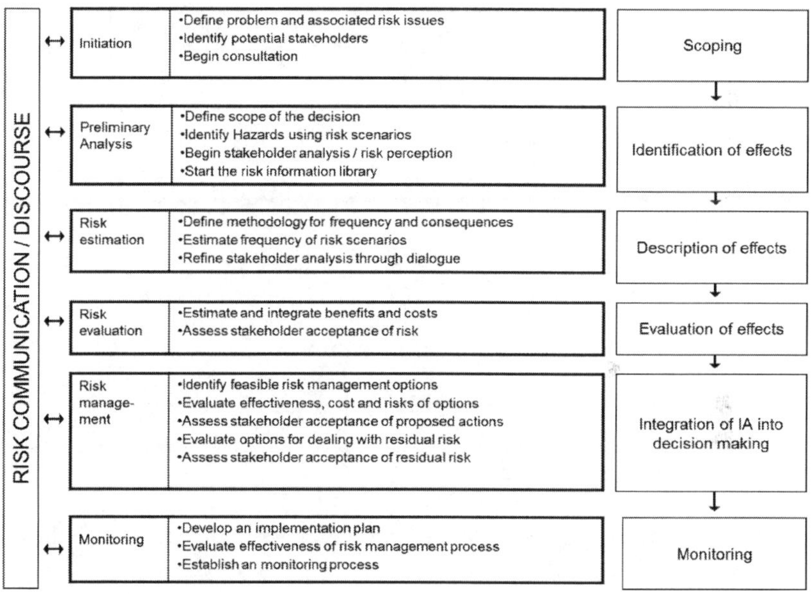

Source: Greiving and Angignard (2013)

The great Elbe and Danube flood in June 2013

In June 2013, large parts of the Elbe and the alpine part of the Danube catchment were affected by a severe flood event (see Figure 70 – the Elbe catchment is outlined in black). The following chapter focuses on the effects of this event on the Elbe catchment. The spring of the Elbe lies in the Czech Republic. The river flows into the North Sea

Figure 70: River catchments in Germany

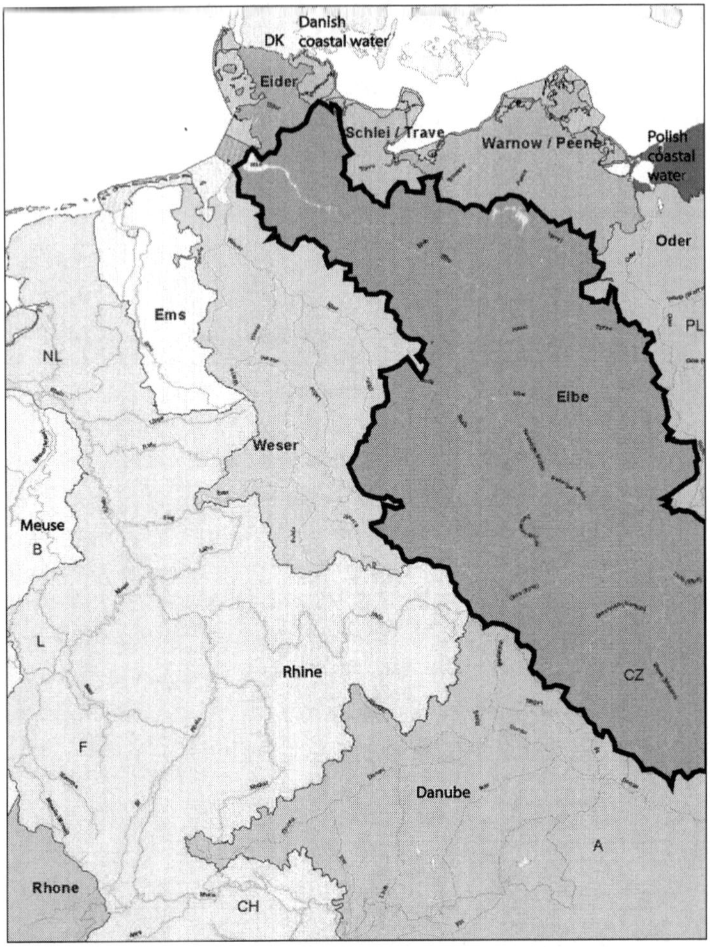

Source: Federal Environment Agency (2000)

close to Hamburg. The length of the river is 1,070km, its medium run-off is 770m^3/s and the discharge of a 1% flood is about 4,600m^3/s (at its mouth).

The return period of this event was partly less than 1:500 years. The flood was caused by a so-called V(5)-track cyclone, which displays a high potential for large summer floods in Europe and also caused previous floods in Eastern and Southern Germany in 1997, 1999 and 2002. Such kinds of cyclone rotate anticlockwise around the Alps and lead to heavy rainfall north of the Alps in Bavaria and also in the Eastern German mountains, as shown in Figure 71.

The estimated damages of the 2013 event were about €6.7 billion (Federal Ministry of the Interior, 2013). As opposed to previous

Figure 71: Weekly rainfall between 28 May and 3 June 2013 in Germany

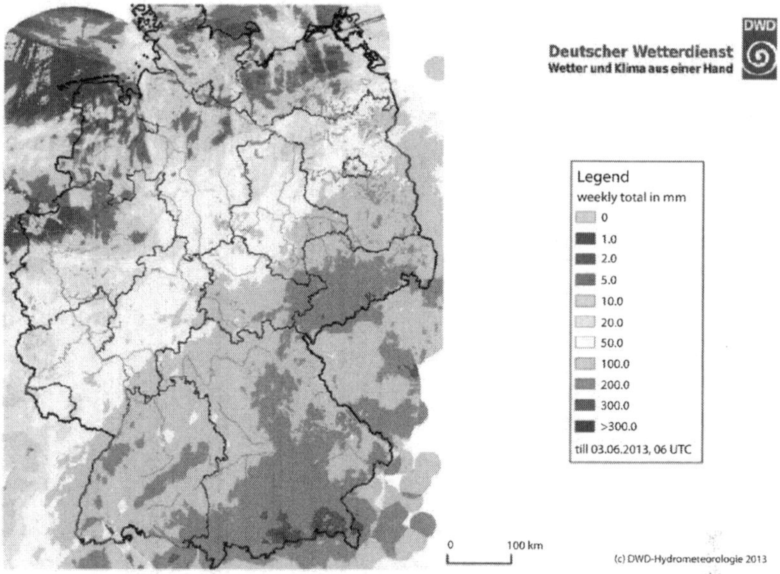

Source: BfG (2013)

events, a major dam failure at the Elbe River in Saxony-Anhalt took place on 9 June (marked by the black circles in Figure 72). About 200m^3/s flew out the river channel for nine days. In consequence, the flood wave downstream was cut by 40–50cm. In other words: this dam failure saved several towns and villages downstream of the river. This exceptional event had different direct and indirect consequences: an area (fortunately only sparsely populated) of about 450km^2 was inundated (see Figure 72) and 25,000 people had to be evacuated for weeks to a month.

The even more relevant, but indirect, effect of this dam failure was the impact on the complex high-speed railway network in Germany. Parts of the high-speed railway connection from Hannover to Berlin, used by an average of 90,000 passengers a day, were flooded. This led to severe damage to its complex infrastructure (power supply, signal technology, etc), but also to the fundamentals of the railway itself. In consequence, the railway connection was closed for almost six months. The established alternative route caused delays of about an hour in both directions; many major towns like Wolfsburg (head office of the Volkswagen Company) were totally isolated from the high-speed network for months.

This event could have been a benchmark for establishing a more resilient way of dealing with major catastrophes in Germany. The

Figure 72: Evolvement of the inundation over time (9–18 June 2013)

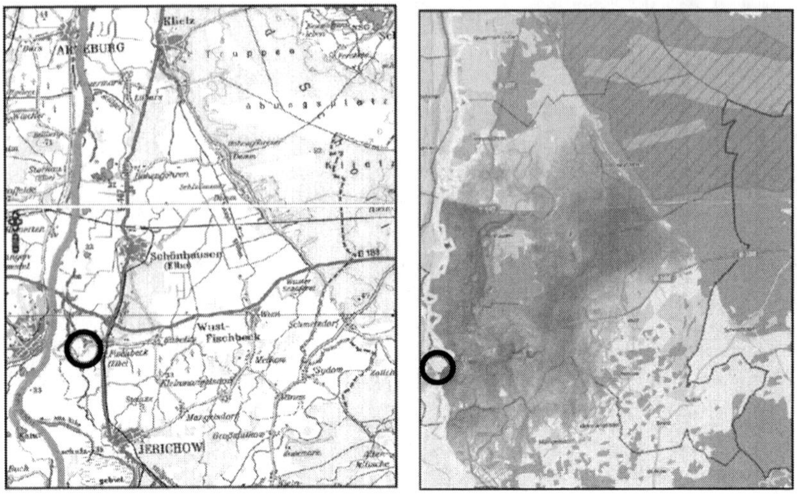

Source: BfG (2013, p 44)

affected region is not only sparsely populated, but also severely affected by ageing and outmigration, having lost one third of its population since the reunification of Germany in 1990. Thus, this catastrophe created the option for a readjustment of the scattered settlement structure in order to maintain services of general interest (public services and goods like health care, education, retail) at central places located outside of the flood zone by retreating from the flooded area. In doing so, an integrated response strategy to both flood risk and demographic change would have been possible.

In reality, all landowners and households received full compensation for their losses from the Federation and the Federal State of Saxony-Anhalt (Federal Ministry of the Interior, 2013). There was not sufficient time in the aftermath of the disaster to fundamentally rethink the settlement structure as the people wanted to reoccupy their homes as soon as possible.

Nonetheless, there is a visible shift from improving technical flood control (as opposed to the aftermath of the 2002 event) to comprehensive, planning-based approaches. More attention has to be spent on non-structural measures, including those described earlier in this chapter. In addition, the early involvement of municipalities and the public is an indispensable requirement for acceptable flood risk reduction measures (Thieken, 2014). This is also clearly underlined by a new two-day audit that assesses the performance of local flood risk management, developed and implemented by the German Association for Water, Waste Water and Waste (Gfrörer, 2013). Figure 73 indicates how this audit works.

Figure 73: Results of the audit

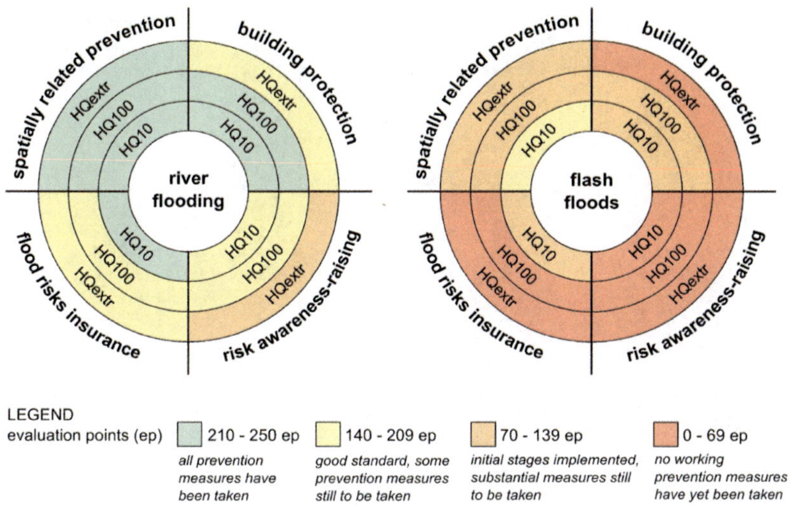

Source: Gfrörer (2013, p 138). The green colour indicates a good, the red a weak performance of a municipality (based on experts' judgement for three return periods).

This audit differentiates between four elements of flood risk reduction: 'Flächenwirksame Vorsorge' (spatially related prevention, see earlier), 'Bauvorsoge' (building protection, see earlier), 'Risikovorsorge' (flood risks insurance) and 'Verhaltenswirksame Vorsorge' (risk awareness-raising). The latter two elements are not relevant for spatial planning and are therefore not discussed in this chapter. This audit indicates a remarkable shift towards a more relevant role for spatial planning in flood risk management but still does not reflect the importance of retreat (see earlier).

Nonetheless, the 2013 flood event has to be seen as a push effect for initiating some first research projects on retreat – particularly from areas that are at risk and whose populations are shrinking (Janssen et al, 2014). Finally, the new national government intends to set up a nationwide flood risk management plan for spatial planning.

Conclusions

Table 9 points at the strength and weaknesses of German spatial planning in the context of (flood) risk assessment and management.

Table 9: Strengths and weaknesses of spatial planning

Task	Milestones	Potential of spatial planning	Description
Risk assessment	Assessment and appraisal of risk and its impact on the human-environmental-system	fair	Possible based on regional impact studies, planning has to have at hand. A strength of comprehensive planning is the traditionally integrated view on different change processes (demography, economy, environment, climate)
	Assessment of frequency and magnitude of extreme events	poor	That is clearly a task for specialised authorities like water management where spatial planning does not have any competence at hand
Change proofing	Identification of interaction between land-uses and the changing risk	good	Such assessments can easily be integrated in the strategic environmental assessment which is obligatory for any spatial plan or program
	New guiding principles (such as "resilience"), suitable for the ongoing global change	good	The concept of resilience is almost in line with existing planning principles like decentralised concentration and could therefore easily adopted in planning practice
Risk mitigation	Avoidance of non-adapted developments	good	This is in focus of planning which is very much about future developments. The effectiveness of actions depends partly on the existing regulatory framework (zoning instruments)
	Adaptation of existing spatial structures (settlements, infrastructure)	poor	Any adaptation of existing structures is hardly possible through regulatory planning due to the given private property rights. What is needed are incentives and good practices aiming at convincing private householders
	Keeping disaster prone areas free of further development	good	At least conforming planning systems have regulatory zoning instruments at hand. Keeping free areas prone to extreme events is thereby possible
	Differentiated decisions on land-use according to the given vulnerability	fair	Almost possible, but not effective with regard to existing settlement structures

Task	Milestones	Potential of spatial planning	Description
	Relocation/retreat from threatened areas	poor	In conflict with property rights. Full recompensation is needed, which fails mostly due to the lack of financial resources. Possible in the aftermath of a disaster or in areas with shrinking population where the existing building stock will be (partly) deconstructed based on planning strategies (see Eastern Germany)

Adapted from Greiving and Angignard (2013)

Spatial planning is theoretically able to mitigate flood risks to a particular extent. Planning authorities are key actors in risk management. However, the quality of planning-based response strategies depends on the quality of risk information. The outcomes of risk assessments have to be tailor-made to the needs of their users, such as planning authorities. They also have to be explained and visualised in a way that meets the educational background of the general public as planning processes are based on the participation of the public.

In practice, the effectiveness of on-paper disaster risk management also depends on the level of trust in public decision-making. As already explained, many hazard zones (ie flood zones) have been settled for centuries. Both problems call for discourse-based strategies and inclusiveness. There is a growing need to involve all stakeholders from the very beginning of the risk assessment and management process in order to improve the effectiveness of disaster risk response, particularly in already built-up areas. Future catastrophes should be understood as chances for rethinking obsolete settlement structures that are no longer sustainable with regard to demographic change.

Notes

[1] See also the website of Humanitarian Aid & Civil Protection, available at: www.ec.europa.eu/echo/index_en

[2] See the website of Humanitarian Aid & Civil Protection, available at: www.ec.europa.eu/echo/index_en

References

BfG (Bundesanstalt für Gewässerkunde) (ed.) 2013, 'Länderübergreifende Analyse des Juni Hochwassers 2013', BfG Report 1797, Koblenz.

European Commission, Directorate-General for Environment 2013a, *Guidance on Integrating Climate Change and Biodiversity into Environmental Impact Assessment*, viewed on 23 May 2013, http://ec.europa.eu/environment/eia/pdf/EIA%20Guidance.pdf

European Commission, Directorate-General for Environment 2013b, *Guidance on Integrating Climate Change and Biodiversity into Strategic Environmental Assessment*, viewed on 23 May 2013, http://ec.europa.eu/environment/eia/pdf/SEA%20Guidance.pdf

Federal Ministry of the Interior (ed.) 2013, *Bericht zur Flutkatastrophe 2013 – Katastrophenhilfe, Entschädigung, Wiederaufbau*. Berlin. https://www.bmi.bund.de/SharedDocs/Downloads/DE/Broschueren/2013/kabinettbericht-fluthilfe.pdf?__blob=publicationFile

Fleischhauer, M., Greiving, S., Flex, F., Scheibel, M., Stickler, T., Sereinig, N., Koboltschnig, G., Malvati, P., Vitale, V., Grifoni, P. & Firus, K. 2012, 'Improving the active involvement of stakeholders and the public in flood risk management – tools of an involvement strategy and case study results from Austria, Germany and Italy'. *Nat. Hazards Earth Syst. Sci.*, 12, 2785–2798.

Gfrörer, J. 2013, 'Audit – "Hochwasser wie gut sind wir vorbereite"'. Proceedings of the 36. Dresdner Wasserbaukolloquium 2013: 'Technischer und organisatorischer Hochwasserschutz', in *Dresdner Wasserbauliche Mitteilungen*, vol. 48, pp. 135-140.

Greiving, S. 2004, 'Risk assessment and management as an Important Tool for the EU Strategic Environmental Assessment'. DISP 157, pp. 11-17.

Greiving, S. 2006, 'Dealing with natural hazards in Germany's planning practice', in M. Fleischhauer, S. Greiving, S. Wanczura (eds.), *Natural Hazards and Spatial Planning in Europe*. Dortmunder Vertrieb für Bau- und Planungsliteratur: Dortmund.

Greiving, S. & Angignard, M. 2013, 'Disaster Mitigation by Spatial Planning', in T. van Asch, J. Corominas, S. Greiving, J.P. Malet & S. Sterlacchini (eds.), *Mountain Risks: From Prediction to Management and Governance. Advances in Natural and Technological Hazards Research*, vol. 34. Springer, pp. 287-302.

Greiving, S. & Lindner, C. 2011, 'Assessment of flash flood risk in a continuous urban fabric by the example of the City of Dortmund', in G. Zenz & R. Hornich (eds.), *Urban Flood Risk Management – Approaches to enhance resilience of communities*, Verlag der Technischen Universität Graz, pp. 257 – 262.

Greiving, S., Pratzler-Wanczura, S., Sapountzaki, K., Ferri, F., Grifoni, P., Firus, K., & Xanthopoulos, G. 2012, 'Linking the actors and policies throughout the disaster management cycle by "Agreement on Objectives" – a new output-oriented management approach', *Nat. Hazards Earth Syst. Sci.*, 12, 1085-1107, doi:10.5194/nhess-12-1085-2012

Hendler, R. 2012, '§69 Bau-, Raumordnungs- und Landesplanungsrecht im föderalen System der Bundesrepublik Deutschland', in I. Härtel (ed.), *Handbuch Föderalismus – Föderalismus als demokratische Rechtsordnung und Rechtskultur in Deutschland, Europa und der Welt.* Springer-Verlag: Berlin, Heidelberg.

Janssen, G., Keimeyer, F., Rubel, C. & Schulze, F. 2014, *Siedlungsrückzug – Recht und Planung im Kontext von Klima- und demografischem Wandel*, 2nd Interim Report, Dresden.

Krahe, P., Nilson, E., Carambia, M., Maurer, T., Tomassini, L., Bülow, K., Jacob, D. & H. Moser 2009, 'Wirkungsabschätzung von Unsicherheiten der Klimamodellierung in Abflussprojektionen – Auswertung eines Multimodell-Ensembles im Rheingebiet. Hydrologie und Wasserbewirtschaftung', vol. 5/2009. pp. 316-331.

Maurer, H. 2010, *Staatsrecht I. Grundlagen, Verfassungsorgane, Staatsfunktionen*, 6. überarbeitete und ergänzte Auflage, Verlag C.H. Beck, München.

Moser. H., Cullmann, J. Kofalk, S., Mai, S., Nilson, E., Rösner, S., Becker, P., Gratzki, A. & Schreiber, K.-J. 2012, 'An integrated climate service for the transboundary river basin and coastal management of Germany', in *World Meteorological Organsation (2012) Climate ExChange*, pp. 88-91. Tudor Rose, http://www.wmo.int/pages/gfcs/tudor-rose/index.html#/88/

PLANAT (National Platform for Natural Hazards) (ed.) 2008, *Strategie Naturgefahren Schweiz. Projekt B.2 – Schutzziele*, Bern.

Rivolin, U. J. 2008, 'Conforming and Performing Planning Systems in Europe: An Unbearable Cohabitation', in *Planning, Practice & Research*, vol. 23, No 2, May 2008, pp. 167-186.

Slovic, P. 1987, *Perception of Risk, Science* 236: 280-285

Thieken, A. 2014, 'Hochwasservorsorge in Deutschland – Impulse für das nationale Hochwasserschutzprogramm (Draft from 18 July 2014), Verbundprojekt 'Untersuchungen zur Bewältigung des Hochwassers 2013', Potsdam.

Thiele, A. 2012, '§ 54 Katastrophenschutzrecht im deutschen Bundesstaat', in I. Härtel (ed.), *Handbuch Föderalismus – Föderalismus als demokratische Rechtsordnung und Rechtskultur in Deutschland, Europa und der Welt*, Springer-Verlag: Berlin, Heidelberg.

Turowski, G. 2002, *Spatial Planning in Germany: structures and concepts. Studies in spatial development*, Akademie für Raumforschung und Landesplanung (ARL), Hannover.

Turowski, G. 2005, 'Raumplanung (Gesamtplanung)'. In: *Handwörterbuch der Raumordnung*, ARL, Hannover, pp. 893-898.

UBA (Federal Environment Office) (ed.) 2014, *Siedlungsrückzug - Recht und Planung im Kontext von Klima- und demographischen Wandel.* 2nd Interim Report. Dessau.

UN-ISDR 2015, *Sendai Framework for Disaster Risk Reduction 2015–2030*, www.wcdrr.org/uploads/Sendai_Framework_for_Disaster_Risk_Reduction_2015-2030.pdf

CHAPTER A5b

Major-accident hazards in spatial planning

Nadine Mägdefrau

Introduction

Continually, tragic accidents in technological establishments that deal with dangerous substances are proof that they are jeopardising the health and safety of the people who live in nearby residential areas. For instance, the explosion of a fireworks depot in Enschede (Netherlands) in May 2000 caused the death of 23 people, damaged 1,500 houses and left 1,250 people homeless. This incident showed once more how important it is to establish certain conditions that ensure the safest coexistence of establishments dealing with dangerous substances and their surrounding land uses. To establish these conditions, the European Council enacted Council Directive 82/501/EEC on the major-accident hazards of certain industrial activities in 1982 (also known as the Seveso Directive). This directive mainly focused on technological system safety (Louis and Wolf, 2007). With the replacement of the directive in 1992, it not only approaches the problem of major-accident hazards at the organisational level, but also calls spatial planning into account. Based on the directive, the task of spatial planning is to examine and establish certain distances between establishments that deal with dangerous substances and sensitive land uses (eg residential uses). This requirement is based on spatial planning's obligation to consider the population's good health within the planning process – which can be threatened by (technological) hazards. In German spatial planning, hazards are generally determined in accordance with the order of optimisation. This means that its concerns must be considered within the planning process but can be outweighed by other reasons, which might be more important for certain cases (for further information about the German spatial planning system and its approach to hazards, see Chapter A5a).

Major-accident hazards are not always solely based on technological failure. Technological accidents can be caused by natural events (NaTech accidents are a significant category of technological accidents;

see Cozzani et al, 2014). The most frequent natural hazards that cause technological accidents are floods and lightning. Floods mainly cause structural damage and the failure of electrical equipment, which can lead to water dispersion and the reactions of released chemicals with water. One example of this is the release of toxic chlorine at the Spolana chemical plant in Neratovice (Czech Republic) during the summer flood of 2002 (Cruz et al, 2004). The water lifted tanks that contained liquid chlorine, which resulted in the damage of pipes and the subsequent release of 80 tonnes of liquid chlorine and 10 tonnes of chlorine gas. Although no people were harmed, the incident resulted in the destruction of the surrounding crops and fields (Cozzani et al, 2014). An additional example of a technological hazard triggered by a natural event – although not covered by the Seveso Directive discussed in this chapter – is the damage to the Fukushima Daiichi Nuclear Power Plant in Japan after the Great East Japan Earthquake and the subsequent tsunami, which showed once again how important it is to prepare for multiple hazards.

This chapter explains the establishment of the Seveso Directive and its relevance for spatial planning. Following this, the Seveso Directive's implementation into German law and its consideration at the regional and local level is discussed. The example of the expansion of the Frankfurt Airport and the connected relocation of a plant covered by the Seveso Directive illustrates how the application of the directive works in reality. The chapter then concludes with a summary of possible points of transferability.

Major-accident hazards in the European Union

As a response to two major chemical accidents that occurred in the European Community in 1974 (Flixborough, UK) and in 1976 (Seveso, Italy), the European Community established Council Directive 82/501/EEC on the major-accident hazards of certain industrial activities in 1982 (Renn, 1989). The aim of the directive, which became known as the Seveso Directive, is to prevent major-accident hazards that involve dangerous substances in the future and to improve the handling of such accidents should they nonetheless occur (European Commission, 2015).

Directives are legislative acts issued by the European Union (EU). In comparison to regulations, they are not directly applicable in the member states, but only set a certain result that has to be achieved. The directives include a timeline for the attainment of the intended result but leave the form and method to achieve this result to national authorities. This guarantees a certain degree of freedom to the member states. If a member state fails to comply with a directive, the European Commission (EC), which represents the interests of the EU, may initiate legal actions in the European Court of Justice (European Union, 2014).

Since the Seveso Directive came into effect, it has been repeatedly amended and replaced, mainly to broaden its scope (see Figure 74). On 9 December 1996, Council Directive 96/82/EC on the control of major-accident hazards (aka the Seveso II Directive) was adopted, replacing the original Seveso Directive. Besides some other changes, this version of the directive added the necessity for land-use planning to oversee the settlement of new establishments that deal with dangerous substances in the vicinity of protected areas under Article 12 (Schoppengerd, 2015). This article aimed at ensuring appropriate distances between hazardous establishments and sensitive areas when new establishments are planned, existing establishments are modified or new developments near existing establishments are planned. With the commencement of the Seveso III Directive (see later), the requirements for land-use planning were moved to Article 13.

Figure 74: Evolution of the Seveso Directive

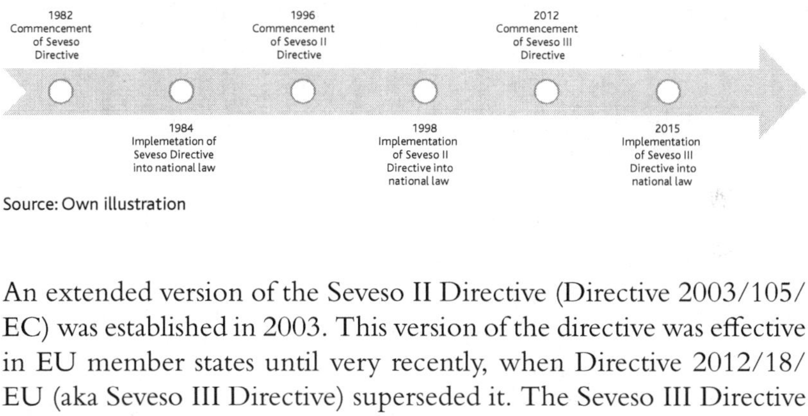

Source: Own illustration

An extended version of the Seveso II Directive (Directive 2003/105/EC) was established in 2003. This version of the directive was effective in EU member states until very recently, when Directive 2012/18/EU (aka Seveso III Directive) superseded it. The Seveso III Directive entered into force on 13 August 2012 and had to be implemented into the national law of EU member states by 1 June 2015 (European Commission, 2015).

The major reason for the replacement of the directive was the required updating of the list of chemical substances in appendix 1 to the EU's new Regulation on the Classification, Labelling and Packaging (CLP) of substances and mixtures. The CLP Regulation adapts the EU system to the United Nations (UN) international chemicals classification (Globally Harmonised System [GHS]), coming into effect in June 2015. Other changes in the directive include the improvement of access to information for citizens and their ability to get justice if this access is denied. Additionally, stricter standards for inspections ensure the precise enforcement of safety rules in establishments (European Commission, 2015).

The relevance of the Seveso Directive for spatial planning

The Seveso Directive intends to tackle the problem of major accidents that involve dangerous substances on two levels: first, it intends to prevent those accidents; and, second, it aims to improve the handling of such incidents if they nonetheless occur. The content of the Seveso III Directive can be divided into obligations for operators, obligations for member states and citizens' rights. The main obligations of the operators focus on the prevention of major accidents and the limitation of the consequences in case they still occur. The articles of the Seveso III Directive directed at them are as follows (European Commission, 2015):

- Notification of all establishments covered by the directive (Article 7).
- Development of a major-accident prevention policy (Article 8).
- Development of a safety report (Article 10).
- Development of internal emergency plans (Article 12).
- Provision of information in case of accidents (Article 16).

In addition to the operators, member states are required to take care of the following articles:

- Development of external emergency plans (Article 12).
- *Land-use planning (Article 13).*
- Information for the public (Article 14).
- Guaranteeing that necessary actions are taken after an accident (Article 17).
- Reporting accidents to the EC (Article 18).
- Prohibiting the unlawful use or operation of establishments (Article 19).
- Conducting inspections (Article 20).

As mentioned earlier, the Seveso III Directive strengthened citizens' rights and the requirements for citizen participation in order to ensure that they are sufficiently informed about existing risks and can prepare themselves accordingly. The articles of the directive that guarantee citizen participation are as follows:

- Consultation of the concerned public (Article 15).
- Relevant information must be available (Articles 14 and 22).
- Access to justice must be granted for the cases listed in Article 23.

Most of the addressed points lie within the responsibility of the operator of the establishment (eg major-accident prevention policy, internal emergency plans, accident reporting) and the emergency response units (eg external emergency plans, taking necessary actions after an accident). However, to manage spatial risks through land-use planning, planning competence is needed. Due to this, spatial planning is responsible for dealing with Article 13 of the Seveso Directive.

According to its federal structure, Germany's planning competence rests with the federal states (see Chapter A5a). Hence, they are responsible for handling the regulations of Article 13 of the Seveso Directive. How the federal states deal with this topic is based on the implementation of the Seveso Directive into German law, which is outlined in the following section.

The implementation of the Seveso Directive in Germany

In Germany, the requirements of the Seveso Directive that deal with land-use planning were mainly implemented by the revision of section 50 of the Federal Control of Pollution Act (*Bundesimmissionsschutzgesetz* [BImSchG]) and the supplement of section 9 I 24 of the Federal Building Code (*Baugesetzbuch* [BauGB]) (KAS, 2010, p 3), when the establishment of the Seveso II Directive required these changes. The paragraphs of the directive that deal with the management of the security of the establishments and the provision of information were implemented in an amendment of the Hazardous Incident Ordinance (*Störfallverordnung*). With the establishment of the Seveso III Directive, the Hazardous Incident Ordinance and the Federal Control of Pollution Act were adjusted accordingly. However, the plans and examples presented in this chapter still follow the requirements of the Seveso II Directive.

To simplify the implementation of section 50 of the Federal Control of Pollution Act, the Commission on Process Safety (*Kommission für Anlagensicherheit* [KAS]) developed a guideline. In accordance with the Seveso Directive, the guideline covers (KAS, 2010, p 8):

- the designation of land-use zones for new establishments dealing with dangerous substances;

- the designation of land-use zones for the extension of establishments; and
- the approach of sensitive land uses to existing establishments.

Besides distance recommendations that should be met between establishments covered by the Seveso Directive and sensitive land uses, the guideline also gives information on how those distance recommendations can be implemented by the instruments of spatial planning (KAS, 2010).

The guideline's recommended distances differ based on the existing knowledge about the planned establishment ('establishments with detailed knowledge' and 'establishments without detailed knowledge'). Establishments where the substances that will be processed and/or stored are known in advance require individual case assessment. If the substances are unknown, the distances between the establishments and sensitive land uses are much broader. The guideline sets up four classes that are composed of the necessary distances for certain chemical substances.

Class I intends a distance of 200m for establishments that deal with the substances benzene, propane, methanol, acrylonitrile and/or ethylene oxide. With an intended distance of 500m, Class II contains fluorine, hydrogen fluoride, ammonia, bromine and sulphur trioxide. Class III includes hydrocyanic acid, formaldehyde, hydrogen sulphide and sulphur dioxide, with an intended distance of 900m. The furthest distance of 1,500m is intended for the substances of Class IV: chlorine, hydrogen chloride, phosgene and acrolein. If the processed or stored substances are still unknown, the greatest distance of 1,500m has to be satisfied in order to keep all future utilisation options open.

When section 50 of the Federal Control of Pollution Act is considered, it is always in accordance with the order of optimisation. This means that the distances between establishments covered by the Seveso Directive and residential and other sensitive land uses might be outweighed by other – in this case – more important reasons (Louis and Wolf, 2007, p 4). Thus, there is no general priority of sensitive land uses over other concerns; the priorities have to be weighed up for each case (KAS, 2010, p 8).

Major-accident hazards in regional plans

The level of regional planning is not entitled to make legally binding designations for the distances between establishments covered by the Seveso Directive and residential or other sensitive land uses. However, because of its superior level, regional planning is able to recognise and reduce conflicts between different land uses at an early stage. Section 9 of the Federal Spatial Planning Act demands an environmental assessment for all spatial structure plans (eg land development plan, regional plan). The content of the environmental assessment must refer to current knowledge and generally accepted methods of validation; its level of detail must correspond with the level of detail of the plan (§ 9 I ROG). It considers possible impacts on human beings, animals, plants, soil, water, air and so on. As the failure of an establishment covered by the Seveso Directive is a possible threat to people and nature, it is important to consider the location of such an establishment as part of the environmental assessment. However, only few regional planning authorities actually include the impacts that establishments covered by the Seveso Directive might cause in their plans. This chapter introduces as examples the regional plan for North Hesse (*Regionalplan Nordhessen*) and the regional plan for Düsseldorf (*Leitlinien Regionalplanfortschreibung Düsseldorf*).

The regional plan for North Hesse includes a very accurate examination of the distance between existing establishments covered by the Seveso Directive and newly planned residential areas. This examination includes all 15 establishments on the regional territory that are covered by the Seveso Directive and checks whether any residential areas or federal highways are planned within the maximum distance of 1,500m. This analysis resulted in four case-by-case reviews for residential areas that were planned within the 1,500m distance of an establishment covered by the Seveso Directive. The case-by-case review considered the actual existent substances in the establishments, which resulted in the reduction of the required distance between the establishments and the planned residential areas, according to the Commission on Process Safety's guideline. In this way, all conflicts concerning the Seveso Directive could be resolved (Regionalversammlung Nordhessen, 2009, pp 110–11).

In North Rhine-Westphalia, the guidelines for the updating of the regional plan for Düsseldorf form the only regional document that mentions the Seveso Directive. These guidelines aim to enable the future settlement of establishments covered by the Seveso Directive in commercial and industrial zones. Consequently, the settlement of sensitive land uses in commercial and industrial zones is prohibited, and in the case of sensitive land uses approaching these areas, the possible impact of this development on the future settlement of establishments covered by the Seveso Directive has to be considered. The benchmark for this consideration is 1,500m, the maximum distance that is recommended by the Commission on Process Safety. This strategy ensures that there are sufficient areas with tolerable distances to sensitive land uses available to locate establishments covered by the Seveso Directive in the future (Bezirksregierung Düsseldorf, 2012, pp 41–3).

Major-accident hazards in land-use plans

The instruments of the Hazardous Incident Ordinance cannot prevent the location of sensitive land uses in the vicinity of establishments covered by the Seveso Directive. Only spatial planning can take these preventive measures, which emphasises the importance of spatial planners in this field (KAS, 2010, p 10). Since urban land-use planning is responsible for the allocation of land uses, this is the main planning level that deals with the requirements of section 50 of the Federal Control of Pollution Act.

To enable the future location of establishments covered by the Seveso Directive in certain areas, it can be useful to include sufficient distances between this area and existing or planned residential areas in the preparatory land-use plan (Louis and Wolf, 2007, p 4). Additionally, planned areas for residential and other sensitive land uses must be checked for tolerable distances to establishments covered by the Seveso Directive in order to offer a maximum of security. In most cases, the principle of separation, according to section 50 of the Federal Control of Pollution Act, is only considered in binding land-use plans (*Bebauungsplan*). The reason for this is that binding land-use plans are the only kind of plan that are legally binding for both the administration and private individuals. Nevertheless, the environmental report of the regional preparatory land-use plan for Frankfurt (*Regionaler Flächennutzungsplan*) considers this topic and states that the distance requirements for establishments covered by the Seveso Directive have to be considered during the preparation of a regional preparatory land-use plan if the scale permits it (Regionalverband FrankfurtRheinMain, 2010, p 504).

Like at the superior level, urban land-use planning must implement an environmental assessment in the course of the planning process. As section 2 IV of the Federal Building Code states, the content of this environmental assessment must refer to current knowledge and generally accepted methods of validation; its level of detail must correspond with the plan's level of detail (§ 2 IV BauGB). This means that the demands for environmental assessment in land-use planning equal the demands at the superior level of regional planning. The only exception in this regard is the level of detail, which varies with the scale of the plan. This chapter presents as examples the regional preparatory land-use plans of the Frankfurt Region and the Ruhr Region.[1]

The current regional preparatory land-use plan for Frankfurt (*Regionaler Flächennutzungsplan Frankfurt Rhein/Main*) comprehensively considers the Seveso Directive, whereby the main contents are included in the environmental report (*Umweltbericht*) of the plan. The plan addresses establishments covered by the Seveso Directive combined with other environmental impacts of spatial relevance. Besides the new designation of sensitive land uses in the vicinity of establishments covered by the Seveso Directive, the plan additionally indicates already-existing land uses within a potentially dangerous distance to existing establishments covered by the directive. However, in this case, the plan does not state any further need for action. Concerning newly planned sensitive land uses, the plan rejects some designations by former land-use plans within the region because they were planned within the distance of establishments covered by the Seveso Directive. All in all, 294 hectares of designated areas were discarded. Still, the plan intends for 118 hectares of newly planned residential building/mixed-use areas and areas for commercial purposes within the distance of establishments covered by the Seveso Directive. In addition to this, the plan reduces areas for public good and special building areas within the distance of establishments covered by the Seveso Directive by 28 hectares. An environmental assessment, with individual examination, was conducted for all newly planned areas. In the case of the Seveso Directive, this assessment included the investigation of possible negative effects for public health in residential areas, transport infrastructure and green spaces. The environmental assessment resulted in a data sheet that presents relevant restrictions and conflicts for each planned residential area, traffic infrastructure, utility area and area for agriculture and forestry within the region. For each of these areas that are planned in the vicinity of an establishment covered by the Seveso Directive, a case-by-case review was conducted. The distances for each establishment were determined based on the suggestions of the Commission on Process Safety and the processed and or stored substances. As a result of this analysis, the report contains a map of the required distances for the establishments covered by the Seveso Directive, and the statement that planned sensitive land uses within these areas are not inevitably infeasible, but can result in a conflict. Further information on how to approach these possible conflicts is missing (Regionalverband FrankfurtRheinMain, 2010).

The environmental report (*Umweltbericht*) of the regional preparatory land-use plan for the Ruhr Region (*Regionaler Flächennutzungsplan*

Städteregion Ruhr) also approaches the specifications of the Seveso Directive. This plan includes land-use planning for the cities of Bochum, Essen, Gelsenkirchen, Herne, Mülheim (Ruhr) and Oberhausen. Similarly to the environmental report of Frankfurt, the Seveso topic is included within the environmental assessment. The basis for the analysis was a list of all establishments covered by the Seveso Directive in the region. Based on the standards of the Commission of Process Safety and the processed and or stored substances of each establishment, specific required distances were defined (Planungsgemeinschaft Städteregion Ruhr, 2009, p 70). Based on this, the distances between newly designated sensitive land uses[2] and the 49 establishments in the planning region, as well as the seven establishments in the neighbouring municipalities, were determined (Planungsgemeinschaft Städteregion Ruhr, 2009, p 75). In 39 cases, the analysis revealed an intersection of the two zones. For these cases, an individual case profile was put together and summarised the ascertained conflict for the binding land-use plan. Any further examination of those cases, as well as the final decision about the land uses, is transferred to this lower level of spatial planning (Planungsgemeinschaft Städteregion Ruhr, 2009, pp 100–1).

These examples show that legally binding designations can only occur at the level of binding land-use planning. Superior planning levels can only prepare important information for the lower level and work out possible conflicts in advance. In cases where a conflict between newly planned land uses and an establishment covered by the Seveso Directive is too big to be outweighed, this could save time and money – as the example of the Frankfurt Airport proves.

The example of the Frankfurt Airport and Ticona plant

One example that shows how an establishment covered by the Seveso Directive can affect the resilience of its neighbourhood is the case of the relocation of the Ticona plant at the Frankfurt Airport. At the beginning of the century, the Frankfurt Airport demanded a new landing runway due to increased traffic. Fraport AG, the company that runs the airport, chose the north-west runway as the best option out of three different locations (Konersmann, 2009, p 233). The regional planning statement describes the north-west runway as 'compatible with regional planning and development aims and ... also preferable to the other alternatives' (European Commission, 2004, p 2). However, this statement did not consider one major problem: the distance between the new north-west runway and a chemical plant run by Ticona Ltd[3] would have been only 700m, and the altitude of airplanes flying over the plant would have been only 60–100m (Konersmann, 2009, p 233).

The Seveso Directive covers the airport as a sensitive land use and the chemical plant as an establishment that deals with hazardous substances (European Commission, 2004, p 2). The vicinity of those land uses and the fact that 'accident statistics clearly show that the takeoff and landing phases are the most risky flight phases' (Konersmann, 2009, p 233) called attention to the necessity of a further investigation into the impacts of an airplane crash into the Ticona plant.

Nevertheless, it was not until the local initiative *Zukunft Rhein-Main* (ZRM) – which pursued the target to prevent the expansion of the Frankfurt Airport – made a complaint to the EC that the problem was considered. The complaint resulted in a written warning from the EC to the German government. It demanded the consideration of the aims of the Seveso II Directive in this case. Hence, the then Federal Ministry for the Environment, Nature Conservation and Nuclear Safety (*Bundesministerium für Umwelt, Naturschutz und Reaktorsicherheit* [BMU]) took on the problem and asked the former Major-Accident Hazard Commission (now Commission on Process Safety [KAS]) for an analysis of the situation.

The Major-Accident Hazard Commission formed a working group for this purpose, which came to the conclusion that the intended expansion of the north-west runway of the Frankfurt Airport was not compatible with the operation of the existent Ticona plant (AG Flughafenausbau Frankfurt/Main, 2004, p 3). The statement was based on various expert assessments, where the one elaborated by the Technical Control Board Palatinate (*TÜV Pfalz*) was of special importance (AG Flughafenausbau Frankfurt/Main, 2004, p 2). This assessment came to the conclusion that every plane crash with effects on the Ticona plant would have resulted in its total loss. The incident risk amounted to about $4 \cdot 10^{-5}$ a year (1 in 25,000 years). Due to the fact that about 1,000 people were working at the Ticona plant during the day and 60–70 people at night, it was expected that the total loss would have resulted in the death of more than 100 people (AG Flughafenausbau Frankfurt/Main, 2004, pp 6–7).

The values were compared to the Swiss limits for incident risks because such limits do not exist in Germany. The Swiss limit for incidents that affect 10 people is 10^{-5} a year (1 in 1000,000 years); the one for incidents that affect 100 people is $< 10^{-7}$ a year (1 in 10mio. years). This means that the calculated incident risk for the Ticona plant was unacceptably high (AG Flughafenausbau Frankfurt/Main, 2004, p 7).

As a consequence, Ticona Ltd relocated its chemical plant from its former location in Kelsterbach to the industrial park Höchst. The new plant opened in 2011. Fraport AG paid Ticona €670 million for the relocation (Winter, 2011). On 20 October 2011, the new north-west runway of the Frankfurt Airport was opened ('Vierte Piste', 2011).

This example shows how the relocation of a chemical plant can enhance the resilience of a region. Even though the Ticona plant still possesses a certain risk at its new location, the risk that was connected with its location in the vicinity of the airport and its new north-west runway was unlikely higher.

Concluding remarks

As the example of the relocation of the Ticona plant at the Frankfurt Airport shows, risks that involve dangerous substances have one advantage over natural hazards: not only vulnerable land uses, but also the potential hazards threatening them, can be controlled by spatial planning. However, this alleged advantage can lead to an erroneous feeling of controllability. Past events indicate that natural hazards have a higher degree of foreseeability than technological ones since they usually have a long history and follow certain patterns. The existing knowledge about natural hazards also enables better preparation. Additionally, the sense of inevitability of disasters caused by natural hazards leads to a higher degree of acceptance. This is different for technological hazards: in many neighbourhoods, establishments that deal with hazardous substances remain undetected until an accident happens. Due to this, few people are prepared for such a situation – which stresses the importance of proper information. Furthermore, because man-made hazards are always connected with failure, they can generate the feeling that the incident would have been avoidable.

To keep the risk of an establishment that deals with hazardous substances as low as possible, land-use planning, as a form of spatial-oriented risk management, is an appropriate instrument. Land-use planning is usually carried out by spatial planning authorities. The relevance of spatial planning for the protection from harmful environmental effects or other hazards is proven by section 9 I 24 of the Federal Spatial Planning law, which is the legal foundation for urban land-use planning in the municipalities. With its ability to make legally binding designations for urban development, local development planning holds all competences to control the development of establishments covered by the Seveso Directive and residential or other sensitive land uses. This point is also emphasised by the guideline of the Commission on Process Safety (KAS, 2010, p 10).

The implementation of appropriate distances between establishments covered by the Seveso Directive and residential areas and other sensitive land uses is an important step to increase the resilience of urban areas. Since incidents like the one in Seveso happen unannounced, it is important to prepare beforehand and keep the risk to a minimum.

As the example of the Seveso Directive in Europe shows, setting up an overarching framework can help to speed up the implementation of security standards. However, the Seveso Directive leaves many points open, like required distances between establishments and sensitive land uses, which can result in insecurities in application. Setting up such distances or similar standards could solve this problem. Of course, it has to be considered that the fast implementation of the directive can only be assured if a superior authority (like the EC) exists and is able to pressure national governments to consider certain points even if they might not lead to the easiest or most attractive solution. Finally, having legally binding land-use designations, like in the case of Germany, helps to actually implement a plan once it has been established.

Notes

[1] In Germany, preparatory land-use plans usually only cover one municipality. The preparatory land-use plans of the Frankfurt Region and the Ruhr Region are exceptional because they cover several agglomerated cities. However, the reason that they were selected for this chapter was solely their consideration of the Seveso Directive, not their wide scope.

[2] Besides sensitive land uses, the analysis also includes commercial development zones since these areas are potential locations for new establishments covered by the Seveso Directive. The reason for this is the prevention of a domino effect, where two establishments affect each other during an incident.

[3.] Ticona Ltd is a leading manufacturer of technical plastics worldwide.

References

AG Flughafenausbau Frankfurt/Main (2004), *Ergebnis der Beratungen der Arbeitsgruppe "Flughafenausbau Frankfurt/Main"*, Updated 30-January-2004.

Bezirksregierung Düsseldorf (2012), *Leitlinien Regionalplanfortschreibung*.

Cozzani, V., Antonioni, G., Landucci, G., Tugnoli, A., Bonvicini, S. and Spadoni, G. (2014), 'Quantitative assessment of domino and NaTech scenarios in complex industrial areas', *Journal of Loss Prevention in the Process Industries*, vol. 28, pp. 10–22.

Cruz, A.M., Steinberg, L.J., Vetere Arellano, A.L., Nordvik, J.-P., Pisano, F. (2004), *State of the Art in Natech Risk Management.*

European Commission (2004), Press Release IP/04/422 'Commission takes legal action against Germany on breach of EU environmental law'. 30 March 2004.

European Commission (2015), 'Industrial accidents. The Seveso Directive – Summary of requirements', 19 November, viewed on 15 December 2015, http://ec.europa.eu/environment/seveso/legislation.htm.

European Union (2014), *How the European Union works: Your guide to the EU institutions, The European Union explained*, Publications Office of the European Union, Luxembourg.

KAS (Kommission für Anlagensicherheit) (2010), *Leitfaden. Empfehlungen für Abstände zwischen Betriebsbereichen nach der Störfall-Verordnung und schutzbedürftigen Gebieten im Rahmen der Bauleitplanung – Umsetzung § 50 BImSchG.* 2. Überarbeitete Fassung.

Konersmann, R. (2009), 'Chemical Plants in the Vicinity of Airports – Assessing the Risk', in, *Chem. Eng. Technol., 2009*, vol. 32, no. 2, pp. 232-240.

Louis, H.W. & Wolf, V.A. (2007), 'Die erforderlichen Abstände zwischen Betrieben nach der Störfall-Verordnung und Wohngebieten oder anderen schutzwürdigen Bereichen nach §50 S. 1 BImSchG', in, *Natur und Recht*, vol. 29, pp. 1-8.

Planungsgemeinschaft Städteregion Ruhr (2009), *Umweltbericht zum Regionalen Flächennutzungsplan (RFNP) der Planungsgemeinschaft Städteregion Ruhr (Bochum, Essen, Gelsenkirchen, Herne, Mülheim an der Ruhr und Oberhausen).*

Regionalverband FrankfurtRheinMain (2010), *Umweltbericht Regionaler Flächennutzungsplan.*

Regionalversammlung Nordhessen (2009), *Regionalplan Nordhessen 2009.* Umweltbericht.

Renn, O. (1989), 'Risk Communication at the Community Level: European Lessons from the Seveso Directive', *JAPCA*, vol. 39, no. 10, pp. 1301-1308.

Schoppengerd, Johanna (2015), *Umsetzung rechtlicher Anforderungen in der Bauleitplanung am Beispiel der Seveso-II-Richtlinie*, Dortmund.

'Vierte Piste' (2011): 'Vierte Piste am Frankfurter Flughafen eröffnet. Merkel durfte als Erste landen', FAZ Rhein-Main, 21 October, viewed 15 December 2015, FAZ.net.

Winter, Thorsten (2011): 'Ticona feiert neues Werk und lobt Standort', FAZ Rhein-Main, 26 September, viewed 15 December 2015, FAZ.net.

CHAPTER A6

Cross-case analysis: lessons learned and overview of case examples

Stefan Greiving, Nadine Mägdefrau and Teresa Sprague

Introduction

The previous chapters covered five different countries (Japan, Indonesia, the US, Slovakia and Germany) and addressed a manifold of spatially relevant risks that are attributable to disasters in the past. The foci of the chapter contents include earthquakes and tsunamis (Japan and Indonesia), hurricanes (USA) and river floods (Slovakia and Germany). Additionally considered are major-accident hazards that can potentially be caused by natural hazards. This chapter will give an overview of the legal frameworks of the five countries and point out similarities and differences between them, as well as important learning points.

Legal framework and instruments

In Part A, a great deal of information can be learned about the efforts of countries to address disaster risks with various spatial planning instruments. Within the present chapter, a short overview concerning the most important legal facts that set the frame for spatial planning and disaster risk management in the five countries will be given.

As Table 10 shows, all of the five countries have a legal basis for disaster risk assessment and management. Furthermore, in four of the addressed countries (namely, Japan, Indonesia, Slovakia and Germany), legally defined coordination mechanisms between disaster risk management and spatial planning exist. The only exception to this is the US, mainly because its strict federal structure transfers the legislation for spatial planning to its 50 states. This does not necessarily mean that, on the ground, no coordination between hazard mitigation planning and local land-use planning takes place. However, the establishment of a coordination mechanism is up to the individual states and depends on the existence of both the mitigation plan and the land-use plan as neither of them are legally required nationwide.

Table 10: Overview of the legal bases and existing instruments

Country/Element	Japan	Indonesia	USA	Slovakia	Germany
Hazard profile addressed by part A chapters	Earthquake Tsunami	Earthquake Tsunami	Hurricanes Storm Surges	River floods	River floods Major accident hazards
Legal framework for disaster risk assessment and management	Disaster countermeasure basic act (1961), revised in 2013	Law 24/2006 on Disaster Prevention Law 24/2007 on National Disaster Management	Disaster Mitigation Act (2000) 2004 National Response Plan (NRP) National Disaster Recovery Framework (NDRF)	Directive 2007/60/EC on the assessment and management of flood risks, implemented into national law by Act No. 7/2010 on protection against floods	Council Directive 2007/60/EC on the assessment and management of flood risks, implemented in national law by Federal Water Act (WHG)
National legal framework for spatial planning	National Land Use Planning Act of 1974 City planning law (1968)	Law 24/1992 on Spatial Planning Law 17/2007 on national long term development planning Law 26/2007 on Spatial Planning	No legally binding legislative framework Legal bases vary between states	Building Code (1976)	Federal Regional Planning Act (ROG, 2008) Federal Building Code (BauGB, 2013)
Hazard and risk maps	Hazard maps exist for flood, inland inundation, storm surge and tsunami	No scientific standards for risk assessment and mapping Some hazard maps exist for coastal areas	To be developed within hazard mitigation plan by means of the GIS based tool HAZUS MH (multi-hazard)	To be prepared in accordance with Art. 6 Council Directive 2007/60/EC for floods with high, medium and low return period	To be prepared in accordance with Art. 6 Council Directive 2007/60/EC for floods with high, medium and low return period
Disaster plans	Disaster management plans on various spatial levels laid down in the Disaster Counter-measure Basic Act	National Master Plan of Rehabilitation and Reconstruction (2005) in particular book 2 on "Spatial Plan and Land affairs" of the aforementioned Master Plan Informal Village Spatial Plans (community recovery plans)	Hazard mitigation plans (on state and local levels) (Informal) local recovery plans	No comprehensive plans, but Flood Risk Management Plan (Art. 7 Council Directive 2007/60/EC) for water basin districts	No comprehensive plans, but Flood Risk Management Plan (Art. 7 Council Directive 2007/60/EC) for water basin districts Flood prevention strategies on federal state level
Land-use plans	National plans, prefectural plans, municipal plan Local land-use basic plan for different city planning areas	Long Term Development Plan Medium Term Development Plan (national level) Local Development Plans	No obligatory national or sub-national spatial plans Regional and local land use plans as well as zoning ordinances vary between states Legally binding coastal management plans in some states	National Spatial Development Perspective (2011) Regional territorial plan Municipal territorial plan Zoning Plan	Regional plan (Art. 8 ROG) for entire planning regions Preparatory land-use plan (Art. 5 BauGB) on municipal level Legally binding land-use plan (Art. 9 BauGB) for a development site

Table 10 provides a brief glimpse into the legal bases and relates to existing instruments for disaster risk assessment and management, as well as spatial planning. For reference purposes, the hazards addressed in the Part A chapters are listed. This does not include the full range of hazards experienced within these countries; however, it at least establishes what kinds of hazards the legal framework attempts to address. For a more detailed comparative analysis of these frameworks, the reader is encouraged to see Chapter B1 in Part B of this volume. A summary statement for each country is provided in Box 1.

Box 1: Overview of the legal frameworks

Japan

Article 39 of the Building Standard Law enacted in 1950 provides the framework for regulating land uses. The article allows local governments to designate areas that are prone to landslides, water surges, floods and tsunamis as hazardous. This means that the areas are expected to be kept free from further settlement development in the future.

Indonesia

Disaster risk reduction has only been applied very recently in Indonesian spatial law (Law 26/2007). Clause 'a' of Article 6 explicitly states the importance of spatial planning for dealing with hazards (eg earthquakes, volcanic eruptions, landslides, land movement and floods). The law also transfers the responsibilities for disaster management to local authorities.

USA

Due to the constitutional power of its states, the federal government of the US does not hold legislative power for land-use planning. Therefore, the requirements for land-use plans differ widely between the federal states. This means that the extent and requirements of land-use planning depend on the respective state legislation. While many local-level entities (eg municipalities, districts and counties) have zoning plans with prescribed land uses (eg residential, commercial and industrial), risk mitigation efforts are rarely included into land-use plans.

Slovakia

Since 2010, the legally binding Act No 7/2010 Coll governs flood mitigation in Slovakia. It is the Slovakian implementation of the European Flood Risk Directive (Directive 2007/60/EC). The Act requires all municipalities to include the content of flood risk maps into their municipal territory plans with its next revision. In flood-prone areas, certain activities, as well as the development of particular buildings, are prohibited.

Germany

In Germany, Article 78 of Section 1 of the Federal Water Act (WHG) prohibits the passing of land-use plans that allow the development of new settlements in established flood hazard zones. It is also not possible to grant building permissions in these areas. In addition to this, Section 8 VII 1 of the Federal Spatial Planning Act (ROG) allows the designation of certain areas as priority zones, for example, for flood prevention.

The summary statements describe the characteristics of the legal frameworks for both disaster risk assessment and management and spatial planning. The focus of these statements is to pinpoint how these fields are connected in a regulatory sense. The next section identifies some of the similarities and differences observed across these regulatory frameworks.

Similarities and differences

Three primary themes are found that highlight the similarities and differences among the analysed cases. These include the reactive tendency of actions, the role that supranational bodies play in establishing commonalities at the implementation level (eg in Germany and Slovakia) and the importance of visualisation instruments for hazard and risk mapping.

Reactive development of disaster-related legal frameworks

In considering the reactive nature of post-disaster actions, it is also important to pay attention to the fact that most of the case countries are prone to a variety of disaster risks. This disaster profile makes a multi-hazard perspective and, consequently, the development of a comprehensive legal basis indispensable. One such example is found in Hurricane Sandy in the US, an event comprised of not only a hurricane, but also a severe storm surge and inland flooding. This event illustrates a (multi-hazard) disaster-driven effect in which the occurrence of a disaster leads to the development of a legal basis to address this disaster risk (see Chapter A3a in Part A).

The reactive tendency in the development of disaster-relevant legislation is also shown by the example of Banda Aceh in Indonesia. Although many residents agreed to move to a safer place directly after the disaster, in the end they rebuilt their houses in their former location. This occurred because the pace of the official relocation process was too slow, while, at the same time, non-governmental organisations (NGOs) immediately offered to help residents with the reconstruction of their buildings. By the time the government revealed its plan for relocation, most residents were already living in their newly constructed houses – in the hazardous area (see Chapter A2a in Part A). In this regard, the goal to increase resilience was not met due to hasty and uncoordinated NGO intervention. Of course, this development could also have been prevented by a better coordination between the Indonesian government and the NGOs. However, it would have been additionally helpful to initially inform the residents about the relocation process and to begin with the relocation earlier on. The tendency to act after a disaster and the importance of the implementation of measures in a certain time period will be further examined in Chapter B2 in Part B.

The influence of a superior body

As part of the European Union (EU), Slovakia and Germany have to follow the 'Framework Directives' of the supranational body of the EU. In accordance with the Maastricht and Lisbon Treaties (which established the European Union), the European Commission can issue directives to harmonise national standards in different policy areas. Some environmental directives also set standards for disaster risk assessment and management. The directives only set a certain aim and a date until it has to be achieved. In this way, the EU member states have some flexibility to implement the directives into national law. In the context of the assessment of flood hazard and risks (for predefined return periods), this freedom includes the definition of responsible actors, as well as the methods used. As the case of the expansion of the Frankfurt Airport in Germany (see Chapter A5b in Part A) shows, a superior body like the European Commission can oversee the efforts of its member states and intervene if circumstances require it. This can ensure that states follow the law even if it might not be the easiest or most attractive solution. The other countries that are addressed in this book are not part of a supranational body, but they are all member states of the United Nations (UN); and as such, all have agreed to the Sendai Framework for Disaster Risk Reduction, which provides international guidelines for disaster risk reduction.

The importance of visualisation instruments for hazard and risk mapping

As all examples show, the visualisation of risk is an important factor in risk assessment. This can happen with the help of maps or simulations. In the case of the Hornad River in Eastern Slovakia, an entire simulation was developed to show which places along the river were especially endangered (see Chapter A4b in Part A). The importance of sharing the developed information with residents is discussed in Chapter A1a in Part A. In Japan, making hazard maps open to the public had been taboo for a long time and the sharing of such information has started only very recently. However, the revelation of such information did not result in the expected fall of land prices in endangered areas. This example shows that the withholding of information should not be justified by an assumed but unproven fear of economic drawbacks. Access to information for citizens is a basic factor in risk governance and should not be underestimated. Moreover, hazard and risk maps need to be understood by their addressees. This is not always the case as they are normally made by experts for other experts (including land-use planners) but not the general public. Nevertheless, the visualisation of risk is not explicitly addressed by Part B of this book as it is mainly a responsibility of sectoral planning divisions like water management and geological surveys. This book is written from a spatial planning perspective, one among many users of this visualised information.

Important topics and patterns from case examples

The examples from Japan, Indonesia, the US, Slovakia and Germany offered a wide overview regarding the experiences, problems and advantages when spatial planning deals with disaster risk. Some overarching points that are especially noticeable in the case study examples are discussed here.

The importance of coordination

The example of Indonesia shows that coordination between stakeholders on different levels needs improvement. This would increase the opportunity for a more resilient recovery after a disaster (see Chapter A2b in Part A). This problem was of special importance during the development of the village plans in Banda Aceh. The lack of coordination between neighbouring communities resulted not only in plans with divergent qualities, but also in missing links for road networks between villages, among other issues (see Chapter A2a in Part A).

Since the implementation of risk mitigation measures is not always an economically optimal solution, it can also make sense to have a superior authority that reminds the states to consider certain important factors. This is shown by the case of the expansion of the Frankfurt Airport in Germany (see Chapter A5b in Part A). In this case, the European Commission demanded the consideration of security factors set by the Seveso Directive that had been ignored at the early stages of expansion. The importance of coordination for reaching a comprehensive planning approach after a disaster are explained in Chapter B3 in Part B.

The importance of participation and enforceability

Of course, it is desirable to include citizens' views and ideas in the recovery process and offer them a forum for participation. One example regarding this point is the development of village plans in Aceh, Indonesia (see Chapter A2a in Part A). Another important factor is the certain degree of enforcement that is used in the designation of hazardous areas. As the example from Nobiru District in Sendai City shows, the residents' opportunity to rebuild their houses where they used to be or (alternatively) to move to another area led to areas with very low population densities (see Chapter A1c in Part A). This cannot be considered desirable from a spatial planning point of view.

To achieve this enforcement, legally binding designations can be very helpful, as the example from Germany shows (see Chapter A5b in Part A). They help to effectively implement the designated spatial plan. Such designations also exist in Japan, though they were rarely used before the Great East Japan Earthquake (GEJE) in March 2011, as Chapter A1a in Part A explains. After the GEJE, the area designated as hazardous increased by about 84%. In addition to this, the number of collective relocation programmes increased notably (see Chapter A1a in Part A). However, the effectiveness of such designations depends on the extent to which a society is aware of a given plan and is willing to follow the law. In developing countries like Indonesia, this is not always the case and explains why particularly informal settlements are located in hazard-prone areas. A further elaboration of the different types of participation exemplified within the country cases will be given in Chapter B4 in Part B.

The problem of uncertainty

When speaking of risks, there are always uncertainties. First, there is the issue of uncertainty regarding the frequency and magnitude of a hazard. A typical approach in all five countries to address this problem is the application of hazard lines or zones, whose demarcation depends on predefined thresholds. However, uncertainty can be found in cases where there is a lack of knowledge of past events or an obvious change in the environmental conditions due to climate change. Despite the importance of this topic, none of the Part A contributions incorporated strategies for dealing with uncertainty. This obvious lack of attention on the ground to this emerging problem is therefore a major weakness of the state of the art of disaster response worldwide.

Second, regarding appropriate response actions, there is often uncertainty about the number of housing units that have to be constructed after disasters. For example, the recovery process in Japan shows that the number of people who wanted to live in certain areas sharply decreased over time. The longer the time until the completion of a certain settlement, the greater this decrease was. Reasons for this might be that families had already built a new house in another area of the town or that they had decided to move to another area of the country. To predict the demand in advance is very complicated. As Chapter A1c in Part A explains, the degree of utilisation of a settlement in Watari Town is about 90%, while some settlements in Ishinomaki City only have a utilisation degree of 50%. The problem of uncertainty, which is deeply connected with dealing with risks, will be discussed in Chapter B5 in Part B.

The importance of the consideration of different factors in creating structural measures

As the example of the seawall construction along the coast of Tohoku Region shows, it is also important to consider additional reasons aside from the (physical) security factor. Spatial planning requires the weighing up of different concerns and interests and is therefore more able to find a balanced solution than authorities that are only in charge of risk management. After the experience of such a devastating disaster as the GEJE, calls for security of course grew louder. Nonetheless, it is advisable to consider additional factors that might cause further problems for the region if they are not acknowledged. Chapter A1b in Part A gives two examples of areas where the height of the planned seawall did not consider any local factors: in Nonoshima Island in

Shiogama City, the residents demanded the decrease of the designated levee height because of their reliance on the fishing industry and, thus, the importance of visibility beyond the seawall; in Kesennuma City, the construction of the high wall will most likely influence tourism, which is an important factor for local industry. Aside from these economic considerations, there are also other factors that have been completely ignored, including: the fact that construction of the seawall will also cut off any ecological connection between the land and sea along the Tohoku coast, and consideration of the social relationship between the local residents and the sea.

Conclusion

This chapter provides some of the main observations from Part A of this book. As one can see, some topics are not only relevant, but also important, for the five presented countries (eg participation, coordination, uncertainty). Part B of this book will look into some of these topics in a more detailed way. It will introduce the duties and capabilities of spatial planning for risk reduction before (Chapter B1) and after mega-disasters (Chapter B2). It will further explain the importance of the role of coordination in the recovery process (Chapter B3). Part B will also look into the relevance of residents' participation for recovery (Chapter B4) and illustrate ways for how to deal with the uncertainties that are associated with future disasters (Chapter B5). At the end of the book, a conclusion will summarise the most important findings and develop a concept for tackling the key challenges of spatial development in order to make cities and regions more resilient, if not change-proof.

Part B

CHAPTER B1

Planning systems for risk reduction and issues in pre-disaster implementation

Kanako Iuchi

Introduction: planning to mitigate hazards

Efforts to plan for disaster mitigation have increasingly gained attention only recently. Even countries that have a reputation for progressive mitigation efforts only implemented systems that urge planning to play a central role in disaster mitigation within the past few decades. Examples include the US Federal Disaster Mitigation Act 2000 (DMA 2000), Indonesia's 2007 Disaster Management Law (*Tentang Penanggulangan Bencana*) and the 2007 European Commission's directive on the assessment and management of flood risk that is legally enforceable and conceptually the closest to managing risk through land use control. Japan has a longer history of possessing disaster risk reduction systems that include laws and legislation related to spatial control, starting with the modern Disaster Countermeasures Basic Act 1961. Otherwise, mitigation concepts – proactively taking measures to reduce possible disaster impacts – were not centralised in disaster management schemes. In the US, for example, disaster relief was the major response to disasters until the latter half of the 20th century, and was managed mainly by humanitarian and volunteer organisations (Rubin, 2007).

Since the mitigation concept became more widespread, planners and policymakers have largely emphasised this phase among others of disaster management – that is, preparedness, response and recovery – in planning practice. It addresses the importance of evaluating different types of hazards, including seismic (see Olshansky, 2001; Nelson and French, 2002), floods (see Burby and French, 1981; Morris, 1997) and landslides (see Varnes, 1984; Fella et al, 2008). Studies on mitigation via land use control have also interested planners for quite some time (Olshansky and Chang, 2009). Although empirical research on this topic is sparse, disaster mitigation efforts have proven to be beneficial if enforced appropriately at the local government level (Schwab et al, 1998). For instance, studies found that planning measures taken at the local governmental level in Los Angeles prior to the Northridge Earthquake of 1994 significantly contributed to mitigating losses. In particular, the culture of information sharing via hazard maps created an environment that made it easier to advance engineering measures, and land use plans prepared with appropriate mitigation measures contributed to reducing damage (Olshansky, 2001; Nelson and French, 2002). Mileti (1999) lays out several tools for hazard mitigation: land use planning, building codes, insurance, engineering and warnings. Among these, land-use planning plays an important role in spatially managing risks as it guides urban development in a way that minimises risk. Expanding urban areas can theoretically be controlled through land-use decisions that incorporate the idea of disaster risk mitigation. Enforcing these types of land-use controls could deter developments from defined hazardous zones, but can also enforce residential relocation out from hazardous to less hazardous areas in order to reduce future risk. To this end, planning communities have developed techniques and tools that aim to reduce vulnerabilities.

In the process of incorporating disaster risk information into land use, risk assessment is widely suggested as a basis for land-use decisions. It starts with looking at hazards and exposure in order to identify vulnerability, and then uses this to assess risk and concerns within the local context (see Renn and Walker, 2008; RCC, 2011). Risk assessment could also include ecological footprint analysis, assessment of sustainability indicators and environmental impact statements to influence land use decisions (see Mileti, 1999).

Incorporating disaster mitigation efforts into planning is becoming increasingly critical to fully addressing rapid global urbanisation. Urbanisation management is a lingering concern for planners and policymakers in international development. In 2008, for the first time in history, half of the world's population resided in urban areas (UNFPA, 2007). Although industrialised countries have a higher proportion of their population living in urban areas, emerging countries – with Latin American countries leading the way – are urbanising at an unprecedented speed. Angel et al (2011) cautioned that the world's built-up areas – those that are paved or developed – would double in 17 years and triple in 27 years at the current speed of population growth and built-up area expansion. This rate of expansion has never been observed in the development of industrialised countries and no procedures exist to manage harmonious development. What makes this more threatening is that such speedy urbanisation is almost always accompanied by vulnerability: procedures to obtain proper land rights involve lengthy processes; there is a lack of available land in already urban areas; and so new developments are most likely to occur in areas more vulnerable to disasters, such as river banks, gullies and hillsides (see Figure 75).

Figure 75: Development on hillsides: development without legal permission on hillside, Bogota, Colombia

Source: Photo by author

Second, extremely rapid urbanisation is almost always accompanied by malfunctioning urban planning systems. Illegal settlements are likely to happen in hazardous areas and in areas ill-suited for further development (eg extremely dense areas with substandard buildings) because the pace of actual development and population inflow exceeds that of planning procedures. Due to this, local governments often lose control over land use enforcement and lack adequate funding for urban infrastructure to keep up with the needs of the urban population (see Figure 76).

Third, to make matters worse, climate change is causing sea-level rises, affecting coastal cities and development. The rises will directly impact the severity of inundation, flooding and surges affecting coastal settlements. The predictions vary due to the different models and scenarios adopted. For example, the Inter-governmental Panel on Climate Change's (IPCC's) fifth assessment report (AR5) in 2014 predicts a worst-case scenario mean sea-level rise of between 0.52

Figure 76: Development without adequate urban infrastructure: illegal settlements along a river bank, Manila, Philippines

Source: Photo by author

and 0.98 metres by the year 2100, though more aggressive calculations project increases of over three metres during this time period (Oliver-Smith, 2009). Regardless of the actual amount, sea-level rises will impact a significant portion of the population; although only 2% of land is located between 0–10 metres above sea level, 13% of the urban population has settled these areas, with 13 out of the 20 largest cities located along the coast (McGranahan et al, 2007). Researchers have estimated that, with continued population growth and increases in greenhouse gas emissions, the number of people affected by flooding will reach 21 million per annum by 2030, 55 million by 2050 and 370 million by 2100 (Oliver-Smith, 2009; Pachauri & Meyer, 2014).

Even though reducing coastal risk is not the sole objective of disaster mitigation from natural events, effects from hastened urbanisation, population growth and climate change, represented by the sea-level rise and other environmental change, can severely affect large portions of the population at any moment (see Figure 77). Thus, the way to control and manage the use of land in order to mitigate disasters is increasingly central to disaster management discussions.

In line with this, the Sendai Framework for Disaster Mitigation 2015–30 (UN-ISDR, 2015, p 35) aims 'to guide the multi-hazard

Figure 77: Vulnerable development along the coast: development along the coast is likely to be severely affected with the rising sea level, Padang, Indonesia

Source: Photo by author

management of disaster risk in development at all levels as well as within and across all sectors'. It also identifies the participation of relevant stakeholders – including national, regional and local governments – as critical to strengthening disaster management governance. The Sendai Framework includes planning – in particular, land use, building codes and information sharing – as an important driving force to secure and promote disaster risk reduction. It also acknowledges that planning must be advanced with collaboration among stakeholders and include the disadvantaged, both genders and people with differing financial power and political strength. As in Sendai Framework there are various efforts undertaken in different parts of the world, however, mainstreaming disaster management with land use has not been an easy task. As a first step to mainstreaming disaster mitigation with land use control, this chapter seeks to: (1) understand the current situation of mitigation efforts in countries encountering different types of disasters and that have historical backgrounds on disaster management; (2) identify the issues and challenges faced by each case study country; and (3) interpret current issues and challenges of land use and disaster mitigation to advance research and effective practice.

Disaster reduction efforts across countries

Understanding disaster management efforts across different countries is necessary to highlight opportunities and limitations in order to further advance practical disaster mitigation through planning. There is little available literature that describes variations in land-use control to mitigate disaster impacts. This section thereby explains the disaster reduction efforts through planning utilised by different countries considered as disaster-prone. These include Japan, Indonesia and the US. In addition, examples from Germany and Slovakia will be referenced to aid in explaining planning efforts, as well as the opportunities for and constraints upon implementing land-use control.

Mitigating efforts targeting multi-hazards since 1961: Japan

Japan is one of the most developed countries in terms of disaster mitigation and planning systems. The island nation is surrounded by the Pacific Ocean and is located along one of the most seismically active zones in the world. It has suffered various natural disasters, including earthquakes and tsunamis, volcanic eruptions, typhoons, torrential rain and heavy snowfall (for details addressed in this section, see Chapter A1a in Part A). Sea-level rises due to climate change will also threaten a significant proportion of the Japanese population. Due to its geographic and topographic conditions, 50% of the entire population and 75% of all assets are agglomerated in 10% of land within the flood plain (Ministry of Land, Infrastructure, Transport and Tourism, 2007). Population-wise, approximately 5.4 million people live on land below sea level with 75% found in the Tokyo, Osaka and *Ise* (Nagoya region) bays. The country has suffered from many natural disasters and continues to be threatened by potential future events.

The first disaster management Act suggesting the use of land-use control was enacted in 1961. The *Disaster Countermeasures Basic Act* holistically systemised planning for disaster mitigation and was a reaction to the *Isewan* typhoon. The Central Disaster Management Council, established by the Cabinet Office of Japan in 1962, is responsible for national management. The typhoon was one of the largest in terms of damage since the late 19th to early 20th century, and caused more than 5,000 casualties. The storm surge that developed from the typhoon played a major role in the unprecedented damage and devastation of Tokai region (Nagoya being its main city), accounting for 80% of the total damage and loss assessed (Cabinet office, Government of Japan, 2008). Other relevant laws for mitigation, to a lesser extent in relevance to land use, include the *Flood Control Act 1949* and the *River Act 1964*.

With the Disaster Countermeasures Basic Act in effect, all levels of governments at the national, prefectural and local levels are mandated to have a Disaster Management Plan and to implement them. The plan's structure is similar at different levels of government, which is typically organised by disaster type (eg earthquake, tsunami, wind and floods, volcanoes and snow storm, and nuclear hazards) and addresses the different phases of mitigation and preparedness, response, rehabilitation and recovery. Plan contents are often revised after major disasters; for example, the 1995 Kobe earthquake triggered major revisions in planning systems for disaster management, including legislative aspects on mitigating disaster impacts through land use. The 2011 Great East Japan Earthquake (GEJE) and tsunami underscored the inclusion of land use reflecting tsunami impact mitigation. Different levels of government then update their plan contents, coordinating with higher-level governments. For example, the *Regional disaster management plan* of the Tokyo Metropolitan Government (2014a) promotes spaces resilient to earthquakes through land-use control, together with other appropriate urban infrastructure and building control. It additionally refers to land-use change in case the areas are deemed inappropriate. The Tokyo Metropolitan Government (2014b) further states that local governments must implement hazard mapping to guide land use for flood control and mitigation. Thus, local governments systematically conduct potential damage assessments for earthquake and hazard mapping for water-related disasters, reflect this in their local disaster management plans, and share the information with citizens. Since the Tokyo Metropolitan Government revised their *Regional disaster management plan* in 2014, the majority of wards in its jurisdiction also revised their plans in 2014 to align. Plans are often updated after a major disaster to incorporate the most updated information on disaster management policies and lessons in order to mitigate future disaster.

The sharing of hazard maps has advanced drastically over the past decade. As of June 2015, local governments have developed six types of hazard maps – flood, inland inundation, storm surge, tsunami, landslide and volcanic eruption – which are available to the public (see Table 11). Of these, flood hazard maps are the most advanced; as of April 2014, 1,308 out of 1,718 local governments (76% of the total) have developed and published their results. Additionally, 92% of the publicised maps are available online.

From a practical perspective, local governments are the ones that make decisions on and permit the use of land based on the City Planning Law

Table 11: Number of nationally available hazard maps by type

Type of disaster	No. publicized	No. on website	% on website
Flood	1,308	1,207	92.28
Inland innundation	288	265	92.01
Storm surge	114	90	78.95
Tsunami	546	469	85.90
Land slide	842	683	81.12
Volcanic eruption	82	68	82.93

Source: Table prepared by author using data from the MLIT (see: http://disaportal.gsi.go.jp/publicate/index.html?code=1)

1968. The law aims to promote orderly urban development by defining urbanisation promotion areas and urbanisation controlling areas within the planning boundary. In theory, hazardous land should be included in the urbanisation controlling area; yet, due to a constrained amount of land in most cases, development still takes place in areas that are potentially hazardous for some types of natural disasters.

There are several hazard mitigation programmes that encourage relocation, represented by the 'Collective Relocation Promoting Program for Disaster Prevention' (*Bosai shudan iten sokushin jigyo: Boshu*) and the 'Relocating Program for Hazardous Residential Buildings Adjacent to Cliffs' (*Gakechi kinsetsu tou kiken jyutaku iten jigyo*). These programmes are financially supported by the national government to help local governments relocate communities. However, few areas practised relocation prior to disasters. It has also been used primarily after disastrous events, including the *Isewan* typhoon of 1959 (see Figure 78) and the GEJE of 2011. Thus far, large-scale pre-disaster relocation has typically not been considered for implementation, even in Japan.

Emphasising land use in disaster reduction efforts since 2007: Indonesia

Indonesia is a country comprised of over 13,000 islands and extending more than 5,000km east to west. It is located in a seismically active zone and experiences earthquakes, tsunamis and volcanic eruptions. The country also faces droughts and forest fires during the dry season, as well as flooding due to rain during the wet season. Natural disasters occur frequently in Indonesia. In the past decade alone, the country has experienced several large-scale disasters, including: the 2004 Indian Ocean earthquake and tsunami; the 2006 Java earthquake; the 2006

Figure 78: Land use control after the Isewan Typhoon in Nagoya City

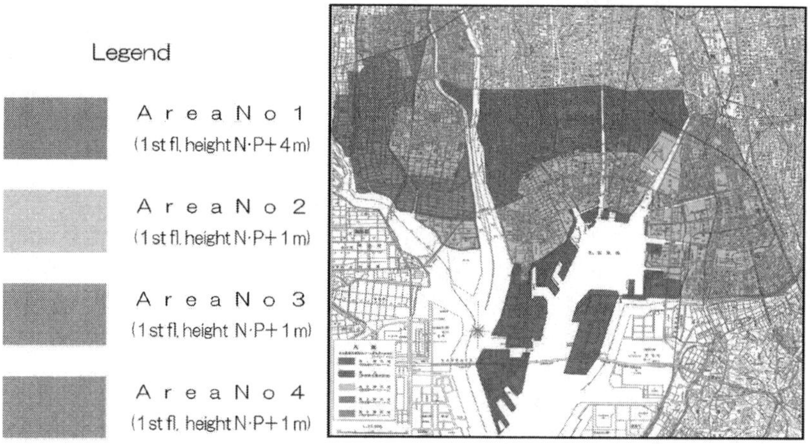

Nagoya City Coastal Disaster Management Area
(Area partially modified, Aug 2007)

Source: http://www.city.nagoya.jp/jutakutoshi/cmsfiles/contents/0000011/11898/jourei_kaisetu.pdf

Pangandaran earthquake and tsunami; the 2007 Sumatra earthquakes; the 2010 Sumatra earthquake and tsunami; the 2010 Mount Merapi eruption; and the 2014 Mount Sinabung eruption. The 2004 Indian Ocean earthquake and tsunami, which caused more than 170,000 deaths, was a significant turning point for the Indonesian government to transform its disaster management system from reactive to proactive.

In 2007, the Disaster Management Law (24/2007) (*Tengtang Penanggulangan Bencana*) was established to protect communities from disasters and establish comprehensive disaster management. With this law, national and regional governments are in charge of disaster risk reduction and recovery activities, and must secure and allocate necessary funds. Requirements to establish national and regional disaster management agencies are articulated in this law. In 2008, Presidential Regulation No 08/2008 was enacted to establish the National Agency for Disaster Management (BNBP) (*Badan Nasional Penanggulangan Bencana*) in order to coordinate disaster management planning and implementation at the national level. This replaced a prior organisation, the Indonesian National Board for Disaster Management (BAKORNAS) *Badan Koordinasi Nasional Penanggulangan Bencana*), which was responsible for relief and emergency response activities.

The Disaster Management Law highlights the role of spatial planning as a tool for disaster mitigation, together with 'regulation development, infrastructure development, building lay-out' and 'conventional and modern education, counseling, and training' (President of the Republic of Indonesia, 2007, p 20). Spatial planning should be implemented together with safety standards (such as building codes), and penalties are considered for violators. Furthermore, it requires that disaster management policy be aligned with regional development policy so that regional governments can establish local disaster management and the actual implementation of land control through developing, deciding and sharing maps of disaster-prone areas.

The year 2007 was also a turning point for spatially related laws to incorporate disaster management aspect. Two key influential ones include the Law on Spatial Plan (26/2007) (*Tentang Penataan Ruang: Nomor 26 Tahun 2007*)[1] and the Law on Management of Coastal Areas and Small Islands (27/2007) (*Tentang Pengelolaan Wilayah Pesisir dan Pulau-Pulau Kecil: Nomor 27 Tahun 2007*). These new laws regulate the need to include the appropriate use of land for risk reduction purposes. Development on land deemed hazardous for future disastrous events is actively discouraged. To begin this process, the revised Law on Spatial Plan of 2007 assigned greater responsibility to local governments to control their land (Rachmawati et al, 2010). As a result, local governments are mandated to adopt land-use decisions that reflect mitigation purposes. Depending on the scale of spatial management, provincial and municipal governments develop mitigation-considered plans and adopt tools that advance risk reduction. While the Municipal General Plan (*Rancangan Tata Ruang Wilayah* [RTRW]), developed at a scale of 1:50,000 to 1:100,000, aims to establish the spatial patterns between conservation and development, the Municipal Detail Spatial Plan (*Rencana Detail Tata Ruang Kota* [RDTRK]), developed at a scale of 1:5,000 to 1:25,000, aims to develop hazard mapping for a more detailed understanding of land use (for details of Indonesia, see Table 12). These are then translated into strategies, policies and regulations for actual implementation, including zoning regulation, building permission, development control and infrastructure development to increase resistance to different types of disasters.

Merging hazard reduction efforts and land-use control is under way in practice. According to the Law on Spatial Plan of 2007, the revised plan is supposed to be implemented within three years after the law enters force. Some local governments, as represented by Semarang City and Padang City (for details, see Chapter A2b in Part A), have adopted land-use or spatial development policies that integrate mitigation into local governments' regulations. Efforts to assess vulnerability are also under way, despite the fact that the results are not always incorporated into the land-use plan. Nevertheless, enforcing hazard-controlled land use prior to large-scale disasters is another necessary layer. Land-use

Table 12: Indonesia's newly defined land use approach on risk mitigation at different scale

		Municiapl spatial plan		
Risk mitigation approach	**Provincial spatial plan (RTRWP) 1:250,000**	**Municipal general plan (RTRW) 1:100,000-50,000**	**Municipal detail spatial plan (RDTRK) 1:25,000-5,000**	**Municipal technical spatial plan (RTBL/RPP)**
Prevention-oriented	Incorporating vulnerable area to land use restriction	Establishing spatial patterns: distinguishing areas to conserve and develop	Hazard mapping	Developing local action plans for disaster risk reduction
Non-structural	Keeping conservation areas away from development to absorb and reduce possible hazardous impact	a) Information management b) Land use planning c) Local codes	a) Zoning regulation b) Building permission c) Development control, etc	a) Adopt land development b) Identify evacuation route
Structural	–	Secure availability of space for protective infrastructure	Planning infrastructure resistant to disastrous events/hazards	Detailed engineering planning to develop infrastructure resistant to hazards

Source: Modified by author from Rachmawati, Deguchi, Yoshitake, & Wijaya, 2010

implementation that aims to adopt risk reduction aspects has only been visible in post-disaster recovery thus far.

Two recovery examples worth noting are those after the Indian Ocean tsunami of 2004 in Aceh province and after the Merapi mountain eruption of 2010 in the Special Region of Yogyakarta. In the post-Aceh recovery, tsunami-risk-mitigating land-use planning was not implemented, even though the government had planned for it (for details, see Chapter A2a in Part A). Within the first three months, a recovery master plan that included a spatial plan of the devastated areas had been developed. Five zones were proposed with land uses: zones I, II and III fall within three kilometres inland from the coast and are development controlled areas; zone IV is located inland and is a development promotion area; and zone V encompasses conservation areas. This spatial plan lawfully came into effect with Presidential Decree 30/2005, two weeks after the master plan became available to the public. However, actual reconstruction had already started two months after the tsunami. With the support of (international) non-governmental organisation (NGOs), local residents favoured a rapid recovery process that provided visible progress and led to reconstruction onsite, but which did not consider future tsunami risk. Six months after the tsunami, the government decided to modify the rebuilding policy from the proposed spatial plan considering future risk in land use to a rebuilding plan developed on-site.

After the Merapi eruption of 2010 recovery also addressed land-use implementation reflecting hazard mitigation. To rebuild with reduced risk after the Merapi eruption, one of the first steps was to revise an already-existing volcanic hazard map (from 2010). The Republic Public Works Department led this activity with other departments and agencies, including the National Development Planning Agency (BAPPENAS), and a revised map was published in 2011 (see Figure 79). Three types of hazard zones were classified: (1) KRB (Kawasan Rawan Bencana) 3, a high hazard zone (zone ATL [area terdampak langsung] included, the directly affected area with the highest alert); (2) KRB 2, a moderate hazardous zone; and (3) KRB 1, a low hazardous zone. Using this hazard map, residents from the directly affected area (ATL) identified resettlement sites distanced from Merapi eruption-related hazards through the community-driven rebuilding programme REKOMPAK (Rekonstruksi Masyarakat dan

Figure 79: Revised Mt. Merapi eruption hazard map in 2011

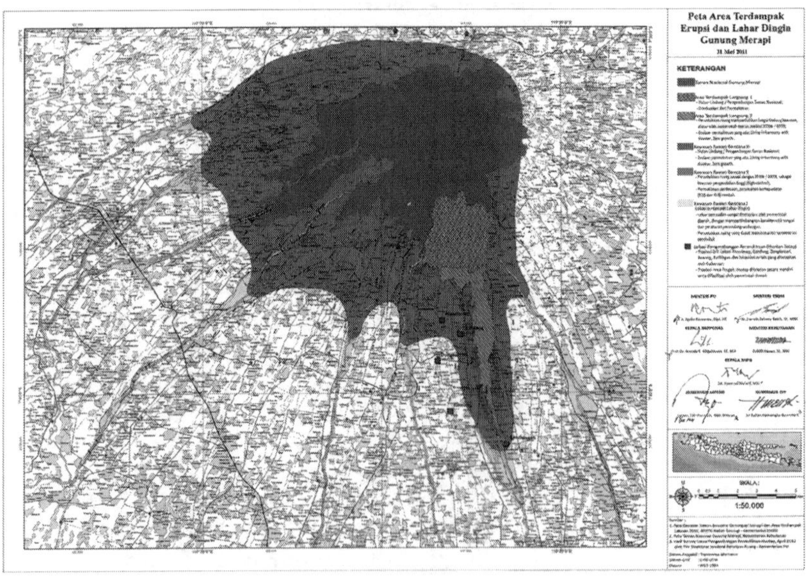

Source: Cipta Karya

Permukiman Berbasis Komunitas). In 2015, people who relocated with the REKOMPAK programme have continued to reside outside the ATL areas, suggesting that there is a certain degree of success in controlling land use (Iuchi et al, 2015). Nevertheless, the hazard map has no lawful power to keep people out. The revised land-use plan is being formalised with the government of the Special Region of Yogyakarta, and one of their land-use strategies was to transform the ATL zones into the Merapi National Park, legalised in 2014 by the President of the Republic of Indonesia (President of the Republic of Indonesia, 2014).

Newly introduced law on mitigation after the long-term operation of disaster management in 2000: USA

The US is the third-largest country by land area and has varied climatic and geographic conditions. Disaster types therefore differ by location. For instance, coastal regions in the Gulf of Mexico often face hurricanes, while the west coast experiences earthquakes, volcanic activities and wildfires due to its geography and the dry climate. The Midwest, by comparison, is often plagued by flooding due to wetter weather and the relatively flat low-lying terrain. Climate change impacts have widely affected coastal communities; parts of the Gulf Coast and the Mid-Atlantic Coast are two areas that experienced the largest sea-level rise. Over the last 50 years, the Gulf Coast has experienced an estimated rise of eight inches or more, occurring at a rate twice as fast as the rest of the globe (EPA, 2014). The threat to coastal cities is acute as 23 out of 25 of the most densely populated are along the coast (CCSP, 2009).

The US disaster management system slowly evolved during the 20th century, though there were no laws directly relevant to mitigation efforts until the Disaster Mitigation Act of 2000 (DMA 2000). Prior to its adoption, the US disaster management system focused on managing emergencies, and was more about response rather than proactive mitigation (Rubin, 2007). Key laws related to US emergency management before this 2000 Act include the Disaster Relief Act 1974, which enables the president to declare disasters and thus provide relief, and the Stafford Disaster Relief and Emergency Assistance Act 1988, which amended the Disaster Relief Act 1974. The Stafford Act sets the operational procedures for responsible entities (eg local, state and the federal governments) when responding to disasters and identifies relief programmes, including the Hazard Mitigation Grant Program (HMGP), for long-term use to mitigate repetitive disaster losses (Johnson, 2013). The Federal Emergency Management Agency (FEMA), established in 1979,[a] oversees the HMGP.

The result of 10 years' effort by hazard mitigation specialists who closely examined hurricanes, earthquakes, floods, tornadoes and blizzards in the 1990s,[3] the DMA 2000 was legalised to mitigate recurrent disaster impacts and reduce losses. State and local governments were required to prepare Hazard Mitigation Plans by 2004 in order to receive federal aid through the Hazard Mitigation Assistance (HMA) programme. FEMA currently manages three types of programmes: the HMGP, the Flood Mitigation Assistance (FMA) programme and the Pre-disaster Mitigation Program (PDM). The HMGP continues to operate in the same manner as pre-DMA 2000, in that the grants can only be utilised in disaster-affected areas. FMA is used for projects that aim to reduce impacts on buildings in flood-risk areas, and PDM funds can be secured for hazard mitigation planning activities and projects (FEMA, 2015b). FMA includes a property acquisition (buyouts) programme that supports voluntary pre-disaster relocation.

In developing a Hazard Mitigation Plan, FEMA encourages local governments to assess community capabilities and risk when developing a mitigation strategy. These plans must be continuously updated, involve the public for implementation and be formally adopted (FEMA, 2015b). FEMA also encourages the inclusion of hazard mitigation into local-level planning (FEMA, 2013). While some local governments successfully incorporated a Hazard Mitigation Plan into their land-use plans,[4] only 33% of local governments had approved Hazard Mitigation Plans four years after the 2004 law went into effect (Jackman and Beruvides, 2013). This reveals that developing a Hazard Mitigation Plan appropriate at the level of adoption requires effort and time.

Implementing land-use decisions for mitigation is not easy. As evidenced by FEMA-related cases (FEMA, 2013), local governments with mitigation plans that included land-use were actually ones that had recent disaster experiences. Even disaster-affected regions requiring land-use decisions upon rebuilding did not include mitigation in their land-use procedures. For instance, in post-Hurricane Katrina New Orleans, various planning stakeholders initially attempted to develop a hazard-reflected rebuilding master plan to mitigate land use in long-term flood-prone areas. However, the first plan – the Bring New Orleans Back (BNOB) Plan, which emphasised hazard mitigation through land use (see Figure 80) – was largely criticised by residents and thus dropped due to its ill-reputation. It took a third round to finally accept the United New Orleans Plan (UNOP), a rebuilding master plan led by the State of Louisiana, in 2007 (Olshansky and Johnson, 2010).

Figure 80: Parks and Open Space Plan in areas with highest hazards

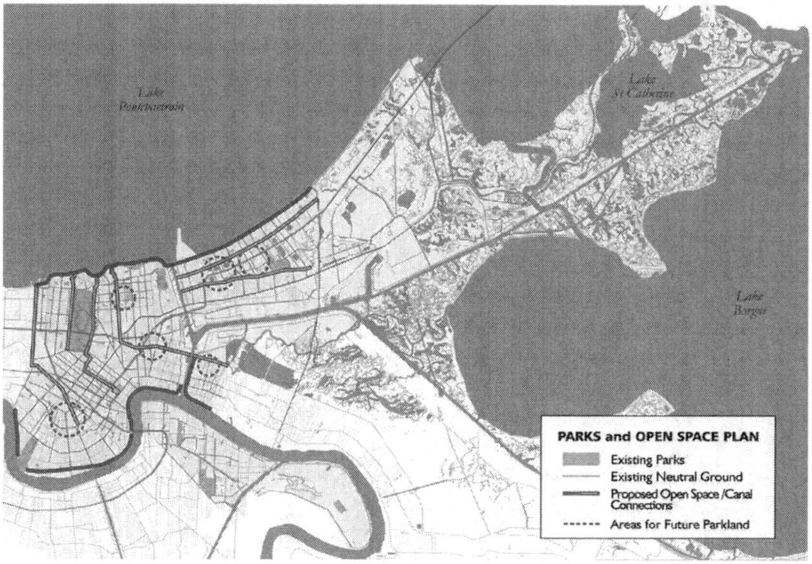

Source: Prepared by WRT

Actual implementation, however, did not directly follow the approved plan. Instead, a private effort by Project Home Again (PHA) has continued with a small-scale land swap programme, targeting homeowners that rebuilt in original, high-flood-risk areas post-Hurricane Katrina New Orleans. These homeowners are relocated into less hazardous areas with support from the PHA by swapping their former land for new land in higher locations, which are also strategic redevelopment areas (Nelson, 2014). This project has continued to move forward because the operation is micro-scale and enables face-to-face consultation. Similarly, the implementation of long-term land-use that incorporates mitigation is still unfolding in 2015 after Hurricane Sandy of 2012 in New York. The city aims to achieve spatial resiliency through: (1) strengthening coastal defences; (2) upgrading buildings; (3) protecting infrastructure and services; and (4) neighbourhood efforts (City of New York, 2014). As land is so expensive, controlling hazardous areas from urban use is almost a non-option and land buyouts though the federal programme, for example, are rare (eg only in Staten Island in New York City at an early stage of rebuilding). Instead, the New York City 'Built it Back' programme has prepared a package that supports families to repair or rebuild homes with elevation, among other options (see Figure 81).

Figure 81: Houses elevated in rebuilding

Source: Photo by author

Mitigation efforts across regions

What does in-depth assessment of three countries tell?

Narratives of the planning systems for disaster reduction, planning actions and enforcement status of mitigation-based land-use systems in Japan, Indonesia and the US lead to an important question: how effectively have mitigation-based land-use management systems been implemented? In brief, each country is at a different level of achievement with a correspondingly varied history and governmental structure, but all are putting greater emphasis on mitigation through land-use control (see Table 13). All three countries have nurtured some form of planning system for disaster mitigation, with established legislative systems that require a hazard-controlled spatial plan and disaster management plan to be in place. Conducting risk assessments, sharing results with the public and linking both with actual implementation is underscored.

The level of actual planning and its implementation, however, varies. For instance, in Japan, more than 75% of the local governments as of 2015 have developed flood hazard maps that are available to the public. On the other hand, as of 2008, only 33% of US local governments have developed and adopted Hazard Mitigation Plans after conducting

Table 13: Overview of mitigation efforts by three hazard-prone countries

	Japan	Indonesia	US
Planning system for disaster mitigation	Established legislation Existing disaster management plan and land use plan	Established legislation Existing land use plan and disaster management plan	Established legislation Existing hazard mitigation plan and land use/zoning plans
Actions related to mitigation plans	Scientific assessments (majority) and concern assessments are in place Assessment results shared with public 76 percent of the total local governments publicize some sort of hazard map	Scientific assessments are minimally done in progressive local governments In local governments conducting such assessment, mitigation aspects are reflected in spatial plan	Scientific assessment and concern assessments are being conducted Assessment results shared with public Results incorporated into hazard mitigation plans comprehensively Plan adoption rate was 33 percent in 2008
Land use control to reduce risk	Implemented after major disasters Post-Isewan typhoon Nagoya Post-GEJE and tsunami Tohoku Approached top-down, with allocated plan, programs and budgets	Attempted to implement after major disasters Post-tsunami Aceh Post-Merapi eruption Yogyakarta Top-down planning approach was not favored Community-driven land use decisions more successful	Attempted to implement after major disasters Post-hurricane Katrina New Orleans Post-hurricane Sandy East Coast Top-down planning approach was not favored Individual, small-scale, private driven land use decisions are being more successful

risk assessments. Nevertheless, the Hazard Mitigation Plans developed in the US are more comprehensive than the Japanese ones as they include scientific data as well as social dimensions, such as ways to mobilise local capacity. Indonesia, on the other hand, is at an early stage of development – at the time of publication, only a few progressive local governments have begun reflecting mitigation aspects in their comprehensive land-use plans.

Attempts to control land use to reduce risk have been observed in all countries, but these efforts are significant only after major disasters. Two cases that adopted major land-use control in Japan are in Nagoya City and Tohoku region after a devastating typhoon and tsunami; two cases that attempted to adopt land-use control in Indonesia are post-tsunami Aceh and post-Merapi eruption Yogyakarta; and two cases that considered about controlling hazard upon land use in the US include post-Hurricane Katrina New Orleans and post-Hurricane Sandy New

York. Notably, all cases approached land-use control differently – Japan carried out top-down implementation, systematically following the plans, programmes and budgets developed and allocated. Meanwhile, Indonesia and the US experienced difficulties in implementing their respective plans and programmes. In both countries, a small-scale, community-driven approach – exemplified by the Yogyakarta and the New Orleans cases – were a more realistic approach to considering hazards in land-use development.

How do Germany and Slovakia fit into this dialogue?

Germany and Slovakia are examples of countries with a planning system for disaster mitigation that is also supported by the legislative system. Their approach to the disaster mitigation effort is particularly comparable because they are both member countries of the European Union (EU), and the EU established a directive that manages flood risks (for details on Germany, see Chapter A5a in Part A; for details on Slovakia, see Chapter A4a in Part A). One unique similarity is that they also assess risk and share the results with the public for appropriate spatial control. However, while Japan, Indonesia and the US aim to mitigate the damages of different types of disasters, systems for disaster impact mitigation in these two countries are largely tailored towards single risks – for example, flood risk reduction in Germany, and both flood risk and landslides in Slovakia. This tailoring system is effective as disasters such as floods and landslides occur on a much more regular basis than other disasters, as opposed to other countries.

The EU established Directive 2007/60/EC to assess and manage flood risks in response to such recurrent flood disasters. Since its establishment, hazard map development continues to advance in both countries. Germany has been proactively developing flood risk maps to identify high-risk areas and to incorporate these results into spatial plans. For instance, the Cologne region has assessed river flood hazards with different return periods of low, medium and high (URBIPROOF, 2015). However, these hazard maps are not legally enforceable; as a result, incorporating the information into land-use plans or implementation has been challenging (for more details on Germany, see Chapter A5a in Part A).

Slovakia has also been expanding its scientific assessment of flood risk under a concerted effort of the directive. The preliminary results are currently being translated into flood risk management plans to identify detailed mitigation efforts (for details, see Chapter A4a in Part A). Besides flood risk, Slovakia has also devoted resources to landslide assessment over the last 35 years due to its geographic conditions and recurrent landslide disasters, the results of which are publicly available. However, spatial consideration is mainly lacking due to an absence of integration between risk and land use.

Efforts to incorporate hazard components into planning are evident in both Germany and Slovakia to differing degrees. As seen in Japan, Indonesia and the US, the adoption of hazard-related information into spatial plans has accelerated after disasters or due to recurrent disasters. In Germany, attention to flood risk management increased over the last two decades due to a series of floods and the related impacts. The most recent significant flood, the Elbe and Danube summer flood of 2013, finally triggered the federal government to develop a plan for spatial planning that incorporates technical hazard assessment results. In Slovakia, the Geological Act (Act No 596/2007 Coll On Geological Works) was established in 2013 after an extraordinary volume of rainfall generated massive landslides in the eastern portion of the country and affected residential areas (URBIPROOF, 2015). This Geological Act underscores the need for regional governments to prepare new spatial plans that consider landslide hazard zones.

Holistic hazard-controlled land use is a recent endeavour and the large-scale implementation of land-use control, including relocation, remains to be seen. Although the German example of Roderau-Sud was able to successfully relocate 140 housing units after a 2002 flood, this was only possible due to substantial political compensation by the state government, which was responsible for developing the new residential site (URBIPROOF, 2015). Reflections on hazard-considered land use in both the German and Slovakian cases suggest that the dialogue on hazard-mitigation planning developed similarly to the other countries examined. That is, both countries have strengthened hazard-considered planning systems via shaping legislation due to repeated disaster impacts, and significant advances tend to occur after major impacts are observed.

Reflections

There are three important aspects that are drawn from the cases observed:

- First, enhanced laws and regulations related to the promotion of hazard-controlled land use are likely to be implemented after major disasters, at a time when planners and policymakers feel their importance under urgency. However, different constraints – such as opposition to the proposed land use – hinder such control even after disasters. Even under normal circumstances, lack of interest, concern or knowledge impedes the planning, adoption and implementation of hazard-controlled land use. In most cases, hazard mapping and information sharing efforts have shown the most progress across all countries.

- Second, having a land-use control system and disaster management system working in parallel, as opposed to being integrated, impedes the adoption of hazard-controlled land use. Typically, land-use control systems are older and are strongly embedded into local governance. On the other hand, disaster management systems are new and mandates a group or an agency as responsible for coordination at all disaster phases (ie mitigation, preparedness, response and recovery). Effectively embedding these two largely depends on local governments being responsible for land-use management and whether or not they will proactively incorporate hazards into land-use planning and decisions. However, not all local governments across the countries examined are prepared or are open to begin such efforts.
- Third, local governments with disaster experience or those that are highly concerned about future disaster impacts are proactively assessing potential hazards to be integrated into land-use decisions. Nevertheless, implementing spatial control – particularly to limit the use of a significant amount of valuable lands – has been slow to gain traction. Cases that illustrate potential hazards being integrated into land use include: post-Tohoku, where central decisions and funding are strong; post-Merapi, which successfully involved and negotiated with the affected residents to consider risk; and post-Katrina, where private efforts at negotiating with local governments are proceeding on a small scale but in a promising direction. Although participatory processes are not often considered in the current system of hazard-controlled land-use implementation, the importance of valuing local residents' involvement has been clear.

Final thoughts

The 2000s saw disaster-prone countries establish planning systems for risk reduction in order to mainstream disaster mitigation. Avoiding the usage of high-hazard lands is considered the first step towards creating a resilient space through land-use control. Such efforts are increasingly important for several reasons, including: addressing the hastened urbanisation that often results in substandard urban development in hazardous areas; sea-level rises that are predicted to affect many coastal cities; and water-related disasters that unpredictably inundate various inhabited areas. Due to an urgent need for immediate action regarding smart and resilient growth against increasing risk, the development of planning systems for risk reduction and the actual planning of hazard-controlled land use have advanced robustly in different locations across these countries. Countries examined in the case studies highlighted that these efforts are occurring at various levels of government.

Challenges lie ahead, however. One key issue is that utilising plans for mitigation is almost always neglected prior to disasters. There are several reasons for this: first, disasters are unpredictable in magnitude and timing; second, measuring pre-disaster investments objectively is difficult in an unpredictable environment; and, third, reaching an agreement with citizens to implement pre-disaster risk reduction efforts in exchange for giving up their normal habits is almost always impossible. The cases analysed, and also the German and Slovakian cases referenced, suggest that drastic advances in planning systems are likely to occur more urgently after major disasters, and although multi-level governments (including on a global scale) are putting effort towards advancing the idea of land control for mitigation through hazard mapping and understanding, developing a plan in a system that links to actual implementation continues to be difficult. To address this, recommended procedures in the long run include: (1) having the hazard map integrated into the land-use plan and adopting it for implementation; and (2) implementing spatial transformation based on the approved plan.

Although creating hazard maps is challenging, this information is vital and should be generally incorporated into local land-use planning. It is however important to understand that the hazard map does not have lawful power to implement land control. With this in mind, the first step for local governments to integrate hazard mitigation efforts into land-use planning is to inform residents of possible disaster impacts *without* any spatial control, and how plan making and implementation can contribute to reduce the impact.

As for implementing the adopted plan, there are several key tools and efforts that may help faster advancement. First are the programmes that discourage the use of hazardous land and encourage the use of lands that are already developed in less hazardous locations. One example is the collective relocation programmes and landslide programmes being utilised in Japan. Another is the flood insurance/buyouts that have been used in the US. Such programmes can be implemented with the support of higher levels of governments (eg national, regional) to gather the scientific information needed for relocation, and to fund the relocation. Second is the promotion of public understanding and participation. The Yogyakarta case and the land swap aspect of the New Orleans case highlighted how important it was for local residents to understand the hazards and risks in order for them to play an active role in long-term realisation. In many cases, especially in developing regions, the government will not have ample funding to conduct risk assessments and pursue relocation. Consequently, and with minimal resources available, valuing local information and residents' knowledge is extremely critical. To this extent, participation by local residents is indispensable, and can only be materialised through cooperation with local governments.

Acknowledgement

This chapter was partially supported by JSPS Grants-in-aid number 26420597.

Notes

[1] Revised based on the former law of 1992.

[2] FEMA's current mission is inclusive of all four phases of disaster management: 'FEMA's mission is to support our citizens and first responders to ensure that as a nation we work together to build, sustain and improve our capability to prepare for, protect against, respond to, recover from and mitigate all hazards' (FEMA, 2015a).

[3] The US experienced recurrent natural disasters in the 1990s, including Hurricane Andrew in 1992, the Midwest flood in 1993, the Northridge earthquake in 1994, the Chicago heatwave in 1995, the Eastern US blizzard in 1996 and the Oklahoma Tornado in 1999.

[4] FEMA (2013) showcases the effort of Cedar Rapids (flood), Miami-Dade (sea-level rise), New Orleans (flooding), Oregon (all hazards), Augusta-Richmond (all hazards), Kings County (earthquake, fire, flood) and Greensburg (tornado, thunderstorm, high windstorm hazards). Many local governments that are taking this approach have experienced some disasters in the past.

References

Angel, S., Parent, J., Civco, D.L., & Blei, A. M. (2011). In Lincoln Institute of Land Policy (Ed.), *Making room for a planet of cities*. Cambridge, MA: Lincoln Institute of Land Policy.

Burby, R. J., & French, S.P. (1981). Coping with floods: The land use management paradox. *Journal of the American Planning Associations*, 47(3), 289-300. doi:10.1080/01944368108976511

Cabinet office, Government of Japan (2008). *Report by special committee for transferring knowledge learnt by the past experience*. Tokyo: Cabinet office.

CCSP (Ed.) (2009). *Coastal sensitivity to sea-level rise: A focus on the mid-atlantic region. A report by the U.S. climate change science program and the subcommittee on global change research. Titus, J.G. (coordinating lead author), K.E Anderson, D.R. Cahoon, D.B. Gesch, S.K. Gill, B.T. Gutierrez, E.R. Thieler, and S.J. Williams (lead authors).* Washington DC: US Environmental Protection Agency.

City of New York. (2014). *PlaNYC: Progress report 2014.* New York: City of New York.

EPA (2014). Climate change: Coastal areas impacts. Retrieved from http://www.epa.gov/climatechange/impacts-adaptation/coasts.html#impactssea

Fella, R., Corominas, J., Bonnardc, C., Cascinid, L., Leroie, E., & Savagef, W. Z. (2008). Guidelines for landslide susceptibility, hazard and risk zoning for land use planning. *Engineering Geology*, 102(3-4), 85-98. doi:10.1016/j.enggeo.2008.03.022

FEMA (2013). *Integrating hazard mitigation into local planning.* Washington DC: FEMA.

FEMA (2015a). About the Agency. Retrieved from https://www.fema.gov/about-agency

FEMA (2015b). Hazard mitigation assistance. Retrieved from https://www.fema.gov/hazard-mitigation-assistance

Iuchi, K., Matsumaru, R., & Maly, E. (2015). Community-driven post-disaster rebuilding policy and its relocation patterns: A case study of resettlement after the volcanic eruption of Mt. Merapi in Yogyakarta, Indonesia. *Journal of City Planning Institute of Japan* (in Japanese)

Jackman, A.M., & Beruvides, M. G. (2013). Local hazard mitigation plans: A preliminary estimation of state-level completion from 2004 to 2009. *Journal of Emergency Management*, 11(2), 121-32. doi:10.5055/jem.2013.0131

Johnson, L.A. (2013). Land use planning in the USA: Its role in natural hazard mitigation and post-disaster recovery. *Symposium on Spatial Planning Following Disasters,* Sendai. D-1.

McGranahan, G., Balk, D., & Anderson, B. (2007). The rising tide: Assessing the risks of climate change and human settlements in low elevation coastal zones. *Environment and Urbanization*, 19(1), 17-37. doi:10.1177/0956247807076960

Mileti, D.S. (1999). *Disasters by design : A reassessment of natural hazards in the United States.* Washington, D.C.: Joseph Henry Press.

Ministry of Land, Infrastructure, Transport and Tourism (2007). Summary of river projects 2007. Retrieved from http://www.mlit.go.jp/river/pamphlet_jirei/kasen/gaiyou/panf/gaiyou2007/

Morris, M. (1997). *Subdivision design in flood hazard areas (PAS 473).* Chicago IL: American Planning Association.

Nelson, A.C., & French, S. P. (2002). Plan quality and mitigating damage from natural disasters: A case study of the northridge earthquake with planning policy considerations. *Journal of the American Planning Association, 68*(2), 194-207. doi: 10.1080/01944360208976265

Nelson, M. (2014). Using land swaps to concentrate redevelopment and expand resettlement options in post-hurricane Katrina New Orleans. *Journal of American Planning Association*, 80(4), 426-437. doi: 10.1080/01944363.2014.988167

Oliver-Smith, A. (2009). *Sea level rise and the vulnerability of coastal peoples: Responding to the local challenges of global climate change in the 21st century.* (No. 7). Bonn: Germany: UNU-EHS.

Olshansky, R.B. (2001). Land use planning for seismic safety: The Los Angeles county experience, 1971-1994. *Journal of the American Planning Association, 67*(2), 137-185. doi: 10.1080/01944360108976227

Olshansky, R.B., & Chang, S. (2009). Planning for disaster recovery:Emerging research needs and challenges. *Progress in Planning, 72*(4), 200-209. doi: 10.1016/j.progress.2009.09.001

Olshansky, R.B., & Johnson, L.A. (2010). *Clear as mud.* Chicago; Washington DC: American Planning Association.

Pachauri, R. K., & Meyer, L.A. (Eds.) (2014). *Climate change 2014: Synthesis reort. Contribution of working groups I, II and III to the fifth assessment report of the intergovernmental panel on climate change [core writing team, R.K. Pachauri and L.A. Meyer (eds.)].* Geneva: Switzerland: Intergovernmental Panel on Climate Change (IPCC).

President of the Republic of Indonesia (2007). Disaster Management Law (Tentang Penanggulangan Bencana, Nomor 24 Tahun 2007)

President of the Republic of Indonesia (2014). Spatial strategic plan for mount merapi national park areas (*Tentang rencana tata ruang kawasan strategis nasional taman nasional gunung Merapi*).

Rachmawati, T.A., Deguchi, C., Yoshitake, T., & Wijaya, I.N. (2010). Changees of spatial planning law after national disasters in Indonesia. *Asian Pacific Planning Review*, 6(1), 73-80.

RCC (Regional consultative committee on disaster management) (2011). *Promoting use of disaster risk information in land-use planning.* Bangkok: RCC.

Renn, O., & Walker, K.D. (Eds.). (2008). *Global risk governance: Concept and practice using the IRGC framework* (1st edn). Geneva: Switzerland: Springer.

Rubin, C.B. (Ed.) (2007). *Emergency management: The American experience 1990–2005* (Second Edition). Virginia: Public Entity Risk Institute.

Schwab, J., Topping, K.C., Eadie, C. C., Deyle, R.E., & Smith, R. A. (1998). *Planning for post-disaster recovery and reconstruction.* Chicago IL: American Planning Association.

Tokyo Metropolitan Government. (2014a). *Regional disaster management plan - Earthquake.* Tokyo: Tokyo Metropolitan Government.

Tokyo Metropolitan Government. (2014b). *Regional disaster management plan - Water disaster.* Tokyo: Tokyo Metropolitan Government.

UNFPA (2007). In UNFPA (Ed.), *State of world population 2007: Unleashing the potential of urban growth.* New York: UNFPA.

UN-ISDR (2015). Sendai Framework for Disaster Risk Reduction 2015-2030, http://www.wcdrr.org/uploads/Sendai_Framework_for_Disaster_Risk_Reduction_2015-2030.pdf

URBIPROOF. (2015). *Handbook: Increasing resilience of urban areas.* Dortmund, Germany: TU Dortmund University. Sendai framework for disaster risk reduction 2015-2030.

Varnes, D. J. (1984). *Landslide hazard zonation: A review of principles and practice.* Paris: United National Educational, Scientific and Cultural Organization.

CHAPTER B2

Land-use planning after mega-disasters: between disaster prevention and spatial sustainability

Michio Ubaura

Introduction: 'After the disaster is before the disaster'

This chapter aims at the comparison and analysis of disaster management after mega-disasters through land-use planning described in Part A. This aim differs in treating the cases 'after' the disasters from the previous chapter, which deals with the cases of the 'pre'-disaster phase. In the process of reconstructing in the aftermath of serious disasters, the biggest challenge is how to prevent similar disasters that may occur in the future.

Needless to say, disaster risk management efforts should be undertaken before the disaster to minimise the damage. In most cases, however, these efforts face a harsh reality. In ordinary situations, the existence of built-up areas is taken for granted and people's interests also lie in a variety of topics other than disaster prevention; it is thus not often recognised that the biggest challenge is raising such disaster prevention capabilities. However, since urban areas face outflows or devastating blows in the wake of serious disasters, the enhancement of these disaster prevention capabilities becomes a primary concern, and the possibility of their realisation is higher where the degree of damage is greater, which is to say where urban areas are becoming vacant.

As tools for disaster risk management, we may cite so-called 'soft' measures, which include practices such as the securing of evacuation routes, the regular implementation of evacuation drills and the provision of early warning information. However, in addition to these, it is also important to manage risks through the control of how space is used by utilising the fact that hazards and their intensity differ according to their spatial location, as does the extent of exposure to such hazards, or vulnerability, according to the way in which land is used.

On the other hand, a disaster prevention perspective cannot encompass everything that must be considered when formulating spatial plans. Even ensuring that disaster prevention is taken into account, where this has a major impact on other elements, such as lifestyle or means of subsistence, for example, it will eventually lead to a decline in the local community. In order to accomplish reconstruction in the normal sense of the word, what is required is sustainable local development that will return local lifestyles and industry to the state that they were at prior to the disaster, and, moreover, to use the disaster as an opportunity to overcome earlier constraints. In other words, what is required is an overall consideration that takes sustainability and feasibility into account from a variety of perspectives, including those of economic development, societal consensus and environmental impact. Of course, needless to say, the perspective of disaster prevention occupies an important position at this juncture. People have no desire to live where they are not safe, and nor should they be expected to. However, this is merely one element that must be given comprehensive consideration when thinking in terms of how space is used.

Thus, while spatial planning and risk management are merely positioned relative to one another, they are also seen to be mutually closely related; in particular, it can be seen that both are given important positions in the context of reconstruction planning.

So, how are these factors given their mutual weightings? In the first place, this will differ in different regions and countries. In addition, reconstruction plans are formulated under the extraordinary circumstances of the aftermath of serious disasters, and are thus given different weightings than in normal situations. Spatial planning in the context of reconstruction is determined as these various elements intertwine, and new urban areas are gradually formed on this basis, though this is not always the case.

In what follows, I summarise the realities and implementation statuses of spatial planning responses that have taken place in various regions and countries in the wake of such serious disasters so as to reveal an overall picture and individual elements through a comparative study to indicate points of similarity and difference. In doing so, particularly in view of the problems described earlier, I will focus primarily on examining how disaster prevention, as well as sustainability elements in addition to disaster prevention (hereinafter 'spatial sustainability'), are handled spatially at the planning and implementation stages, as well as how sustainability is being secured.

Plans relating to spatial use

Tohoku (Japan)

In the Tohoku region, the risk of a tsunami that may occur only once in a hundred years or even longer is defended against by constructing seawalls. In addition, with regard to areas still facing the risk of tsunamis at the level of the one that struck in March 2011, which may occur once in several hundred years, it has been decided that residential structures will either not be allowed or only allowed on the condition that they are able to withstand a tsunami of that level. Also, it has been prescribed that residents of these areas relocate to higher ground or inland areas where there is a lower tsunami risk. Thus, here, planning has placed a strong emphasis on disaster prevention in protecting residential areas from the risk of tsunamis that might occur only once in several hundred years.

Although spatial plans have been formulated for each district, in concrete terms, these are basically formulated by the municipalities concerned. However, because the planning and implementation for each project, including the construction of seawalls, is carried out by the respective project agencies, whether national, prefectural or municipal, the challenge of their overall synthesis has been left behind. The formulation of plans takes place on the basis of certain scientific grounds, such as the results of tsunami simulation. However, in terms of their assessment, in some cases, this was not carried out based on social evaluation criteria on a village-by-village basis. On the contrary, especially where a prefecture is in charge of the project, this was sometimes carried out according to a uniform standard for the entire prefecture, which has also been seen to generate occasional opposition.

Given the context of a large number of communities in which population decline has been progressing dramatically, a pillar of planning that may be regarded as comparable to disaster prevention is 'compact' community development, which has a certain level of density and scale. Specifically, by conscientiously carrying out opinion surveys with disaster victims on a number of occasions, attempts have been made to create relocation housing complexes on a scale consistent with demand, and to increase sustainability at the district level. In addition, macro-level intensification – in other words, the concentration in one resettlement site of multiple scattered settlements, or else the integration of adjacent built-up areas – has featured relatively frequently in planning for farming villages on the plains. However, particularly in fishing villages along Ria coastlines, because of the need to secure connection with livelihoods, there have been few cases of such consolidation, and relocation housing areas on higher ground basically adjacent to existing villages have been planned on a small and individually distributed scale. In these cases, although safety from natural hazards is assured, many problems remain from the viewpoint of the sustainable community.

Aceh (Indonesia)

With regards to the city level, the Master Plan for the Reconstruction of Banda Aceh City formulated immediately after the disaster by the National Development Planning Agency (BAPPENAS) places an emphasis on disaster prevention, specifically by setting 'Creating a safer and better life' as the plan's primary objective. In addition, in terms of urban structure, the indicated intention is to continue maintaining the existing urban structure, that is, to 'minimize changes in the existing structure, hierarchy, density and land use'.

In light of these general principles, physical zoning has been determined and placed into the actual spatial context. Therein, the coastal areas that were devastated by the tsunami are specified as 'Restricted Development Areas', with the intention of carrying out land usage that takes tsunami risk into account so that similar devastation would not take place a second time. However, there was not necessarily any clear scientific basis for the formulation of these plans, and the context of the 2004 disaster was utilised as a basis for reference. For this reason, assessment of the disaster prevention plans is impossible. In addition, the designation of residential areas for absorbing the demand for relocation from the affected areas has also been carried out. This is in a safe and secure area on the south side, and the designation has been carried out in such a way that it will be adjacent to built-up areas, with the intention of avoiding sprawl-type expansion, which is to say that it may be understood as being intended to ensure spatial sustainability.

With regard to the district level, one characteristic feature is that village spatial plans have been formulated for each individual village. Certainly, the fact that the agencies that have carried out (or supported) these formulations are various in nature, as well as the fact that their substance and level vary by their respective districts, has given rise to criticisms that they have only a thin relation to the higher-level Reconstruction Master Plan. However, in the sense that they have basically been formulated through a public participation process, as well as formulated based on their respective social evaluation criteria, the plans may be said to conform to realities. In terms of disaster prevention, although assessment is impossible, as described earlier, the plans may be said to have a certain level of sustainability.

New Orleans (USA)

Damages caused by Hurricane Katrina were the result of the failure of flood walls – damages that, conversely, might have been prevented had the walls been strengthened. For this reason, the strengthening of embankments (either through restoration or partial raising) has become a basic premise in all of the plans seen in the following, thereby making it possible to address the hazard levels found in recent disasters, or, more precisely, to deal with high tides that occur with a probability of once in a century.

In the reconstruction plans for New Orleans, the biggest problem was how to handle the Lower Ninth Ward, the district that had suffered the heaviest damage. Plans formulated immediately after the disaster, which centred on the city's Bring New Orleans Back Commission, proposed the greening of the Lower Ninth Ward. However, the problems of race and poverty that were also involved led to fierce criticisms from city residents, who wanted immediate assurances of housing and a return to the city from evacuation sites across the local region, and, in fact, the plans were withdrawn.

The Unified New Orleans Plan (UNOP), formulated in 2007, which forms the basis of the current reconstruction strategy, sketches a reconstruction strategy that takes into account the dual risks of population decline and flooding. In other words, it proposes strategies such as the concentration of housing through residential relocation, targeting areas with high flood risk, and the associated revival of social service functions as a kind of 'clustering programme'. Specifically, one of the first things highlighted as a 'recovery goal' was to 'promote the integration of multi-level flood protection systems into rebuilding plans' (City of New Orleans, 2007), with an indication of the implementation of a variety of flood measures as part of the reconstruction process. From the perspective of land use, too, the aim of disaster-resilient community development was incorporated as the goal of 'helping residents/businesses relocate from the most vulnerable areas, accommodating additional population in less vulnerable areas' (City of New Orleans, 2007).

On top of that, a second cited goal was to 'foster remedies to address blighted neighborhood conditions throughout the City' (City of New Orleans, 2007), aiming towards clustered development in disaster-resilient areas through strategies to induce heightened neighbourhood sustainability, with the view that 'a more rational pattern of resettlement can be encouraged by concentrating community services and commercial activity in areas of higher elevation, offering incentives to residents/business owners and developers to relocate into a more clustered development' (City of New Orleans, 2007). More specific policies are indicated for each district, categorised by repopulation rate and future flood risk. For example, with regard to areas with very slow repopulation rates and a high risk of future flooding, after first determining the necessity of focusing on stabilisation, an aim towards clustered development to ensure spatial sustainability is demonstrated in the statement that 'the heavy damage to infrastructure must be repaired and residents and businesses will be encouraged to return and rebuild in more sustainable clusters within their neighborhoods' (City of New Orleans, 2007). In this way, clustered development incorporates an aim towards the amelioration of both spatial sustainability and disaster preparedness.

Elbe riverbanks (Germany)

In the areas along the banks of the Elbe River, which principally includes the German state of Saxony, a tremendous amount of damage was caused by the flooding of the river in 2002. Despite this, no comprehensive plan for reconstruction was ever established, and nor does there seem to exist any specific vision of new spatial objectives. Efforts focused on the immediate repair of housing, such that residents would be able to rebuild their lives in the same location as early as possible. Only a single planning initiative seems to have leveraged the disaster as an opportunity to bring about major spatial changes, particularly at the macro-level, through the consolidation of smaller scattered settlements, though this has, in part, been due to budgetary restrictions, as well as light damages. Even this, rather than having been carried out in a planned fashion, was largely accomplished as the result of political influence.

Methods of planning implementation

Tohoku (Japan)

In order to implement planning, districts in the affected areas where it is deemed that a certain level of tsunami risk remains even given the construction of defensive facilities have been designated as disaster hazard zones in a manner similar to the zones targeted by the Disaster

Prevention Group Relocation Promotion Project, and the purchase of the previous residences and the provision of grants to cover relocation expenses has been carried out under the auspices of the same project. Although, in actual practice, such purchases have been carried out by municipalities and the purchased land is now held under municipal ownership, the costs are borne by the country. Moreover, residential usage is basically prohibited in the municipally designated disaster hazard zones. This is intended to prevent the progress of the re-urbanisation of the original areas from which relocation took place, that is, to inhibit the progressive spread of land usage once again into areas at risk from tsunami damage.

With regard to relocation areas, the nationally funded construction of relocation housing estates has taken place under the auspices of the Disaster Prevention Group Relocation Promotion Project, with these units to be sold or leased to victims of the disaster. In principle, the intention is for these buildings to be rebuilt by victims of the disaster on their own. However, public housing has been made available for victims who no longer have the ability or the intention to do so.

Aceh (Indonesia)

In Aceh, as described earlier, while there were initially plans to prohibit residence in coastal regions, the measures necessary to implement this, including the imposition of land-use restrictions by local municipalities, the provision of subsidies for relocation by the national government and the establishment of relocation sites to provide to disaster victims, were never carried out. Causes for this failure include the amount of time that was taken to arrive at decisions by local assemblies, which held regulatory authority over land use, and a lack of financial wherewithal on the part of the national government.

On the other hand, an important role has been played in connection with relocations by international non-governmental organisations (NGOs). Some of the more well-funded organisations have independently developed relocation housing estates on higher ground and elsewhere inland, and provided housing to disaster victims free of charge. While some of this housing was built in areas that were quite conveniently adjacent to existing urban areas, there were also those that were conversely built in suburbs located a half-hour drive or more away from the existing urban areas. Neither of these relocation housing estates has necessarily been developed on the basis of any comprehensive reconstruction plan; rather, they have been developed on the basis of the various plans of their respective agencies.

New Orleans (USA)

In New Orleans, it has been deemed that the implementation of plans must ultimately be strictly voluntary and incentive-based, and there have been no proposals relating to mandatory relocation programmes, such as land-use restrictions or forced relocations. This was due to the fact that, from the failure of the first plan, it was determined that a forced relocation programme would be unable to obtain the consensus of citizens. To implement the plan, land purchases and relocation projects would be carried out by the state government.

Drawing on funding contributed by the federal government, the Louisiana state government has purchased lands in the affected districts under the auspices of its Road Home Program. However, residents' decision to sell their lands is voluntary. In addition, special land-use restrictions have not been applied in districts where land purchases have taken place, or where they are able to take place. This indicates that this programme is being carried out from the perspective of supporting disaster victims rather than for the purpose of risk control based on spatial planning.

In terms of reconstruction, NGOs have been engaged in activities to assist with relocation and rebuilding in the field. Specifically, for those who wish to rebuild on the original site, homes are being rebuilt at no charge (chiefly in clusters), and the relocation of residents is being promoted through the transfer of properties in the Lower Ninth Ward, which suffered heavy damage, and the free acquisition of buildings and land in other, safer districts. Although these activities have no direct relationship with reconstruction activities being promoted by public agencies such as the New Orleans Recovery Agency (NORA), both sides are actively engaged in the exchange of information, and they are thus being carried out in indirect association.

Elbe riverbanks (Germany)

On the banks of the Elbe River, relocation projects have not been carried out as a general rule. Even with flood protection concepts that have been formulated in order to protect against damages similar to those of the recent flooding, while provisions have been made for elements such as the strengthening of embankments and the expansion of floodplains, similar provisions have not been made for relocations.

The sole example where such relocations have been carried out is in a new urban district composed of around 60 residential units and land for industrial use that was formed as the result of development that has taken place since 1994. In this case, the state government, despite the existing flood risks, that is, despite the fact that the area had been designated as a flood zone under the Water Act, with a probability of floods occurring once in a hundred years, had issued permits with regard to local plans to allow development, and the responsibility for this action had become a political issue. For this reason, the state government made full compensation for lands and properties in this new urban district, whereas such compensation would normally have covered only 80% of the damages in other districts.

The lands in question all became state property, and no special land-use regulations have been applied. On the other hand, with regard to existing settlements that have not been relocated, development consistent with the conditions in each settlement has been allowed as 'unplanned inner areas', as set forth in the conventional Building Code. In addition, apart from this, while mitigation measures and the like are required in cases where areas are designated as flood zones under the Water Act, this does not extend to regulation that would represent a total ban on development.

The destination to which victims are relocated is free, yet while a basically scattered model of relocation has been carried out, including individually to neighbouring municipalities, there is also an example of a voluntary group relocation encompassing about 10 households. In any case, the burden of these relocation expenses is borne by the residents, who have already received compensation for their lands and buildings.

Substantive formation conditions in urban areas

Tohoku (Japan)

Although land use for industrial purposes continues to take place in districts still facing the risk of tsunamis, these areas are not used for residential purposes. In this sense, it may be said that disaster safety has been adequately secured.

On the other hand, and as a result, a variety of land-use planning challenges is also apparent. With regard to the original areas, in the context of an already declining population, there is no reason to expect demand in connection with industrial land usage in the first place. In addition, because not all land has necessarily been targeted for acquisition by the municipality, a scattering of privately held properties remains, creating a situation that makes their intensive use difficult. Moreover, even in disaster hazard zones, some of these are subject to provisions that allow reconstruction given the attachment of certain conditions, and there are cases where low-density urban areas can be seen to be taking shape as the result of the mixture of relocation with reconstruction in the original sites.

With regard to relocation housing estates on higher ground or in inland areas, because project plans have been determined while striving to understand actual demand through the careful administration of questionnaires to residents, residential units have basically enjoyed a high residential take-up rate. However, due to difficulties in obtaining an accurate grasp of demand, particularly in the case of large-scale intensive relocation such as that planned for the plains areas, there are also cases where a mismatch has emerged between the scale of development and actual demand. Many of the areas affected by the disaster are experiencing a trend towards population decline, and there is no strong possibility of filling in the resulting vacant areas with new developments in the future. Furthermore, even if these could be filled immediately after relocation, the fact that demographic ageing is already extremely advanced means that there is a strong possibility that vacant lots and empty houses will begin to appear in the near future.

In addition, while there are some situations in which relocation by individual victims' reconstruction of their own homes can be seen to lead to the elimination of vacant lots and vacant houses in existing urban areas, there are also examples where this can be seen to have led to the expansion of sprawl in suburban areas. In this way, while densification is continuing to take place in existing urban areas, from the perspective of spatial sustainability, a dual-faceted situation can be seen in that there are also some cases in which low-density urban areas and settlements are gradually forming in low-lying areas from which relocation has occurred, in relocation settlements on higher ground and in inland areas, as well as in suburban parts of existing urban areas.

Aceh (Indonesia)

In the districts affected by the disaster, no special land-use regulations have been applied. Due to this, in addition to the fact that some of the previous residents of these areas have returned, there has also been an influx of new residents who have come in, drawn by the prospect of inexpensive rents due to the effects of disaster in the area, resulting in an ongoing process of gradual re-urbanisation. However, this does not mean that there has been sufficient development pressure until now for development to have succeeded in covering the entirety of the zone affected by the disaster; rather, this has resulted in the formation of low-density urban areas.

Also, with regard to relocation housing estates, those that have been developed on sites conveniently located adjacent to existing urban areas have been taking shape and maintained at an appropriate density. On the other hand, where they have been created at sites located at some distance from existing urban areas, these have faced challenges. As an example, where disaster victims draw their livelihoods from the fishing industry, they now have to spend a considerable amount of time commuting to the fishing ports on the seashore. Also, the lack of nearby commercial facilities in relocated communities means that residents must also spend more time shopping, or else rely on vendors who sell at a relatively high premium. Relocation housing estates developed in such inconvenient areas have, in some cases, seen their residents returning to existing urban areas, resulting in a gradual process of hollowing out. Conversely, a high level of density has been maintained in existing urban areas that managed to escape large-scale disaster. Thus, the perspective of spatial sustainability shows that a dual-faceted situation can also be seen in the case of Aceh.

New Orleans (USA)

The focus of ongoing reconstruction, even in areas that bore the brunt of the catastrophic disaster damage, like the Lower Ninth Ward, has been in highly convenient areas such as along major highways and in the vicinity of public facilities such as schools and libraries. This could likely be evaluated as an example of the implementation, and, to a certain extent, success, of planning aimed at clustered development.

Beyond this, however, while no special land-use regulations have been applied, no more than a portion of the previous residents has had the desire to rebuild, leading to the formation of urban areas of extremely low density. Furthermore, despite their ultra-low density, even urban areas such as these have been provided with public facilities, including parks and schools, which, in some areas, constitutes an apparent inconsistency with the conceptual aim of improving the relative attractiveness of clusters through 'selection and concentration', which is also a means of promoting their further development. This is likely the result of inadequate collaboration and cooperation in terms of foundational spatial planning among different departments in the public sector. On the other hand, new developments are being carried out and prices are rising in existing urban areas in districts that are more highly secure, such as the slightly elevated areas formed by natural levees.

Elbe riverbanks (Germany)

Changes to urban areas on the banks of the Elbe River have been relatively simple. Nonetheless, with regard to the new urban area where relocations have taken place, the original sites have become the property of the state, and these lands have been cleared through the removal of all buildings. Also, even in the relocation areas, because these are subject to development regulations according to the Building Code, development was carried out in existing urban areas or adjacent areas, with no apparent evidence of adversely affecting urban formations.

On the other hand, with regard to other urban areas and settlements where relocations have not taken place, there has been a return to the previous status quo. In addition, 80% of the costs of restoring damages have been covered, and because, unlike with a tsunami, the inland flooding did not necessarily cause catastrophic damage to the buildings themselves, neither was there any particular requirement for residents' voluntary and collective relocation. In that sense, while urban formations have basically become fixed, it might be said with respect to the relocation example that improvements have been seen in terms of both disaster preparedness and spatial sustainability.

Discussion

Disaster preparedness and spatial sustainability as desired objectives

What all of the plans described here have in common is their positioning of disaster-resilient community development as an objective. In other words, the goal of avoiding a repeat of similar disasters in the future lies at the heart of their understanding of the problem. Spatial planning has been situated as an important tool for this purpose. While plans in Tohoku, Aceh and New Orleans have also featured the active promotion of relocation, except for a single exception, plans along the banks of the Elbe River have emphasised rebuilding in situ. A conceivable reason for this difference is that whereas the disasters that occurred in the former locations caused catastrophic damage accompanied by houses being washed away, the Elbe riverbanks presented a continental river-style flooding hazard in which floodwater levels rise more slowly, thereby resulting in less damage.

However, there was also a difference in terms of the presence or absence of a scientific basis regarding the risk of disaster. While there has been some discussion in Tohoku, New Orleans and along the banks of the Elbe as to how to respond to disasters that occur in terms of the probability of their occurrence, in Indonesia, plans have been formulated in the absence of any such discussion. It is assumed that this is due to the fact that developing countries lack the necessary basic data for carrying out such investigation, and that they lack sufficient analytical techniques. However, ultimately, appropriate planning cannot take place without scientific data relating to disaster risk, even though they contain a certain amount of uncertainty. Accordingly, it may be said that what is called for in such cases is the development of simple data creation techniques that can be taken advantage of over a short period until plans can be formulated in the context of insufficient basic data.

Here, we encounter the further problem of data evaluation. This is the problem of how to evaluate risk data so that they can be reflected in spatial planning. First, it is important to build an awareness that this kind of scientific data has a margin of error in two ways: an error of assumption and an error of simulation or calculation. It is, therefore, dangerous to rely too much upon uncertain data. (The theme of 'uncertainty' is thoroughly discussed in chapter B5.)

Here, citizen participation takes an important role. Generally, assessment should take place on the basis of social criteria, on the basis of values rather than evidence, which have the potential to vary by urban areas or settlement units. Particularly with regard to Japan's seawalls and Indonesian's land-use planning, examples have also been seen in which attempts to carry out plans to a uniform standard have been problematic. In such cases, even though the importance of indicating certain uniform standards can be seen from the standpoint of efficiency, it could also be said to suggest the need for the discretion to take local conditions into account.

In addition, alongside improvements to disaster prevention capabilities, the improvement of spatial sustainability – namely, the formation of compact urban spaces that meet a certain standard of density – constitutes another shared planning objective. It might be said that there is also a shared recognition of the necessity to secure not only safety, but also factors such as lifestyle convenience and economical efficiencies, in order to reconstruct sustainable cities. Japan, in particular, appears to hold a sense of crisis with regard to this point. This is a result of the fact that, unlike the other countries, it is experiencing a striking population decline. Thus, even if its cities and settlements continue to shrink in future, there is little prospect of expansion.

The changing reality of urban areas

From a disaster preparedness perspective, the formation of urban areas has been most aggressively promoted in the Tohoku region. Rather than simply building embankments, the creation of housing in at-risk districts has been suppressed through the implementation of land-use regulations in areas that were formerly residential districts and residential relocation projects that seek to address the risk of tsunamis that might occur only once in several hundred years. Along the banks of the Elbe River, the reduction of flood risk has been basically pursued through measures such as the strengthening of embankments and the expansion of floodplains, and little has been done in terms of active resettlement. However, the further expansion of urban areas and settlements that would also broaden risk is also being prevented through regulation by the Building Code. Even in New Orleans, while it has basically been possible to address the current risk of storm surges by strengthening embankments, clustered development along with land-use incentives have been targeted, with the objective of securing an additional degree of both safety and spatial sustainability. However, since such measures have centred on incentives and due to factors such as insufficient coordination among different public agencies, they cannot be said to have had adequate effect.

On the other hand, the case of Aceh represents the opposite scenario. In Aceh, recovery plans have been formulated in the absence of risk characterisation or subsequent risk assessment, with village spatial plans, moreover, having been formulated at the district level without any relation to an overarching plan. Accordingly, neither relocation projects nor land-use regulation to prevent subsequent re-urbanisation were carried out to a sufficient degree. As a result, the affected area is being reconstructed without any countermeasures against tsunamis. For this reason, even though judgement of the extent of risks remains impossible, at least re-urbanisation is taking place in areas that suffered devastating damages at the time of the recent tsunami.

Next, from the perspective of spatial sustainability, how are we to assess the transformation of actual urban areas in each district? From a macro-perspective, relocation can be categorised as leading to either consolidation or distribution. Examples of both of these can be seen in Aceh and Tohoku, which have each actively carried out relocation initiatives. Basically, the former has high spatial sustainability in cases where relocations are made to urban areas or their neighbouring districts, while the latter could be said to have low spatial sustainability in cases where relocations are made in a distributive fashion, moving people to areas removed from existing urban areas.

Then, from a micro-perspective, relocation can be categorised as resulting in either densification in urban areas and settlements or else their hollowing out. Here, in a similar way, examples of both phenomena can be seen in both of these countries. Basically, the former may be said to have positive spatial sustainability, while the latter has a negative effect because of the inefficiency of infrastructure development and maintenance, the inefficiency of providing social services, and so on. Whereas densification progresses in areas relatively unaffected by disasters, a process of hollowing out occurs in areas heavily affected by disasters or in which there has been partial location to higher ground.

So, what accounts for the formation of low-density urban areas that are at odds with planning objectives? The first cause that may be mentioned is the diversity of residents' intentions. In a single affected area, there will be those who hope to rebuild in situ, and there will also be those who would prefer to relocate to somewhere safer. On this point, there will be variation according to factors such as age, whether one feels a sense of belonging to the local area and the type of occupation. In an attempt to reflect this as much as possible, some people will be relocated while others will rebuild in their original location, with low-density urban areas being formed as a logical consequence. Furthermore, the occurrence of such differences is also affected by the fact that the intentions of residents are liable to change over time. The situation on some of the higher-ground settlements to which relocations were made in Indonesia is typical. Housing estates were initially constructed on higher ground in line with the will of residents, who initially focused on safety, resulting in the formation of new urban areas with a certain population density. Nevertheless, over time, residents found that they were unable to maintain lives separate from their jobs and markets, and so relocated once again when they began to place a greater priority on lifestyle convenience, which has occasionally led to decreased density.

Here comes the importance of a plan reflecting long-term objective aspects. Plans formulated shortly after the disaster tend to be subjected to the individual opinion of the affected people, who put weight on a short-run viewpoint. On the other hand, without taking into consideration the adequacy of the spatial structure from a long-term point of view, in reality, the urban and regional structure can be formed in a way different to the initial plan. In this sense, it is important to formulate an adequate plan by taking into consideration both the aspects of the short-term individual opinion of the affected people and the long-term spatial structure. On this occasion, it is difficult to formulate a reconstruction plan only by the administration positioning near citizens. Experts should also participate in the planning process.

A second causal factor that can be mentioned is the absence of land-use planning for the original disaster area after relocations have taken place. Reconstruction and relocation projects are carried out in order to assist disaster victims, not in order to implement development projects in disaster-stricken areas based on a comprehensive vision of the future. From the beginning, elements such as how to make use of the original sites after relocation are not included as planning objectives. Accordingly, in cases where demand is subsequently absent, low-density urban areas will remain unchanged.

Nonetheless, in reality, it is difficult to formulate a comprehensive plan shortly after a disaster, when aiding disaster victims is the first priority. What is much more important is to change the space formed from the aspect of disaster prevention or aiding disaster victims into a space adjusted to the sustainable daily life or industrial activity of the region.

A third cause that may be mentioned, though associated with the first, is the discrepancy between plans, or else problems with planning systems in terms of the absence of any means for their implementation. If what is being attempted is a comprehensive relocation, this will require land-use regulations as well as public subsidies to accompany the relocation projects, being necessary from the viewpoint of property rights issues as well. However, there is little margin for such financial expenditure, especially in the case of developing countries. Moreover, in the US, political issues were also involved, and what could be described as clustered development was still fraught with the difficulty of external land-use regulations. In addition, as the result of insufficient cooperation between public agencies, or, in other words, the absence of a comprehensive coordinating authority for spatial planning, public expenditures were also carried out outside of the areas marked for clustered development. Since this was not accompanied by sufficient implementation measures due to such different backgrounds, the result has been that the individual wishes of various residents have been reflected directly in the realities of land use.

In many cases, such a problem about land-use control did not become an issue in the wake of disasters; rather, it existed before the disaster and first came to the surface in the wake of it. In that sense, it is important to establish the land-use control system during a normal period.

Conclusions

As we have seen in this chapter, every spatial plan for reconstruction from catastrophic disasters was planned and implemented with due consideration to both disaster prevention and the sustainability of the spatial structures. Those reconstruction efforts, however, have taken place in a difficult situation, being hampered by various hurdles.

First, regarding the planning phase, there is a necessity for a scientific basis regarding the risk of disasters and a necessity for risk evaluation based on social criteria. This implies the importance of applying an adequate risk governance process to the spatial planning process, composed of four steps: 'risk pre-assessment', 'risk appraisal', 'risk characterisation and evaluation' and 'risk management' (IRGC, 2008).

Second, with regard to the actual spatial formation of the region and district, some settlements are aggregated on the regional level or the density of some settlements increases adequately on the district level. In other cases, however, especially in the severely damaged areas, the density of the settlements decreases and settlements with ultra-low density form. For these situations, the following three necessities can be pointed out: the necessity to plan from both long-term and subjective aspects; the necessity to transform the settlements to a sustainable form; and the necessity to establish the land-use control system during a normal period. This chapter closes by characterising the reconstruction planning in each region from the aspect of the relationship between disaster prevention and spatial sustainability, as follows.

The land-use planning and implementation process for recovery in the Tohoku region can be characterised as the type 'setting priority on disaster prevention'. What was the most prioritised in the planning process was the safety from tsunami disaster, which occurs theoretically once in several hundred years, and the relocation from those areas with higher risks than that. It was a secondary consideration in the framework as to how to secure the sustainability of the cities or villages spatially. Most of these plans have been or are being implemented on the initiative of national and local governments. This is because of the characteristics of the hazard that caused the catastrophic damage, as well as the capability of government. As a result, the density of some parts of the existing cities and villages has increased because of the demand of the people relocated from the affected areas, which makes the sustainability of those areas higher. On the other hand, however, the density of the other affected areas or relocation settlements is very low as a result of prioritising the support of the affected people, as well as errors in predicting. Considering the tendency towards population decline in the region, it is quite likely that those settlements will never get sufficient density in the future. There is also scepticism as to whether some of the relocation settlements will last into the future due to inconvenient access to shopping and workplaces.

The land-use planning and implementation process for recovery in Aceh can be characterised as the type 'setting priority on the sustainability of daily life'. Due to the limited financial and regulation capabilities of national and local governments, and despite the characteristics of the hazard that caused catastrophic damage to the region, the spatial recovery plan was drastically changed over time and the spatially well-planned relocation projects and land-use regulations in it were not implemented. Relocation projects were implemented mainly at the initiative of international NGOs. As a result, some of the relocation settlements in the outer suburbs have lost their inhabitants because they are inconvenient. On the other hand, re-urbanisation of the affected area proceeded through village spatial plans without taking almost any countermeasures against tsunamis. Settlements with low density are being formed both in the outer suburbs and in the affected area. The density of the affected area, however, is anticipated to get higher in the future because of the population increase in the region and the convenient access to shopping and workplaces.

The land-use planning and implementation process for recovery in New Orleans can be characterised as the type 'pursuing an ideal form that combines disaster prevention and sustainability'. The reconstruction plans have been changed several times, and in the ongoing plan, disaster prevention against the high water shall be achieved by basically strengthening the levee. The reconstruction plan seeks 'clustered development', which treats the density of the areas differently in order to achieve further developments in both disaster prevention and sustainability. This reflects both the hazard that caused partly catastrophic damage and the social situation of the district from before the disaster. In fact, some clusters whose damage was relatively small have been developed further. In other areas, however, even the accumulation in the cluster was not sufficiently achieved, while developments outside of the clusters have been done sporadically because of the lack of land-use regulations and projects, which leads to low-density urban areas. The gaps between the plan, which was changed drastically over time, and the reality are large in the recovery process in New Orleans.

The land-use planning and implementation process for recovery in the Elbe riverbank can be characterised as a 'fixed' type. The hazard of this area is flood, which results in the increase of the water level little by little and thereby causes scarcely any catastrophic damage, with almost no risk for human life; in actuality, the area was not severely physically damaged by the flood compared to other regions hit by, for example, tsunamis. Reconstruction in the original place was, therefore, the basic policy instead of large-scale relocation projects, and aggregation of the scattered settlements was therefore not pursued. However, it should also be noted that the flood risk of the area has not been increased due to strict land-use regulations based on the German Federal Water Act and Building Code.

References

City of New Orleans (2007), *The Unified New Orleans Plan: Citywide strategic recovery and rebuilding plan (UNOP)*, City of New Orleans.

IRGC (ed) 2008, *An introduction to the IRGC Risk Governance Framework*, Geneva.

CHAPTER B3

Role of coordination in building spatial resilience after disasters

Jaroslav Tešliar, Alena Kučeravcová and Ján Dzurdženík

Introduction

Disasters often provide unique opportunities to promote climate-resilient development and to build political will to integrate resilience measures into recovery and development. Recent evidence shows a growing demand for sustained engagement in countries following a disaster to support the implementation of resilient recovery and reconstruction planning. Politicians and donors alike are attuned to the issue, and the general public may be more amenable to the often-difficult trade-offs necessary for risk reduction.

According to the World Bank (2013), a major challenge for post-disaster and climate-resilient recovery support is timely and sufficient access to resources. All climate- and disaster-resilient development actions have an upfront cost. However, if the action is well designed and proportionate to the risk, then the outcome will be cost-effective and save money in the long run.

Disaster resilience is a shared responsibility between governments, communities, businesses and individuals. Enhancing coordination, synergies and linkages among various organisations, institutions and involved actors is fundamental to keeping all stakeholders focused on the goal of reducing vulnerability. Collaboration provides an opportunity to share knowledge, experience and skills with multiple members in order to modify goals and contribute to development. In order to successfully collaborate, there must be sufficient resources, a culture that encourages effective teamwork and cooperation, and clearly defined responsibilities.

The current management of disaster risks is often fragmented due to a lack of coordination between involved actors, for example, civil protection and spatial planning – a phenomenon known as the 'problem of interplay' (Young, 2002, p 23; Greiving et al, 2012). Lack of coordination refers not only to prevention/preparedness measures, but also to preparations for the remediation phase. It should not be forgotten that remediation functions simultaneously as the prevention stage for the next disaster event. Fragmentation and disconnection leads, among other things, to a lack of synergy and mutually subverted or duplicated measures and funding. These problems become more acute in times of limited resources (financial, personnel, etc) (Sapountzaki et al, 2011). Spatial planning is increasingly regarded as one important instrument for disaster risk reduction. Spatial planning is responsible for decisions on the long-term utilisation of land.

Theoretical background

Today, in the 'globalised' world, all areas of human activities are interconnected, and risks 'know no borders'. This is regardless of whether the risk is related to the individual, community, region, state or globe. If individuals are exposed to a particular risk at a certain time, they have to decide and choose a strategy or approach as to how to manage and deal with it. Some of the most prominent examples are regularly repeating floods in Central Europe due to more frequent extreme weather conditions (eg floods in Germany on the Rhine River in Cologne, or in Eastern Slovakia on the Laborec River in 2010).

The more there is interdependence and interconnectedness in a particular environment (either in the organisational units, regions or countries), the more these entities should cooperate to reduce the risk because the risk can often uncontrollably spread to other entities. Each entity (involved) should act responsibly and should take care of its protection. At the same time, all stakeholders should cooperate and coordinate their actions and implemented measures because without that cooperation and coordination, it should be expected that the effectiveness of the individual measures will be only partial.

Horizontal interplay

Coordination at the horizontal level in order to eliminate risks means that parts of society at the same level of organisation that are exposed to risk often communicate and collaborate with their neighbours. This would ideally mean that at the local level, there is lively and well-organised cooperation between neighbouring municipalities. At the regional level, this would mean that various organisations with different responsibilities for tackling the risk inform each other and undertake joint action to solve or minimise the risk.

Horizontal institutional interplay is introduced in three steps (Young, 2013):

1. The first section introduces concepts where the landscape of interplay between and among systems is characterised and identifies several types of interplay. All of them might be similar in some ways but deserve individual explanation in an effort to understand this phenomenon.
2. The next section addresses the politics of interplay. It begins with the observation that the various actors involved carefully follow risk in conjunction with institutional interplay and propose strategies that they adopt in both the formation of regimes and the administration of these arrangements following their creation.
3. The final substantive section is interplay management. It starts from the observation that from much of the interest in the interactions between distinct regimes flows a concern that such interactions may generate harmful consequences with regard to the effectiveness of one or more of the regimes involved. The central question here has to do with the strategies or families of strategies available to those responsible for designing or administering regimes and desiring to maximise the ratio of positive to negative interactions. The objective is not only to identify differentiable strategies, but also to analyse the circumstances under which individual strategies are likely to work best.

Although institutional interplay can become a source of problems, alert actors can often minimise harmful consequences and promote positive interactions. Where such cooperation is successful, there can often be a synergistic effect.

Vertical interplay

In a complex, 'well-established' society, most of the institutions have well-defined missions and they are clearly focused on particular functional concerns, they frequently interact with one another, they complement each other and they coordinate their activities and common approach. No generally accepted recipe exists but it can be said that vertical interplay reflects political design and management. The principal conclusion is that vertical interplay can manage well-functioning interdependencies on a permanent basis. To see how these interactions play out in practice, please see the examples from particular countries later in this chapter. In this context, three general factors for institutional interplay are important – competence, compatibility and capacity (Young, 2002):

- Competence is a matter of political and legal authority.
- Compatibility is a matter of standard practices or procedures for handling governance in individual member states using incentive mechanisms.
- Capacity is a measure of the availability of the social and institutional capital, as well as material resources, necessary to make good on commitments.

The landscape of institutional interplay

The following draws a distinction between horizontal interactions and vertical interactions (Young, 2013). *Horizontal interactions are those that take place at the same level of social organisation*, from local or regional interactions up to national interactions and international or global-level interactions. At the local level, zoning ordinates designed to eliminate or minimise the environmental side effects associated with development often come into conflict with the rights of owners of private property. National regimes created to provide protection for endangered species can generate tensions not only with the property rights of private owners, but also with management regimes created to address mainstream uses of the public domain, such as timber harvesting or hard rock mining.

Vertical interplay, by contrast, occurs when regimes interact across levels of social organisation. Whereas interactions between the international regimes dealing with climate and biological diversity are horizontal in nature, interactions between these international regimes and the regimes dealing with the same or similar issues at the national level, for example, give rise to vertical interplay. Horizontal and vertical interplay can and often do occur at the same time, leading to increasingly complex systems of interactions.

Building spatial resilience across countries

Spatial planning aspires to be an interdisciplinary and cross-cutting coordinator of sectoral policies and decisions with spatial impacts. There are greatly varying forms of organisation on the regional planning level in different countries (Reimer et al, 2014). Taking this into consideration, this chapter also tries to explain the results from the project Increasing Resilience of Urban Planning (URBIPROOF) and its workshop in Sendai and attempts to address the following question: upon creating and implementing more resilient spaces after disasters, how did the coordination, both vertical and horizontal (between departments and agencies, as well as among localities and regions), proceed?

Germany

According to German legislation, the main responsibility in the area of spatial resilience rests with the spatial planning system. The system is well established and responsibilities are recognisable. In Germany, planning occurs within a decentralised decision-making structure and a strong legal framework. Spatial planning is defined as comprehensive, over-sectoral planning. In addition to comprehensive spatial planning, specific aspects with spatial influence are considered within sectoral planning (eg transport, water protection, nature protection).

Vertically, the planning system is stratified into the federal government, 16 state governments, 114 planning regions and 14,000 municipalities, and it interacts with other types of planning (Schmidt and Buehler, 2007). The Federal Spatial Planning Act does not set a statutory basis for a national spatial structure plan that has a binding effect for the entire territory.

The Federal Building Code (FBC) ('*Baugesetzbuch*') is used as a basis for spatial planning at the local level. The FBC sets uniform objectives, instruments and procedures that are binding for urban land-use plans of every municipality. According to the policy provisions of the sectoral planning authorities, the FBC is the only spatial planning tool contributing to risk management through keeping threatened areas free of further development or at least of particular vulnerable land uses (Sapountzaki et al, 2011).

The spatial plans of upper levels are more general and define measurements, corridors and areas at the national and state levels. However, with respect to the scale of plans, they cannot be defined very exactly. The spatial plans of lower levels have to respect higher levels. Nevertheless, the building of spatial resilience is possible particularly in local and regional levels, but national (or regional) spatial plans cannot hinder this effort.

Planning is a shared task among all levels of government. The federal government does not create or implement plans, but rather sets the overall framework and policy structure to ensure basic consistency for state, regional and local planning.

The level of responsibility and degree of plan detail increases with lower levels of government. Specialised sectors (ie ministries for transportation, water, energy) provide input through sector plans, which are formulated independently from spatial plans and then integrated by planning authorities.

Regarding the reduction of vulnerability, there exist several possibilities. From the spatial planning point of view, land-use provisions (eg provisions for the restriction of development on flood plains) ('*Flachenvorsorge*') are issued by the zoning boards ('*Planungsbehorden*') at the regional and local levels. These have the power to influence and control development and land uses through planning and implementation (Grunewald, 2009).

From a long-term point of view, insufficient coordination is visible in the case of Roderau-Sud in Germany, where a former decision (definition of a flood retention area) was not respected and a field was used as a residential area. This inadequate re-decision was possibly influenced by political changes in Germany and is not necessarily a common occurrence. Coordination in spatial planning is influenced by an existing long tradition of state intervention at both the national and regional levels in Germany (in comparison with, eg, the US). State intervention has been accepted, and expected.

Interaction between spatial planning and the assessment of risks can be well realised via strategic environmental assessment, which is obligatory for any spatial plan or programme. An important problem is the adaptation of existing spatial structures (settlements, infrastructure) as it is hardly possible through regulatory planning due to given private property rights. Incentives and good practices aiming at changing the minds of private householders are needed. A second serious problem in Germany is the low possibility of relocation or retreat from threatened areas. This is in conflict with property rights. Full compensation is needed but fails mostly due to the lack of financial resources. Retreat, however, might be possible in the aftermath of a disaster or in areas with a shrinking population, where the existing building stock will be (partly) deconstructed based on planning strategies.

Slovakia

The system of spatial planning in Slovakia is essentially similar to the German system. The main difference is in scale – the regional level in Slovakia can be more detailed than the regional (federal state) level in Germany. However, spatial plans at the local level can be elaborated at similar scales in both European countries. Spatial planners (with the use of expert and scientific data provided by water managers and geologists) are able to define measurements leading to the increasing of spatial resilience after (or before) disasters. Vertical coordination is well ensured by valid legally binding regulations (between 'sub-local', local, regional and national levels). However, horizontal coordination had significant weaknesses in the near past. This led to a state in which spatial plans did not utilise the very relevant data and recommendations elaborated and defined by hydrologists and geologists.

There are few examples of the deterioration of the impact of a catastrophic event caused by a weak respect of recognised threats (horizontal coordination). A very evident example of this can be seen from landslides in the village of Nižná Myšľa, in which case spatial plans did not respect the data generated by the geological survey. The western part of the village was described as a dormant landslide with several earth blocks in 1991. The position of the landslide was put onto the map at a scale of 1:10,000 and has been publicly available in the Geofond since that time. Engineering geological maps are the most important maps for regional planning and comprise the zoning map, a map of land susceptibility to slope movements, a map of significant geofactors and, in some regions, also a map of land susceptibility to collapsibility. Three geofactors (parameters) are evaluated directly: the bearing capacity of foundation soils, the inclination of the land surface and slope stability. Suitability is depicted by three colours (see Figure 82): red (unsuitable area), orange (conditionally suitable area) and green (suitable area). Maps also delineate other important geofactors and legislative restrictions. In spite of the fact that engineering geological maps represent a very good basis for the preparation of territorial/spatial planning documents, their use by planners (architects, designers) is very

Figure 82: The optimisation map for land use planning of the Ľubovnianska vrchovina and Spišská Magura (red colour - unsuitable area, orange colour - conditionally suitable area and green colour - suitable area)

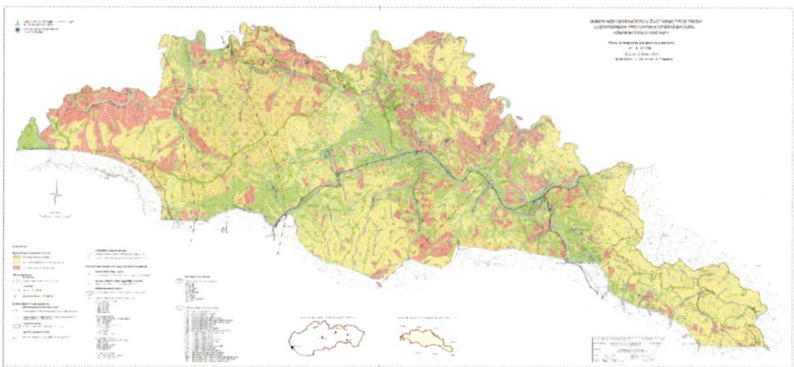

Source: Petro et al (2010)

limited in Slovakia. Until 2010, the construction of buildings and infrastructure proceeded regardless of potential consequences. If coordination had been better, material damages and subsequent expensive technical measurements could have been avoided.

At the local level, several examples of the possible avoidance of damages in the case of better horizontal coordination can be mentioned. Almost all are concerned with flood-prone areas that were occupied by buildings not resistant against flooding. Slovakia does not suffer from such an enormous lack of building plots as Japan and some German regions. For that reason, keeping flood-prone areas free of construction is, in many cases, a feasible solution.

After larger amounts of small-scale catastrophic events (in comparison with Japan, Indonesia or the US), and in the process of the implementation of European Directives into Slovak legislation, a few very important steps in the area of horizontal coordination have been taken. All of them have improved the planning process by establishing a better respect of expert knowledge in the area of geological risk and hydrology during the elaboration of spatial plans.

According to the Geological Act and it latest amendments, after experiences from landslides in 2010, spatial planning has to take into account geological data, especially those connected with possible landslides. The Ministry of the Environment will therefore provide information about potential landslide areas and other geological hazards. Conditions for the utilisation of these areas also have to be defined. This should ensure a so-called precautionary system aiming to eliminate undesirable construction interventions. This can prevent future expenses for the remediation of landslides caused by human activities.

Building spatial resilience in general is a task for municipalities. The municipality coordinates building permits and determines the regulations regarding the spatial arrangement and functional use of land in the local spatial plan or in the local plan of zones, with measures to protect against natural hazards. These competences are transferred from the state administration in the field of spatial planning and building regulations. The municipality also provides a designation of all flood lines shown on the flood hazard maps into the spatial plan of the municipality or spatial plan of zones.

Of course, municipalities (especially in the case of small villages) are not able to 'fight' effectively with threats of a larger scale (floods, landslides, forest fires), which require more expensive measures (dikes, geotechnical constructions). It such cases, the national level is the responsible body (coordinator) and investor for bigger actions.

Japan

Japan is a country very often exposed (eg in comparison with Europe) to various natural hazards. In connection with a high level of organisation of society and economic power, Japan was able to create a very well-developed system of natural hazard management.

Japan has a centralised governing structure in which the national government oversees 47 prefectures and other local governments, including cities, towns and villages. Authorities at the middle and lower levels are based on national delegation. Prefectural governments have an intermediary role between the national and local governments.

Vertical coordination in the area of spatial planning is necessary between three basic levels of planning: national, prefectural (47 prefectures) and municipal. Prefectural and municipal land-use plans have to be based on the national plan. Land-use basic plans are developed and approved by prefectures. These plans define borders for five types of land use: urban, agricultural, forest, national parks and natural preservation. City planning areas are divided into controlled parts and uncontrolled, where zoning is voluntarily enforced. The city planning system is well articulated to function within city areas, being theoretically capable of controlling hazardous lands in order to protect lives and assets from disasters.

Reconstruction and recovery is part of the disaster management plan (one of three different phases). This plan defines the roles of national, prefectural and local governments, as well as residents, for actions at all different phases of disasters. After the Great East Japan Earthquake (GEJE), a revision of the Basic Disaster Management Plan was approved. One of the major focuses in this plan is defining spaces that are resilient to different types of natural hazards. The plan lays out the foundational role of national and local governments, through different ministries, as well as prefectural and local governments. National and local governments, as well as communities are mandated to proactively invest in mitigation efforts by preparing master plans against potential floods and developing flood hazard maps, which are publicly shared. Making decisions on land use is a primary task given to the local governments, through combining different land strategies, such as prohibiting residential use, providing conditions in land use and raising lands in potential inundation areas for low-frequency, large-height tsunamis.

The National Reconstruction Agency was formed after the GEJE and still exists. This agency is directly responsible to the prime minister and sets guidelines for local planning, approves local recovery plans and coordinates the work of national ministries as they implement reconstruction.

The rebuilding of devastated areas gives rise to a large demand for the coordination of all partial projects (roads, levees, buildings). Problems occur if some infrastructure is planned by prefectures and some by the national government. In the town of Minamisanriku, for example, the coastal levee and the mounded national road are planned to run parallel to each other, in close vicinity of about 100m. The road, planned to be behind the levee, is even higher than the levee and the local government is planning to fill the resulting dips between the two infrastructures. Although the construction of a road that also functions as a levee might present an economic advantage, it is hard to coordinate the two projects because the levee is planned by the prefectural government and the road is planned by the national government. Each section is overloaded with such enormous tasks that the active coordination between the two is completely neglected.

USA

In the frame of the US, we can recognise three basic levels: the federal government, individual states and local governments. The established system of urgent response and systematic recovery is in US is tested very often, 'thanks' to hurricanes, heavy storms and floods.

The role of state intervention in the US (in comparison with the analysed European countries), let alone the existence of a structured planning framework, has historically been a very much contested topic. This has produced a heterogeneous planning system at all levels of governance. Spatial or territorial planning is not a part of the US legal planning vocabulary. Land-use decisions have been under the jurisdiction of the state legislatures, which have historically given much power to municipal legislation (thus the coined term 'home rule'). Since property rights are so strongly defined, it is easy to mount a legal defense against any plan perceived as unfair by you (Gawroński et al, 2010). The role of higher levels of government in the planning process is still challenged (Schmidt and Buehler, 2007). Federal intervention is generally confined to indirect means, such as environmental regulation, the management of nationally owned lands, investment in transportation infrastructure, the provision of financial assistance and housing subsidies, and the dissemination of information and technical assistance. It follows that the comparison of coordination in building spatial resilience (mainly in the field of spatial planning) after disasters cannot be generalised for the whole of the US.

However, disaster planning and recovery have long been the task of disaster management professionals, and the role of spatial planning in relation to disasters has been increasing in recent years. Recovery planning in the US happens through the complex interactions of federal, state and local government actions. The federal government has shouldered a heavy burden for short-term response and recovery actions (through the Federal Emergency Management Agency [FEMA]). During the last decades (after some legislative changes), long-term community recovery has increasingly become part of FEMA's mandate. Federal agencies and programmes provide more intensive assistance (in the process of recovery) to local and regional authorities.

Coordination between the federal government and local governments falls under the role of individual states. They act as intermediaries between federal funding sources and local governments (vertical coordination), and help serve a coordinating role among multiple affected local governments (horizontal coordination). Local governments retain that control in the recovery phase (including hazard mitigation and resiliency building) and have the primary role of planning.

Indonesia

The first spatial planning system in Indonesia was established in 1926. Major changes in this system have recently been implemented after 1997 and have been legally released in 2005. Since 2001, the structure of government has shifted from a centralised into a highly decentralised structure. Most administrative affairs, including spatial planning, have been transferred from the central government to provincial and local governments (Hudalah and Woltjer, 2007).

In the recovery phase after the 2004 tsunami, a tool for spatial planning was utilised in Indonesia, similar to Japan. The main coordination task was given to the Rehabilitation and Reconstruction Agency (BRR). The BRR had full authority to coordinate and implement any rehabilitation and reconstruction (Johnson and Olshansky, 2013). After such a large disaster, coordination was the most urgent issue as many individuals and entities have been involved in the recovery process. It was expected that the BRR would have to fully coordinate the rehabilitation and reconstruction process.

After 2008, the coordination role of spatial planning was transferred to the National Development Planning Agency. Local authorities are not able to participate in this recovery process because their capacity to contribute to this process has not been proved. The Master Plan for Rehabilitation and Reconstruction was developed within three months of the disaster. It consists of 12 volumes, and the spatial plan is one of them (Book 2: Spatial Plan and Land Affairs). This spatial plan was used as a reference to revise the existing provincial and local spatial plans. Its goal was to create settlement structures that will be safer and that will create the possibility to relocate citizens exposed to the risk and restore economic activity.

The implementation of this spatial plan was met with many problems. It has not been accepted by communities, mainly because the coastal zone of a width of two kilometres was determined as a restricted zone with no construction allowed. Many landowners did not agree with this 'buffer' zoning. Communities that had been exposed to the tsunami wanted to be quickly resettled, and did not wait for the mitigation measurements of the local spatial plans. In the end, the BRR had to allow the building of new settlements in the restricted zone, which did not respect the spatial plan. The BRR guideline was used as the minimum benchmark. New local spatial plans have been developed in cooperation with local citizens by facilitated discussions and were quickly approved by the village leader. Lack of coordination in terms of vertical interplay (between the local level and the regional or national levels) was another problem.

Conclusion

Comparing the *planning systems*, it is possible to find some similarities between the US and the European countries (industry, democracy, system of governance). Despite all these similarities, the US and Germany have established different planning systems. The US and Germany both have federal systems of government. An examination of the role and function of each level of government (national, provincial, local) as it belongs to planning would uncover important similarities and differences. Although planning operates in the US and Germany within a federal division of powers, German planning is far more integrated between levels of government and consensus-oriented than planning in the US. The US has a more fragmented and often more competitive nature of planning, with a much more decentralised distribution of planning powers.

The system of spatial planning in Slovakia is similar to the German system. The main difference is in scale – the regional level in Slovakia can be more detailed than the regional (federal state) level in Germany. In Slovakia, municipalities are self-governing and fully autonomous within their own territory, as the regional bodies and districts are self-governing on regional issues. Local plans have to comply with the framework given by the regional plans.

The Indonesian spatial planning system is decentralised. This might be one reason why some details suffer from a lack of horizontal coordination in the area of planning. However, in the case of a large disaster, when fast rebuilding is expected by all stakeholders, these mistakes are almost unavoidable. This is also the case in Japan, which is more centralised (in the area of spatial planning as well), but some problems in the recovery phase are also visible.

In Slovakia, legislative conditions for *vertical and horizontal coordination* in the phase of spatial planning are given. There are certainly solutions as to how the problems of the risks of floods, landslides and other catastrophes might be solved. For example, a superior or regional plan should be reflected in the local spatial plan, ignorance of responsibilities at the local or higher levels should be reduced, and the lack of capacity of local municipalities to cover all aspects of life and fulfil all obligations bounded by the wide range of Acts and rules should be addressed. Slovak society still has a lot of work to do in defining and creating the future realisation of coordinated steps for the adaptation of existing structures and for the potential relocation from threatened areas. In the context of ongoing climate change, which is accompanied with an increasing frequency of catastrophic events connected with weather (floods, droughts/fires, windstorms, landslides), there is an urgent necessity to create feasible adaptation strategies at the regional and local levels, with clearly defined measurements.

Coordination is influenced by a long tradition of state intervention. This occurs at both the national and regional levels in Germany (eg in comparison with the US). State intervention has been not only accepted, but expected. The coordinating unit depends upon the scale of the disaster. Representatives of authorities at the local or regional levels are involved in post-disaster management. If the impact of the disaster exceeds the borders of one municipality, or if the local government is incapable of handling the situation on its own, the superior authority takes over the coordination.

In Indonesia, a lack of coordination or vertical interplay (between the local level and the regional or national levels) is observed. Coordination only occurred between the communities where a professional facilitator entered into the process. There was nearly no horizontal coordination with surrounding villages. The results of such a chaotic coordination are visible in Indonesia, for example, the lack of proper continuity of the road infrastructure and so on.

In the case of large disasters in Japan, coordination is problematic due to the extreme overload of offices by various tasks. In the case of difficulties in the coordination of activities at the local, prefectural and national levels, planners may be key actors able to coordinate the activities realised by various subjects.

Coordination between the federal government and local governments is found within the role of individual states in the US. In the case of large disasters, there is substantial interplay between local governments, their state governments and the federal government.

Sometimes, a lack of coordination at the regional level is essentially explained by the dependence of sectoral policies at this level on decision-making at higher levels (national and even supranational), that is, the top-down model of policymaking. In cases of centralised political-administrative systems, where the pattern of top- down policymaking and implementation predominates, the lower regional and local levels are bound by the upper ones to implement the policies of the latter, and enjoy limited discretion in formulating and following their own paths to spatial planning and risk mitigation (Greiving et al, 2012).

This chapter has shown that there are countries in which after a disaster, recovery/reconstruction agencies were established (ie in Japan – the National Reconstruction Agency; in Indonesia – the BRR; in the US – FEMA), which take the lead in the planning process. Lead institutions must have the necessary authority to coordinate powerful sectoral ministries. Experience indicates that this is best done by an agency located at the highest possible government level.

According to the Australian National Strategy for Disaster Resilience (2009), effective partnerships across all areas of society are critical to enhancing disaster resilience. This focus on coordination has also recently been addressed by the Sendai Framework for Disaster Risk Reduction. Priority 2 points at 'Strengthening disaster risk governance to manage disaster risk' and that 'Clear vision, plans, competence, guidance and coordination within and across sectors as well as participation of relevant stakeholders are needed' (UN-ISDR, 2015, p 17).

Many not-for-profit organisations have experience and expertise in areas including community engagement and education, and various facets of service provision. Importantly, their existing networks and structures reach far into communities and can effect real change. Research institutions, for example, have an important role to play in providing advice to federal, state and territory policymakers; worldwide, governments need to engage with academic organisations to provide advice on the need for policy-driven research. Here, the concept of collaborative research is promising for achieving a mutual understanding between science and society and taking the interests, but also the expertise, of the end users into account. Normative decisions on thresholds or the weighting of concerns and interests have to be taken together by scientists and practitioners as users of research results (Greiving et al, 2015). Policymakers at all levels of government need to strengthen their partnerships in order to develop a coordinated response to the changing risk environment. Building better links with the private sector is a particular priority, not least because infrastructure is often owned or managed by private interests, which deliver services that enable communities to function. Businesses, whether large or small, can play an important role in preparing for and dealing with the consequences of a major emergency or event. This role is key in helping the community maintain the continuity of services following a disaster.

So, what can communities do to recover successfully or what can be done towards building spatial resilience? The following transferable conclusions have been drawn from the work in the URBIPROOF project, as well as from the literature (BCLC, 2012):

- **Increase communication between different government entities** – Natural hazards do not contain themselves in one particular political jurisdiction, so it is therefore necessary that different government groups work effectively together. Horizontally, governments from different geographic regions must be able to work well together. Vertically, local, regional and national government agencies must be familiar with how each other operates and know how to work in sync with one another.
- **Increase communication between different sectors** – Similarly, the appropriate individuals from the business community, the government, the media and non-governmental organisations (NGOs) must be familiar with each other's responsibilities and capacities and work towards common goals. This can be especially difficult because different types of organisations have different objectives and dissimilar time frames for achieving those goals in affected communities.

References

BCLC (Business Civic Leadership Center, US Chamber of Commerce) (2012). *Creating Disaster Resiliency: Implementing Regional Coordination and National Disaster Prevention Networks*, A Whitepaper Based on BCLC's February 2012 Systemic Community Resiliency Workshop. BCLC

Gawroński K., Van Assche K., Hernik J. (2010). Spatial planning in the United States of America and Poland. *Infrastruktura i Ekologia Terenów Wiejskich*. Nr 2010/11

Greiving S., Pratzler-Wanczura S., Sapountzaki K., Ferri F., Grifoni P., Firus K., Xanthopoulos G., (2012). Linking the actors and policies throughout the disaster management cycle by "Agreement on Objectives" – a new output-oriented management approach, *Nat. Hazards Earth Syst. Sci.*, 12, 1085-1107

Greiving S., Zebisch M., Schneiderbauer S., Lindner C., Lückenkötter J., Fleisch-Hauer M., Buth M., Kahlenborn W., Schauser I. (2015): A consensus based vulnerability assessment to climate change in Germany., *International Journal of Climate Change Strategies and Management*, 7, 3.

Grunewald U. (2009) Gutachten zur Entstehung und Verlauf des extremen Niederschlag-Abfluss-Ereignisses am 26.07.2008 im Stadtgebiet von Dortmund—einschließlich der Untersuchung der Funktionsfa higkeit von wasserwirtschaftlichen Anlagen und Einrichtungen der Stadt, Emschergenossenschaft und Dritter in den Gebieten Dortmund-Marten, -Dorstfeld und –Schonau, Gutachten im Auftrag der Stadt Dortmund und der Emschergenossenschaft

Hudalah D. and Woltjer J. (2007) 'Spatial Planning System in Transitional Indonesia', *International Planning Studies*, 12: 3, 291–303

Johnson L.A. and Olshansky R.B. (2013). *The Road to Recovery - Governing Post-Disaster Reconstruction*, Lincoln Institute of Land Policy, Land Lines

National Strategy for Disaster Resilience (2009). *Building our nation's resilience to disasters*, The National Emergency Management Committee, Australia

Petro et. al. (2010). *Súbor máp geofaktorov životného prostredia Ľubovnianska vrchovina a Spišská magura, Mapa optimalizácie pre územné plánovanie.* ŠGÚDŠ Košice

Reimer M., Getimis P. and Blotevogel H.H. (eds) (2014). *Spatial planning systems and practices in Europe – A comparative perspective.* Routledge

Sapountzaki K., Wanczura S., Casertano G., Greiving S., Xanthopoulos G., Ferrara F.F. (2011) Disconnected policies and actors and the missing role of spatial planning throughout the risk management cycle, *Nat. Hazards Earth Syst. Sci.*, DOI 10.1007/s11069-011-9843-3

Schmidt S. and Buehler R. (2007). 'The Planning Process in the US and Germany: A Comparative Analysis', *International Planning Studies* 12(1).

UN-ISDR (2015) Sendai Framework for Disaster Risk Reduction 2015–2030, http://www.wcdrr.org/uploads/Sendai_Framework_for_Disaster_Risk_Reduction_2015-2030.pdf

Young, O.R. (2002). *The Institutional Dimensions of Environmental Change; Fit, Interplay, and Scale*, The Massachusetts Institute of Technology Press, USA.

Young, O.R. (2013). *On Environmental Governance; Sustainablility, Efficiency and Equity*, Paradigm Publishers, USA.

World Bank. (2013). *Building Resilience: Integrating climate and disaster risk into development. Lessons from World Bank Group experience.* The World Bank, Washington DC

CHAPTER B4

Residents' participation in rebuilding more resilient space

Nadine Mägdefrau and Teresa Sprague

Introduction

The United Nations International Strategy for Disaster Reduction (UNISDR) has just recently reiterated the importance of participation in its Sendai framework for disaster risk reduction 2015–2030 (UNISDR, 2015). The document states clearly that the 'participation of relevant stakeholders [is] needed' to '[strengthen] disaster risk governance to manage disaster risk' (UNISDR, 2015, p 17). Unfortunately, these terms leave a wide scope for the interpretation of their meaning. This chapter attempts to give an introduction into the various understandings of participation and to explain its importance for risk governance. Within this introduction, the chapter provides an overview of the different forms of participation. For each of these forms of participation, a general example concerning disaster risk reduction is given.

After this general introduction, the chapter will explain the participation processes in Japan, Indonesia, the US, Slovakia and Germany. These case-study examples of participation can be very helpful in identifying and understanding cross-case learning points, specifically in addressing: What are the advantages or residents' participation in risk governance? What are the potential problems? And what can be done to solve these problems? The content of the cross-case synthesis presented in this chapter will attempt to answer these questions.

Background and introduction to participation

There is no common framework for public participation in the world. The UNISDR included participation into the Hyogo Framework for Action and most recently into the Sendai framework for disaster risk reduction 2015–2030, which asks for 'strengthening disaster risk governance and coordination across relevant institutions and sectors and the full and meaningful participation of relevant stakeholders at appropriate levels' (UNISDR, 2015, p 6). In accordance with the International Risk Governance Council (IRGC), risk governance is understood as:

> deal[ing] with the identification, assessment, management and communication of risks in a broad context. It includes the totality of actors, rules, conventions, processes and mechanisms and is concerned with how relevant risk information is collected, analysed and communicated, and how management decisions are taken. It applies the principles of good governance that include transparency, effectiveness and efficiency, accountability, strategic focus, sustainability, equity and fairness, respect for the rule of law and the need for the chosen solution to be politically and legally feasible as well as ethically and publicly acceptable. (IRGC, 2008, p 4)

For successful risk governance to be achieved, participation should include all relevant stakeholders (UNISDR, 2015, p 12), which, according to the Intergovernmental Panel on Climate Change (IPCC) Fourth Assessment Report, includes any person or organisation holding legitimate interest or who might be affected by a given action or policy (Baede et al, 2007, p 87). We interpret this definition within the context of this chapter to also include the general public.

The Sendai Framework appears to support this interpretation and also mentions the special importance of non-governmental stakeholders, such as civil society (especially women, children and youth, disabled people, the elderly, indigenous people, and migrants), academia, businesses and the media (see UNISDR, 2015, pp 20–1). Each one should be enabled to participate in the risk governance process. Although the importance of the participation of each group is mentioned in the document, further explanation as to how the participation process should be organised is not given.

However, what does public participation actually mean? Many definitions of public participation remain unclear. One example is the definition by the International Association for Impact Assessment (IAIA): 'the involvement of individuals and groups that are positively or negatively affected by, or that are interested in, a proposed project, program, plan or policy that is subject to a decision-making process' (André et al, 2006, p 1). Another definition, given by Renn et al (1995, p 2), summarises public participation as 'forums for exchange that are organized for the purpose of facilitating communication between government, citizens, stakeholders and interest groups, and businesses regarding a specific decision or problem'. In both cases, it remains unclear who is supposed to be included (Which stakeholders should be included? Who is affected by something?) and in which situation this involvement becomes necessary (When does a decision require inclusion?). Some intergovernmental bodies, such as the European Union (EU), also attempt to provide an understanding of participation. However, these unanswered questions apply similarly to the understanding of public involvement within EU policies. Concerning flood risk management, the targeted and most appropriate directives are the EU Water Framework Directive (Directive 2000/60/EC) and the EU Floods Directive (Directive 2007/60/EC). These directives provide three types of participation: information supply (required), consultation (required) and active involvement (required to be encouraged). However, neither document clarifies who or what constitutes the 'interested parties'[1] that should be involved, nor what is meant by the term 'active involvement'.

This chapter takes into consideration the aforementioned definitions and types of participation and provides a general working definition of participation as follows: *participation is comprised of the involvement of all relevant and potentially affected individuals, parties and/or organisations in decision-making processes towards reducing disaster risks*. This working definition necessarily includes the public as part of the relevant and potentially affected individuals. This is in light of the importance that public participation has in the creation and reduction of disaster risks. As Glass (1979) states, there are two general purposes behind the need for public participation. The first is provided from an administrative perspective, in that:

> involv[ing] citizens in planning and other governmental processes [can], as a result, increase their trust and confidence in government, making it more likely that they accept decisions and plans and will work within the system when seeking solutions to problems. (Glass, 1979, p 181)

This highlights the need to have implementable solutions that are legitimate in the eyes of affected citizens. Although public participation does not guarantee the acceptance of proposed solutions, it allows for transparency and the potential for trust building. The second purpose is seen through a citizen's perspective, whereby 'provid[ing] citizens with a voice in planning and decision making [is pursued] in order to improve plans, decisions, and service delivery' (Glass, 1979, p 181). The second perspective emphasises a potential benefit in enabling pathways towards understanding the needs of the population and ways in which plans and actions can respond to and address those needs. This draws attention to the importance of allowing pathways through which local knowledge can be incorporated into decision-making processes. This helps to establish a holistic understanding of the local-level context and, consequently, a potentially greater impact, especially at the local level.

These two purposes provide support for the call for public participation in the risk governance process. It is important to involve citizens from the administration's perspective because risk is socially constructed (Slovic, 1999) and therefore the public's view changes what needs to be defined as risk. Consequently, people should be asked about their understanding of risk to achieve a more complete definition of risk within a given context. This is of particular importance as risk is complex and often uncertain. There is inherent uncertainty in environmental and risk-related problems, particularly in that there are scientific limits to what can be known. Given these limits, 'expert' knowledge is not sufficient as the primary base for legitimate decision-making (Evans, 2012, referring to Hajer and Wagenaar, 2003, p 10). Complexity emerges from the 'subjective and contextual nature' (Slovic, 2001, p 23) of risk and how it is defined. Understanding the subjective and contextual aspects necessitates having a public component as an essential part of problem framing. This also helps enable an arena for reducing value conflicts and unknown unknowns, and therefore plays an important role in understanding the perception and acceptance of risks (Slovic, 2001, pp 18–19). Public involvement in problem framing also enables the integration of the deep contextual knowledge of local communities and assists in making decisions more locally relevant, with potential for greater impact (Evans, 2012). In addition to this, a constant dialogue with citizens will improve not only the understanding of their needs, but also the citizens' trust in a given administration (or experts), and can, consequently, improve the acceptance and potential implementation of risk management measures. This issue of trust plays a key role in risk management (Slovic, 1999), especially in terms of establishing a legitimated basis for making decisions (Evans, 2012).

Of course, it is also important to involve the public in risk governance from the citizens' point of view. The ability to define risk is an act of power as those who control this definition also control the solutions that are ultimately selected to reduce disaster risk (Slovic, 2001). Implemented solutions, and, indeed, the problems that they attempt to remediate, may have direct and/or indirect effects on the public.

Taking ethics into consideration, citizens should be able to influence the decisions that affect them (see Evans, 2012, p 192). In this case, public participation can be seen as an act of citizen empowerment. The level of this empowerment depends on the degree to which citizens can influence the decision-making process. These levels can be understood and communicated through different forms of participation.

Forms of participation

There are different forms of participation that can be distinguished according to the type of communication established between the involved parties. In this chapter, we separate the different levels of participation between one-way and two-way forms of communication. One-directional communication pathways can be considered as a passive form of participation. This refers to a one-way provision of information – either from the administration towards the stakeholders/public (eg information events) or from the stakeholders towards the administration (eg by commenting on draft versions of plans). In this form of participation, there is typically very limited to no exchange of information and ideas between different parties, nor a continuous or iterative process of dialogue. Manipulation, therapy and information (as shown in Figure 83), feature one-way processes that do not establish a high degree of involvement and are therefore considered to be below the edge of participation.

Figure 83: Levels of participation

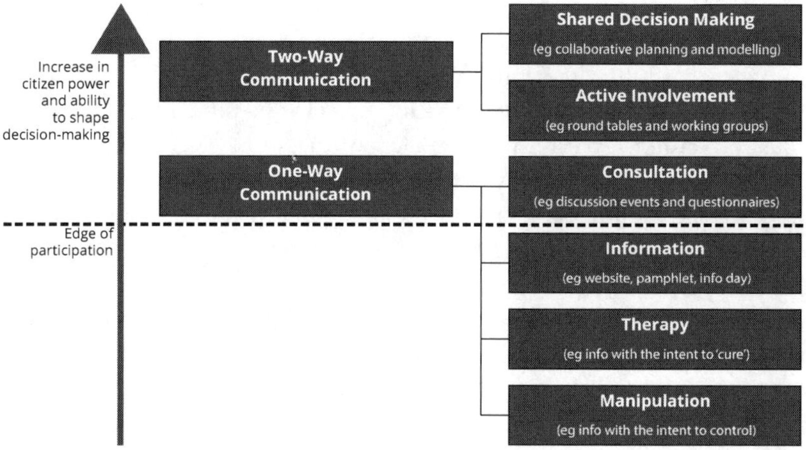

Source: Own illustration, content adapted from Evers (2012) and inspired by frameworks from Arnstein (1969) and Imra (2011)

In Sherry Arnstein's original Ladder of Citizen Participation, the eight rungs (manipulation, therapy, informing, consultation, placation, partnership, delegated power, citizen control) were aggregated into three macro-categories emphasising the increase in citizen power as one climbs the rungs (Arnstein, 1969, p 217). Figure 83 considers the original eight rungs, as well as the escalating power concept, but adapts the number of rungs and adds the edge of participation. The key difference here is the inclusion of 'information' as a level in which no participation occurs. This decision was made as, in most cases, receiving information is comprised of a very passive degree of involvement that constitutes little to no input or influence in the decision-making process.

Manipulation

According to Arnstein, manipulation is a false participation in which citizens are 'educated' and used to gain support, especially for those who already hold power in the decision-making process. This is a form of tokenism and can be seen in the use of advisory bodies of citizens created for the sole purpose of educating the public towards the will of the primary power-holders. This can occur, for example, with the creation of reconstruction committees after a disaster has struck, which exist on paper and potentially also in reality but have zero influence and decision-making power and are used to help support the actions of higher-level decision-makers.

Therapy

This level has the goal of treating citizens in a way so as to ignore any opinion that interferes with the plans of the decision-makers. Rather, it identifies citizens as a problematic group (or possible individual) that must be 'cured' (Arnstein, 1969). Attention is not paid to establishing what the key issues and interests of the potentially affected public are. Instead, the focus is on trying to fix an (often purposefully and falsely) assumed problem through having participants follow the suggested 'cure'. This is also not a form of participation as it does not have as its goal to include and involve input from citizens into the governance process, but rather attempts to keep interference from the public as low as possible. An example of this can be found in the following scenario: a landslide event occurs, causing massive structural damages and casualties in a small town. Regional authorities respond with the decision to construct massive-scale structural mitigation measures and the assumption that the population did not understand or follow instructions for evacuation. The authority does not work to mutually find solutions with the public. On the contrary, the population is distracted with a campaign on how to fix their 'assumed' problem of misunderstanding while the regional authority carries out the decision-making process and final decisions. Had an effort been taken by the authorities to work and communicate with the population, this might have revealed more important information targeting the real source of the problem, for example, a faulty early warning system.

Information

This form includes information events and the provision of information on websites or in brochures. This kind of participation forms the basis for two-way communication and is therefore very important (Evers, 2012). It is, however, only a way to give the citizens the knowledge they need, for example, about their rights, responsibilities and options, as well as details regarding the plans of the decision-makers. This level only consists of a one-way information pathway to the public and does not allow for an exchange or discussion. An example of this can be found in the emergency information provided on the websites of civil protection agencies, environmental protection agencies and municipalities.

Consultation

Consultation allows citizens not only to be informed about the risk governance process, but also to actually give their opinion and general input. This type of participation most often occurs in the form of questionnaires or public hearings. It offers the citizens an opportunity to have their voice heard. However, it is still a one-directional means of communication. It is also possible to neglect comments provided by the public in the further progress of the decision-making process. This can occur, for example, when a comment is provided by an individual or group but the comment is not considered important enough to influence the decision-making process and is therefore disregarded. Although consultation crosses the edge of participation (Evers, 2012, p 8) and provides a means for a degree of involvement, the extent to which this input has influence in the decision-making process may be limited or non-existent. Consultation can commonly be found in the case of public inquiry periods being required for the approval of environmental impact assessments, as well as for the approval of planning projects for the development of public lands. Active involvement, however, requires the potential for a higher degree of influential power as it enables two-way communication and discussion.

Active involvement

Active involvement can be found in the form of round tables and working groups that offer the opportunity for citizens and decision-makers to discuss important topics with each other. This two-way communication pathway enables the potential to directly respond to people's concerns, discuss disagreements and possibly work towards a mutual consensus. It also provides the opportunity for a higher sense of ownership in the decision-making process and outcome, which can, as mentioned earlier, provide a greater sense of trust in the process and a higher level of acceptance of the implementation of the outcome. Active involvement is, at least in the case of the EU, encouraged but not required according to the EU Water Framework Directive and the EU Flood Risk Management Directive. An example of active involvement may be found in a series of round tables set up by the local government that are open to the public in order to discuss plans for relocation and for the discussion of what local-level issues will be integrated into the relocation process. This does not amount to shared (or co-)decision-making, but it can come very close.

Shared decision-making

This level is explained well by Fiorino (1990, p 229) as 'when citizens exercise decision authority or codetermine policies in collaboration with government officials'. Borrowing from Evers (2012, p 8), this means that the decisions are taken in a way in which 'all involved persons or parties have equal rights'. This also holds true in cases where there are no purely government actors involved in the decision-making process. Shared decision-making can be seen, for example, in citizen-led committees that decide and implement changes in their own community. This can also be seen in community mapping activities in which all participants are involved in mapping their community and in discussing and deciding on the implementable measures that should take place. What is important to consider here, and what puts shared decision-making a rung above active involvement, is the equal ownership between the involved parties in the decision-making process. In this case, citizens should retain at least equal power in ruling out options for mitigation measures.

Indirect participation

Although not included in Figure 83, there is an additional form of participation that can be found in the indirect influence of family members relaying their opinions and input to be provided by their head of household in organised town hall meetings. In this example, the individual is not participating directly in the more 'official' mechanisms of participation, but they are, indeed, able to provide an input and to voice their concerns through an intermediary. This form of participation is what Chandrasekhar (2012) termed 'non-participation'. We accept this concept, but choose to use the term 'indirect participation' to communicate a clear connection to and notable degree of influence that one can have even if not directly participating in these 'official' mechanisms. This can, for example, be found in cases where strict gender roles exclude women from decision-making processes, and yet these women influence their husbands to include their needs into formal organisation processes, thereby creating a form of indirect participation through informal influence.

Examples of participation processes

Based on the understanding of participation established earlier, the following section introduces examples of participation that were presented and discussed at the Workshop on Spatial Planning Following Disasters in Sendai, 12–13 December 2013 organised by the project 'URBIPROOF – Improving the resilience of urban planning'. More specifically, the section attempts to address how participation functioned in the aftermath of disastrous events in Japan, Indonesia, the US, Slovakia and Germany. It targets what are the pertinent problems and major issues reflected in the various cases.

Japan

The current form of participation in spatial planning was introduced with the 1992 revision of the Land Use Planning Act. The development of a master or district plan includes a two-stage participation process. First, the public has to be informed about the intention of the plan in a public hearing or other information event. The second stage of participation includes the public display of the draft version of the plan and the opportunity for the public to comment on it. This type of one-way participation can be seen as information and consultation.

The participation within the process of recovery after the Great East Japan Earthquake (GEJE) corresponds with the one-way participation requirements of the spatial planning laws. However, the main type of participation during the recovery phase only belonged to the information category: many municipalities offered basic information via print media, the Internet and email. In addition, they organised either information events where the developed plan was presented – with no opportunity to discuss at all – or workshops that offered the opportunity to review the developed plans and approve the final version. Reasons for this might have been that the time right after the disaster was short and the administrative structure was damaged.

One example of how participation was neglected can be found in the Miyagi prefecture's determination of the height of the levee that is planned to protect the Tohoku coastline against level-1 tsunamis – tsunamis that occur with a frequency of between once every decade to once every hundred years. The prefectural government took this decision without considering the special needs of the different communities along the coast. The only kind of 'participation' was an information event where the designated heights for the seawall were presented. There was no kind of consultation with the public in advance and no chance to object to the plans of the government at this event (see Chapter A1c in Part A).

Of course, this procedure was frustrating for the citizens that depend on connectivity to the ocean. For instance, those involved in the fishing industry need the ability to frequently check sea conditions for their work. Another branch that depends on the view of the sea is the tourism industry, which is also a very important coastal industry. In addition to this, there was also a lack of consideration of the possible ecological impacts that might be caused by the seawall. No environmental impact assessment was considered for the creation of these plans.

In comparison to this, Iwate prefecture decided to open the discussion about the height of the levee to the public. The citizens were allowed to make their requests for a lower or (in some cases) higher than formerly planned levee and stated their reasons for their opinion. In many cases, the discussion led to a compromise and the previously appointed height of the levee was changed.

A general problem with citizen participation in Japan is that most participants are local shop-owners, particularly men over 70 or 80 years of age. Most young people are not interested in involving themselves in participation processes. One approach to improving public participation – that is, participation efforts currently already in application – is the involvement of spatial planners as mediators between the administration and the public. The function of these planners is to present and explain the draft plans to the public, promote the discussions between citizens, help them build a consensus, and facilitate citizens' needs and requirements in cooperation with the administration. A process like this can improve the participation to the level of active involvement. However, all of the two-way participation processes are beyond current legal regulations.

Indonesia

In Indonesia, citizen participation is deeply linked with decentralisation policies, which delegate decision-making from the national to the regional and local levels. The Spatial Planning Law 26/2007 emphasises the importance of citizen participation in the planning process. In addition to decentralisation, participation is connected with democratisation and the empowerment of regional and local authorities. Despite this fact, the active involvement of citizens only happens occasionally and usually has only minor influence on the results of the planning process (see Wever et al, 2012, Birkmann et al, 2014). Participation in Indonesia can therefore be described as one-way.

A broader participation of local stakeholders is still missing, especially because people do not have access to the information that they would need to participate in an informed way. In addition to this, consultants responsible for combining all important knowledge, which should also include the results of the participation process, into the planning documents fail to do so (see Birkmann et al, 2014).

In regard to the reconstruction process after the 2004 Indian Ocean tsunami, the Main Book of the master plan for the rehabilitation and reconstruction of Aceh and Nias Islands states that the recovery process should include 'community members in disaster management' and there should be a 'participatory approach to spatial structuring' (Republic of Indonesia, 2005, s V-11).

After the tsunami, the local governments developed community-based spatial plans at the village level, with a wide range of public participation. This participation process especially involved future leaders like women and younger citizens. This form of participation can be seen as active involvement (two-way participation). However, the most influential people in the village – the heads of the households that earn the primary household income – were unable to participate in most cases due to time restrictions. Additionally, not all villages set up their community plans with the participation of their citizens (see Chapter A2a in Part A).

Throughout the participation process, non-governmental organisations (NGOs) acted as mediators. This means that not only were local people's opinions included into the plans, but also the agendas and mindsets of the involved NGOs. The involvement of the NGOs also resulted in a short-term frame of the village plans because the NGOs only conducted their work in one place for a certain period of time. As a result of this time restriction, the long-term needs of the community were overlooked in many cases. This short-term perspective can also lead to problems after the NGO leaves, particularly as the village's citizens might not have enough knowledge to continue the recovery process without the help of experts.

The integration of the citizens' opinions into the village plan helps to give the decision-makers an overview of what the people actually want. The participation process, with its meetings, discussions and exchange of opinions, was also very helpful for the healing process in the communities. A problem with the participation process in Aceh was the feeling of jealousy within communities and between villages and difficulties with the implementation of the plans. The missing exchange between the different communities led to the problem that the quality of the plans differed strongly and a link between the different village plans was missing. This occurred because the development of the plans happened only from a very bottom-up direction (individual villages) in order to speed up the process (see Chapter A2a in Part A).

USA

Beginning from the 1960s, there have been different federal policy programmes in the US that required public participation. This includes, for instance, the National Environmental Policy Act 1969 (NEPA), the Resource Conservation and Recovery Act 1980 (RCRA) and the Intermodal Surface Transportation Efficiency Act 1991 (ISTEA). Their basic requirements for participation were 'timely notice, public hearings, information, the ability to comment on the information, and the ability to access officials throughout the process' (Cullingworth and Caves, 2014, p 26).

Environmental policy is mainly carried out by the individual federal states. The role of the federal government in this regard is therefore limited. The fact that the states in the US hold constitutional powers results in different regulations for participation all over the country. In consequence, the nature of public participation varies widely between the states. While some states followed the federal government's decision to require participation for their policy programmes, for example, Arizona's Growing Smarter Act 1998, others decided that the 'normal electoral process is considered sufficient' (Cullingworth and Caves, 2014, p 26). The planning process in New Orleans, Louisiana, after Hurricane Katrina is a good example to illustrate how citizens' participation can help increase a city's resilience after a disaster.

After Hurricane Katrina struck New Orleans in August 2005, planning for recovery did not go as expected. After citizens met the developed recovery plans with disapproval and the city was still stuck in chaos one year after the disaster, the decision was made to create a comprehensive plan with a wide range of public involvement. This decision resulted in the United New Orleans Plan (UNOP), a plan that was completely privately funded. The information presented here about the development of the UNOP is taken from Olshansky and Johnson (2010).

The plan included two levels of planning. First, 11 District Plans were developed for all 13 districts of New Orleans (Districts 9, 10 and 11 were combined into one plan). Nationally renowned planning experts developed each of the plans in close cooperation with the citizens of the concerned district. The citizens also helped with the decision as to which planning team would be assigned to which district. In the end, these plans were mainly comprised of the ideas of the people. The second plan combined the ideas of all 11 District Plans into one document called the Citywide Recovery Plan. Eventually, the plan was more like a strategy than an actual plan and included the most important ideas to start the recovery process for the city of New Orleans.

The planning process, which only lasted for about four months from October 2006 until January 2007, occurred in three basic phases. The first phase was a recovery assessment to understand the current conditions in the different neighbourhoods. The second included the work on recovery scenarios, and the final phase put everything together into the recovery plan, which also included a list of priorities. Throughout these phases, different forms of participation were used. The communication strategy included information via the Internet (website and newsletters), flyers and state-wide newspaper ads developed by a media team. To distribute the informational material to the people, each of the planning teams selected one member to be responsible for this job.

As concerns active involvement of citizens, planner attempted to facilitate this involvement by organising different forms of meetings. The content for the District Plans was generated in four neighbourhood meetings per district. On the city level, three Community Congresses were organised in the New Orleans Convention Center. The purpose of these meetings was to discuss topics with citizens from all neighbourhoods. In addition to this, biweekly community support meetings offered another opportunity for the public to interact with the planners and address possible concerns.

The former NGO AmericaSpeaks, whose goal it was to engage citizens in public decision-making, became engaged with the project and used its expertise to invite people to the second and third Community Congress. At the first Community Congress, only 250 to 300 people attended and only half of them were actual citizens. The composition of the group was mainly white, wealthy and from areas that were not highly affected by the disaster. Since AmericaSpeaks had just started its work on the UNOP days before the first Community Congress, they only played a minor role in this first Congress. To get more people involved in the remaining congresses, many invitations were sent out to displaced citizens and offered not only free transportation and catering, but also child care and translation services for Vietnamese and Spanish-speaking people. To simplify the opportunity to participate in the Community Congress for currently displaced people, simultaneous meetings to the one in New Orleans were held in Houston, Dallas, Atlanta and Baton Rouge. At the Community Congress, people were able to discuss their views on certain topics and to add their own options for the questions they were supposed to vote on. People were able to use a keypad voting device to cast their votes. This enabled a way for the planners to find out what the citizens actually wanted.

The keypad was also used to survey demographic data at the beginning of each Community Congress. The survey collected information about each participant's race, income, age and location in the city before the disaster. With this innovative approach, it was possible to get an overview of the composition of participants and therefore the representativeness of the collected data. This data enabled an understanding of what was needed to provide accurate representation of the city's demographic characteristics (ie in terms of attendees) for the second and third Community Congress.

The fact that the first Community Congress was not very well attended but the later ones were shows that it is also possible to still motivate people when the process is already running. The efforts of the UNOP team resulted in the participation of about 2,400 people in the second and 1,300 people in the third Community Congress (in New Orleans and elsewhere), each of them widely representing the demographic characteristics of New Orleans before the disaster.

An extensive overview of New Orleans' planning for recovery after Hurricane Katrina can be found in Robert Olshansky and Laurie Johnson's (2010) book *Clear as mud: Planning for the rebuilding of New Orleans*.

European Union

The EU started to demand citizen participation for plans and programmes relating to the environment with the release of Directive 2003/35/EC on public participation in May 2003. This directive added the necessity to additionally include citizen participation into environmentally relevant plans that were not already covered by the environmental assessment introduced by Directive 2001/42/EC in June 2001. The reason for this directive was to implement the content of the United Nations Economic Commission for Europe's (UNECE) Convention on Access to Information, Public Participation in Decision-making and Access to Justice in Environmental Matters, adopted on 25 June 1998 in Aarhus, Denmark (the Aarhus Convention). Public participation should help to expand 'the accountability and transparency of the decision-making process and [contribute] to public awareness of environmental issues and support for the decisions taken' (European Parliament and Council of the European Union, 2003, Clause 3, L 156/17). Since spatial planning also includes spatially relevant sectoral planning like environmental planning, the content of Directive 2003/35/EC was integrated into the planning laws of the member states.

The EU provides guidance and requirements regarding participation concerning flood risks. With regard to this subject, Directive 2007/60/EC on the assessment and management of flood risks is of major importance. This directive sets the framework for dealing with floods in all member states of the EU. Concerning participation, the directive demands 'active involvement of all interested parties' (European Parliament and Council of the European Union, 2007, Article 9) and therefore 'reinforces the rights of the public to access this information and to have a say in the planning process' (European Commission, 2015).

The term 'all interested parties' is rather vague but can be interpreted to refer to anyone who is interested in participating – including representatives of administrative authorities, NGOs and the general public (see Evers, 2012, p 6). According to the EU Water Directors (2003, p 3f) 'public participation … is recommended at any stage in the planning process'. The EU directive does not state how the member states are supposed to incorporate participation into their planning processes. This is left to the different member states and how they choose to implement the directive.

In Germany and Slovakia, the implementation of the directive incorporates public participation in a similar manner: both countries determined one-way participation (information and consultation) as legally mandatory parts of flood risk management.

Slovakia

In Slovakia, public participation in spatial planning is referred to in the Environmental Impact Assessment Act 1994, which also cross-references the Municipality Act, the Land Use Planning Act, the Building Code and the Environment Act (see Belcáková, 2004). This means that formal participation includes a public hearing in which the citizens are informed about the aims of the plans. During the planning process, the public and other affected stakeholders are allowed to hand in their comments, and planners have to consider them. If a comment is not integrated into the further planning process, the planners have to discuss it with the commenting person and include the comment and the explanation in a written review. The information about the plans and their preparation often happens via Internet pages. Relevant authorities (eg Slovak Water Management Enterprise, Slovak Hydro Meteorological Institute) are integrated into an expert group, which is consulted during the planning process (see Chapter A4a in Part A).

Public participation concerning flood risk protection is similar to the participation in spatial planning: all flood risk management plans are open for written comments that have to be considered in the further planning process. The procedure is based on the EU flood directive outlined earlier. This type of participation mainly consists of information and consultation and is therefore one-way participation.

Germany

In Germany, formal participation within the process of spatial planning has its legal basis in the Federal Spatial Planning Act (Raumordnungsgesetz) and the Federal Building Code (Baugesetzbuch). Public participation has been integrated into section 10 of the Federal Spatial Planning Act and includes the right of the public and affected public bodies to be informed about the establishment of a federal spatial structure plan. They must also be granted the opportunity to give their opinion on the draft version of the plan. All relevant information must be accessible to the public for at least one month. Section 18 of the Federal Spatial Planning Act grants the same rights for the establishment of the spatial structure plans of the Federation. The goal of this participation process is to determine material for the weighting process that forms the basis for every spatial planning decision in Germany (see §7(2) Raumordnungsgesetz) (see Chapter A5a in Part A).

In urban land-use planning, formal public participation is included in section 3 of the Federal Building Code. It constitutes the necessity to inform the public about the aims of the plan and its possible impacts as early as possible. The public must also be granted the chance to comment in this early stage (see §3(1) Baugesetzbuch). Draft versions of urban land-use plans must be publicly displayed for one month. All comments on the plan must be checked. In case a comment is not integrated into the plan, the municipality must explain the reason for its decision in the final version of the plan (see §3(2) Baugesetzbuch). Experts can help as mediators during the planning process, but this only occurs upon request.

Aside from these formal participation procedures, there are also some informal procedures that offer the public a higher level of participation. However, because this kind of participation is not mandatory and takes more time and money, it is not as common as one might wish. Be that as it may, with the public's urge to be heard, there is currently an ongoing debate about the need for additional public participation in spatial planning (see Deutscher Städtetag, 2013; Akademie für Raumforschung und Landesplanung, 2014).

The participation process for flood risk protection in Germany is similar to the one in Slovakia: it consists mainly of informing the public about the purposes and aims of the plans and offers the opportunity to comment. All submitted comments must be considered throughout the weighting process that forms the basis for the final decision. Planning for flood risk protection does not usually include any further, more extensive participation – such as active involvement and shared decision-making. Of course, there are also some rare examples where this takes place, but the main type of participation is primarily one-way, consisting of information and consultation. This is based on the legal regulations that demand very little involvement of the public.

Cross-case-synthesis

The general forms of participation in the five countries that are presented in this chapter are located on the 'edge of participation' (see Figure 83). In each of the countries, forms of participation that are found to be legally binding require the provision of sufficient informational material and the opportunity for the public to state their opinions and possible concerns in the form of comments. Any two-way form of participation (active involvement or shared decision-making) is optional and can only be found in examples where planners and citizens decide to take their own initiative to involve themselves beyond legal requirements.

It is also important to note that participation seems to function better if performed on a smaller scale in more decentralised states. Nevertheless, in these cases, countries also tend to have more difficulties with the actual implementation of the completed plans (eg Indonesia and the US). One reason for this might be that receiving funding is more difficult in such decentralised states.

The presented cases show that participation processes can bring many advantages. The most immediate advantages, like the opportunity for generating new ideas and gaining the people's support for the plan, might be more or less obvious. The creation of a vision for the city that truly represents the people's needs can only be reached if people are empowered in a truly meaningful involvement process. In some cases, participation in recovery and the meeting and exchange with other community members can also support the healing process of individuals or the entire community.

However, the participation of citizens also brings many problems, which are illustrated in the previously presented participation examples. Very general problems include the fact that participation is often expensive and takes an extensive amount of time (often so in the case that it is well done). Additionally, participation cannot guarantee desired results, as opposed to what one can expect when hiring spatial planning experts to develop a plan that reflects the government's expectations. Furthermore, some of the results envisioned by the participants might be unrealistic, which can lead to the disappointment of the participants themselves. Since different neighbourhoods in a city and different people in a neighbourhood have varying interests and views, it is impossible to satisfy everyone who participates in the planning process. In addition to this, participation usually encourages the continued participation of the same people (in many cases, people who are older and male). The aforementioned advantage that participation can help with the healing of a community also has an opposing view that should also be mentioned: after a disaster, many people have lost loved ones and all of their assets. They are traumatised, displaced and trying to get their life back together. To request direct involvement from people in this situation can feel like an extra burden on them that they cannot fulfil. The involvement of NGOs that come to help with the recovery process can also add problems if they are involved in participation processes (which they often tend to be). This can occur, for instance, if the NGOs follow their own agenda and mindset instead of working together for an overarching goal, for example, the goals and priorities set by the government. Another problem can occur after the NGOs leave as the citizens might be beyond their technical capacities in efforts for recovery and are therefore unable to keep up with the work that the NGO started. An overview of the advantages and disadvantages of resident's participation is given in Table 14. How can these problems be solved? Experiences from Japan, Indonesia, the US, Slovakia and Germany can offer some ideas.

Table 14: Advantages and disadvantages of residents' participation

Advantages of residents' participation	Disadvantages of residents' participation
+ Getting new ideas + Getting the people's support for the plan + Empowering people can create a vision for the city that represents the people + Participation can support a community's healing process + Acceptance can enable smoother implementation	– It is expensive and time consuming – It does not offer guaranteed results – Set goals or visions might be unrealistic – It is impossible to reach an overall agreement – people can be disappointed – Usually the same people are encouraged to participate – After a disaster, people are traumatised, participation can be another burden for them – NGOs follow their own agenda and mindset – When NGOs leave, communities might have problems to continue the work they started

The uncertainty of results

Since the results of the participation process depend on the people involved and their interests and needs, it is impossible to ensure guaranteed results from a participation process. Nevertheless, since the main goal of participation is often to achieve a plan that combines the needs of all citizens, there should be a certain amount of time and money spent to reach a demographically representative group of participants. As the case of the UNOP shows, this might take a lot of encouragement, dedication and funds. However, if efforts for achieving and encouraging participation are made, they might as well be made in a way that helps achieve the greatest possible opportunity for successful implementation. Of course, the amount of time and money available to invest in involving a wide variety (or representative group) of people depends on the individual case and it is important to keep in mind that it is never possible to reach everyone. To still reach as many people as possible within the available funding and other resources, it is a good idea to develop a professional communication strategy and to adjust this throughout the process as necessary. In this way, the results of the participation process have a higher potential to be useful for the continued and future recovery process.

The impossibility of an overall agreement

When many people with different needs and interests try to come up with one overall solution, there will always be disagreements. It is impossible to solve this problem. However, what can be done is to increase the people's understanding by keeping the planning process as transparent as possible. It can also help to confront the citizens directly with the fact that it is impossible to implement all of the collected ideas. It is also of great importance to ask them for a list of priorities. In this way, it might be possible to at least implement the most important projects for each neighbourhood and to attempt to encourage public acceptance and reduce disappointment.

Participation of the same (unrepresentative) group of people

As the case of UNOP shows, a tailor-made communication strategy can help to reach a demographically representative group of participants. In order to reach as many different people as possible, it is helpful to know the composition of the population before the disaster and to find the best ways to communicate with them. The involvement of communication experts and the collection of demographic data can help to solve this problem.

Healing versus burden

Only the citizens themselves are able to answer the question of whether participation in the planning process for their future hometown can help them with healing from the disaster or not. Where one person might be traumatised and overwhelmed, another person might see the process as a chance to improve his or her future life. Therefore, every affected citizen should be informed about the content of the planning process in advance. Due to the displacement of many people after a disaster, it might be hard to reach each person. Still, the effort should be made. This can be attempted, for example, through visiting temporary housing units or other areas where the people are currently living, through information meetings in different locations, and, of course, through information via the usual communication channels like newspapers, leaflets or the Internet (especially social networks). When informing people about their ability to participate in the reconstruction process, planners should also consider a community's history of participation. Communities that have always been overlooked in the past can react differently when addressed with an offer to involve them in the planning process, as opposed to communities with a long history of active participation.

Involvement of non-governmental organisations

An onsite governmental representative or coordination station should direct and oversee the assistance provided by all NGOs and other stakeholders right from the beginning of the recovery. It is important to have an overview of the NGOs and other stakeholders who are currently working on recovery, what resources (especially expertise) they are supplying, and what actions they can take. In the best case, the delegate would be able to tell the arriving NGOs and other stakeholders what is needed where and to ask them to help in certain areas based on a plan from the government.

Conclusion

What can be learned from these examples? Including the public through a higher level of participation provides a highly beneficial opportunity as it can enable the combination of the citizens' views into a visionary plan – like the UNOP did for New Orleans. Of course, not every planning process can incorporate participation in a way that demands a huge amount of money and many hours of work by professionals as well as citizens. However, what level of participation is right? Which stakeholders should be included? As learned from this chapter, there is no universally applicable answer to these questions. However, one should note that the requirements for participation depend on different factors, such as:

- The scale of the disaster should match the extent of public participation. For example, when a disaster exceeds a certain size – like Hurricane Katrina – a large-scale and in-depth participation process is more appropriate.
- The complexity of the problem to be addressed should also be taken into consideration. If a problem is simple, with limited impact, then

it often does not demand the integration of knowledge from many stakeholders.

- Complex problems, which are more common in disaster risk, typically require the involvement of a wide variety of (directly or indirectly) affected persons and responsible authorities. This need increases when faced with a multitude of different hazards whose effects are overlapping (eg the triple disaster of earthquake, tsunami and nuclear accident in Japan's Tohoku Region in 2011).

To initiate the recovery process after a disaster, it is advisable to start with a damage assessment and put together a funding plan before moving forward to the development of a management strategy. This strategy should include information about everything that has to be planned, including an agenda on how to incorporate the citizens into the process. Based on this management strategy, tasks can then be distributed to different people.

However, no matter how large the scale of a participation process might be, there are always obstacles it cannot overcome. It is not possible to reach a consensus with everyone. Planners need to know and accept this fact. They also need to communicate the limits of participation to the people. Only with this transparency is it possible to keep the trust of the people and lead the participation process to successfully achieve a common goal.

Note

[1] The term 'interested parties' is not further specified in the EU Floods Directive, but can be understood as everyone who is interested and willing to participate. In other words, this can be interpreted to mean that everyone should be allowed to participate (see also the later section 'European Union').

References

Akademie für Raumforschung und Landesplanung (Ed.) (2014), *Raumordnungsverfahren – Chance für eine frühzeitige* Öffentlichkeitsbeteiligung *bei der Planung von Infrastrukturprojekten,* Positionspapier aus der ARL 99, Hannover, available at: http://nbn-resolving.de/urn:nbn:de:0156-00998 (accessed 29 November 2015).

André, P., Enserink, B. and Connor, D. (2006), *Public Participation International Best Practice Principles, Special Publication Series*, Fargo, USA.

Arnstein, S.R. (1969), "A ladder of citizen participation", *Journal of the American Planning Association*, Vol. 35, No. 4, pp. 216–224.

Baede, A.P.M., van der Linden, P. and Verbruggen, A. (2007), *Annex II. Glossary of Synthesis Report: A Contribution of Working Groups I, II and III to the Fourth Assessment Report of the Intergovernmental Panel on Climate Change, Geneva*, available at: http://www.ipcc.ch/pdf/assessment-report/ar4/syr/ar4_syr_appendix.pdf (accessed 29 November 2015).

Baugesetzbuch. *Version of 23 September 2004 (BGBl. I S. 2414). Last amended by Article 6 of the law on 20 October 2015* (BGBl. I S. 1722): BauGB.

Belcáková, I. (2004): *Strategic Environmental Assessment and spatial planning in Slovakia: Current practices and lessons for practical application of the EC SEA Directive*, available at: http://archive.rec.org/REC/Programs/EnvironmentalAssessment/pdf/seminar2004/Ingrid%20Belchakova_bp.pdf (accessed 29 November 2015).

Birkmann, J.; Garschagen, M. and Setiadi, N. (2014): "New challenges for adaptive urban governance in highly dynamic environments: Revisiting planning Systems and tools for adaptive and strategic planning", *Urban Climate*, Vol. 7, pp. 115-133.

Chandrasekhar, D. (2012), "Digging deeper. Participation and non-participation in post-disaster community recovery", *Community Development*, Vol. 43, No. 5, pp. 614–629.

Cullingworth, B. and Caves, R.W. (2014), *Planning in the USA: Policies, issues, and processes,* 4. ed., Routledge, Abingdon.

Deutscher Städtetag (2013), *Beteiligungskultur in der integrierten Stadtentwicklung*, Arbeitspapier der Arbeitsgruppe Bürgerbeteiligung des Deutschen Städtetages, Deutscher Städtetag, Berlin, Köln, available at: http://www.staedtetag.de/imperia/md/content/dst/veroeffentlichungen/mat/mat_beteiligungskultur_2013_web.pdf (accessed 29 November 2015).

European Commission (2015), *The EU Floods Directive*, available at: http://ec.europa.eu/environment/water/flood_risk/ (accessed 29 November 2015).

European Parliament and Council of the European Union (2003), *Directive 2003/35/EC of the European Parliament and of the Council of 26 May 2003 providing for public participation in respect of the drawing up of certain plans and programmes relating to the environment and amending with regard to public participation and access to justice Council Directives 85/337/EEC and 96/61/EC: Directive 2003/35/EC.*

European Parliament and Council of the European Union (2007), *Directive 2007/60/EC of the European Parliament and of the Council of 23 October 2007 on the assessment and management of flood risks: Directive 2007/60/EC.*

EU Water Directors (2003), *Guidance document on Public Participation: Final version after the Water Directors' meeting, 2 December 2002,* available at: www.eau2015-rhin-meuse.fr/fr/ressources/documents/guide_participation-public.pdf (accessed 29 November 2015).

Evans, J.P. (2012), *Environmental Governance*, Routledge Introductions to Environment Series.

Evers, M. (2012), *Participation in Flood Risk Management: An introduction and recommendations for implementation, Rapportserie Klimat och säkerhet*, Karlstads Universitet, Sweden.

Fiorino, D.J. (1990), 'Citizen Participation and Environmental Risk: A Survey of Institutional Mechanisms', *Science, Technology & Human Values*, Vol. 15, No. 2, pp. 226–243.

Glass, J.J. (1979), 'Citizen participation in planning: the relationship between objectives and techniques', *Journal of the American Planning Association*, Vol. 45, No. 2, pp. 180–189.

Hajer, M.A. and Wagenaar, H. (Eds.) (2003), *Deliberative policy analysis: Understanding governance in the network society, Theories of institutional design*, Cambridge University Press, Cambridge, UK, New York, USA.

IMRA (Integrative flood risk governance approach for improvement of risk awareness and increased public participation) (2012), *Planning and implementing communication and public participation processes in flood risk management: Procedural guidelines and toolbox of methods.*

IRGC (2008), *An introduction to the IRGC Risk Governance Framework*, Geneva.

Olshansky, R.B. and Johnson, L.A. (2010), *Clear as mud: Planning for the rebuilding of New Orleans*, American Planning Association, Chicago, Washington, D.C.

Raumordnungsgesetz. *Version of 22 December 2008 (BGBl. I S. 2986). Last amended by article 124 of the enactment on 31 August 2015 (BGBl. I S. 1474): ROG.*

Renn, O., Webler, T. and Wiedemann, P.M. (Eds.) (1995), *Fairness and competence in citizen participation: Evaluating models for environmental discourse,* Technology, risk, and society, Vol. 10, Kluwer Academic, Dordrecht.

Republic of Indonesia (2005), *Attachment 1 Regulation of the President of Republic of Indonesia Number 30 Year 2005 on Master Plan for Rehabilitation and Reconstruction for the Regions and People of the Province of Nanggroe Aceh Darussalam and Nias Islands of the Province of North Sumatra*, Main Book of Rehabilitation and Reconstruction.

Slovic, P. (1999), 'Trust, Emotion, Sex, Politics, and Science: Surveying the Risk-Assessment Battlefield', *Risk Analysis*, Vol. 19, No. 4, pp. 689–701.

Slovic, P. (2001), 'The risk game', *Journal of Hazardous Materials*, Vol. 86, No. 1-3, pp. 17–24.

UNISDR (2015), *Sendai Framework for Disaster Risk Reduction 2015–2030*, Geneva.

Wever, L., Glaser, M., Gorris, P. and Ferrol-Schulte, D. (2012), 'Decentralization and participation in integrated coastal management: Policy lessons from Brazil and Indonesia', *Ocean & Coastal Management*, Vol. 66, pp. 63-72.

CHAPTER B5

Spatial planning and dealing with uncertainties associated with future disasters

Stefan Greiving

Introduction

The reduction of disaster risk from multiple hazard sources is an explicitly pronounced aim in several international agendas, for example, in the Agenda 21 (UN, 1992), the Johannesburg Plan (adopted at the 2002 World Summit on Sustainable Development) and the *Hyogo framework for action* (UN-ISDR, 2005). Strategies and actions to 'control, reduce and transfer risks' on the basis of risk assessments and analyses can be subsumed under the term 'risk management' (UN-ISDR, 2009).

Priority 2 of the new Sendai Framework for Disaster Risk Reduction points at 'Strengthening disaster risk governance to manage disaster risk' and argues that 'Clear vision, plans, competence, guidance and coordination within and across sectors as well as participation of relevant stakeholders are needed'. At the same time, the framework makes clear that a sound evidence basis is needed in order to ground decisions on risk management (Priority 1: 'Understanding Risk'). Here, the connection between disaster risk and sustainable development becomes clear:

> The development, strengthening and implementation of relevant policies, plans, practices and mechanisms need to aim at coherence, as appropriate, across sustainable development and growth, food security, health and safety, climate change and variability, environmental management and disaster risk reduction agendas. Disaster risk reduction is essential to achieve sustainable development. (UN-ISDR 2015, p 13)

Decisions in the area of so-called 'traditional' hazards like floods or earthquakes are normally based on expert analysis, often combined with results from modelling analysis. Hereby, the calculation of the spatiotemporal probabilities of the natural hazards on the basis of recent field monitoring, but also available historical information, is crucial. However, analysed data are only available for a specific period – and are thus not representative for longer periods. This problem is enhanced when using historical data. These do, indeed, add valuable information, in particular, for the frequency and magnitude analysis of the investigated processes. However, it has to be assumed that historical data are always incomplete information covering, in particular, large-scale events, but not events with smaller magnitudes. Nevertheless, even having the incompleteness of records in mind, the added value is still evident. This principal problem is exacerbated by the observed climate change-related effects on temperature and precipitation, which will certainly lead to new uncertainties about hydro-meteorological hazards, because past events might not be representative anymore. Similarly, other changes in the catchments (eg deforestation, melting of glaciers, surface sealing through settlement development, surface modification by infrastructure, etc) will enlarge these uncertainties. Here, the perspective changes from probabilities to just possibilities (Greiving and Glade, 2013). The role of science in this context is problematic because science cannot provide proof of risk or guarantee safety. In this context, science is inconclusive (Van Asselt, 2005). Many risks, which require societal choices and decisions, are adequately characterised as complex, uncertain and/or ambiguous, but are treated, assessed and managed as if they were simple (Renn et al, 2011).

Uncertainty characterises most assessment, policy and management processes, which have unpredictable consequences. In broad terms, uncertainty can be defined as being any deviation from the unachievable ideal of completely deterministic knowledge of a relevant system (Walker et al, 2003). Uncertainty may have different types of sources, from quantifiable errors in the data to ambiguously defined terminology or uncertain projections of human behaviour (MA, 2003). Uncertainties have to be transparently assessed, honestly reported and effectively communicated (Hill et al, 2013).

Thus, even if it is possible to predict a changing risk with increasing confidence due to advances in modelling technologies and risk assessment methodologies, the prediction of social responses to this changing risk will always remain out of reach. In this context, a robust approach has to put in place tools that can enable processes of institutional learning (Berkhout et al, 2002). Current policies must deal with uncertainty, but they tend to have a lack of flexibility for adaptation to a changing environment (Dessai and Van de Sluijs, 2007; Walker et al, 2013). This means that the future baseline trend is evolving due to climate change and globalisation – mostly independent from decisions taken by local or regional actors (see also EEA, 2013, p 29). In addition to climate scenarios, it is important to consider socio-economic scenarios as this will help assess future vulnerability to climate change. Here, the baseline (given status of the environment) is changing. Within the European Union (EU), the framework directive for Environmental Impact Assessment was recently amended in order to reflect these new uncertainties (EC, 2014).

Within the global change debate, the field of climate change in general, but particularly as a triggering factor for many natural hazards, is of special importance for countries like Germany or Japan, with their existing settlement structures, cultural landscapes and infrastructures that have been developed over centuries. Under these circumstances, mitigation actions, for example, carried out by spatial planning, are less effective than in countries that are still growing rapidly in terms of population and the built environment. Here, disaster-prone areas can be kept free from further development, whereas most of these areas are already built-up in the developed world. However, this calls for authorities to improve public risk awareness and to look for means to mitigate this problem. Moreover, measures based on the mandatory decisions of public administrations, as well as measures that are the responsibility of private owners, need to be understood and regarded as suitable by their addressees for their implementability. This is clearly visible when looking at evacuation orders or building protection measures to be taken by private households, and it calls for a governance perspective.

To conclude, with public decision-making not having any precise information at hand, restrictions on private property rights that are often required when establishing hazard zones and/or protection systems are probably not legally justifiable under uncertainty. Hereby, the justification of actions and consensus about thresholds for acceptable risks and response actions become more important (Greiving and Glade, 2013).

As opposed to the previous contributions of Part B, this chapter refers only to a certain extent to the different country reports presented in Part A. This is justified by the cross-analysis of Part A: it identified the given lack of reflexive strategies in the examined countries that incorporate uncertainty.

The risk governance perspective

The concept of risk governance has been created and evolved in the area of newly emerging, mostly man-made, risks. Nonetheless, it is also of particular relevance for natural hazards, as recently acknowledged by Priority 2 of the Sendai Framework of Action (UN-ISDR, 2015). Actually, the successful management of natural hazards is limited due to the fact that the interactions between individual sectors, disciplines, locations, levels of decision-making and cultures are not known or considered (IRGC, 2005; Greiving et al, 2006).

This policy statement is in the line with the findings of research on the institutional dimensions of environmental change. Young (2002, 2010) referred to the so-called 'problem of interplay': most institutions interact with other similar arrangements, both horizontally and vertically. Horizontal interactions occur at the same level of social organisation. Vertical interplay is a result of cross-scale interactions or links, involving institutions located at different levels of social organisation. Interplay between or among institutions may take the form of functional interdependencies or arise as a consequence of a politics of institutional design and management. The problem of interplay is a consequence of the existence of a multitude of actors.

Linking the relevant actors and policies throughout the disaster management cycle, but also creating an inventory of information on disasters, was propagated as a key objective by the Communication of the European Commission *A Community approach on the prevention of natural and man-made disasters* (EC, 2009). Here, it becomes clear that available knowledge on disasters is currently limited and suffers from a lack of comparability.

Furthermore, the current prevention of risk caused by natural hazards is fragmented, among others, between civil protection and spatial planning (see, eg, Greiving et al, 2006; Sapountzaki et al, 2011). Inadequate publicly available information about risks in terms of societal and natural dimensions, inapprehensible procedural steps, and insufficient involvement of the public in the risk-related decision-making process lead to severe criticism and distrust (Greiving et al, 2012).

Risk governance has become increasingly politicised and contentious. The main reasons are controversies concerning risk that are not about suitable scientific methodologies for hazard and risk assessment (Armaş and Avram, 2009). Rather, risk controversies are disputes about who will define risk in view of existing ambiguity. In many cases, policy discourse is not about who is correct about the assessment of danger, but whose assumptions about political, social and economic conditions, as well as natural or technological forces, win in the risk assessment debate. Thus, the hazard as a potentially damaging physical event is real, but risk is socially constructed.

Scientific literacy and public education are important but not the only aspects necessary to avoid conflicts about risk. Emotional responses by stakeholders to issues of risk are truly influenced by distrust in public risk assessment, as well as in risk management. Due to this fact, those who manage and communicate risks to the public need to understand the emotional responses towards risk and the way risk is perceived by the at-risk population. It is a matter of the definition of risk as to how risk policy is carried out. Moreover, defining risk is an expression of power, Slovic (1999) thereby argues that whoever controls the definition of risk controls risk policy. Within the communication strategies in all approaches, trust, transparency, clarity and confidentiality can be seen as central terms in this respect (Löfstedt, 2005; Greiving, 2009; IRGC, 2009).

To a certain extent, the given distrust is also caused by a lack of communication between science and society. This gap between scientific output and policy demand has often been described (see, eg, Weichselgartner and Kasperson, 2010). In order to solve it, the following major challenges need to be addressed:

- The reduction of data and information, as well as the aggregation of intermediate results, in a scientifically sound and transparent way.
- The communication of the complexity of underlying data sets and the uncertainty associated with the applied methodologies and the results.
- The fulfilment of user requests to provide results from a rather systemic point of view, that is, to take into consideration cross-sectoral interlinkages, the multiplicity of threats or adverse impacts, and inner-systemic feedback loops.
- Agreement on the numerous normative decisions that need to be made for the selection of methodologies, the prioritisation of threats, potential impacts or sectors, the analyses of intermediate and final results, and so on. Due to a lack of existing legally based norms and thresholds, the justification of actions and consensus about thresholds and priorities can be understood as an appropriate way of dealing with uncertainty.
- The difference in future periods of relevance – that is, the usually long-term perspective of natural science (observation periods from hundreds to thousands of years) – needs to be brought in line with

the common time horizon of political decisions, but also with actions to be taken by spatial planning (about 15 years maximum).

Thereby, further attention is paid to this science–policy gap when discussing soft governance strategies (see following section).

Strategies for dealing with uncertainty

Reflexive adaptation

In principle, adaptation strategies aim to reduce vulnerability and increase resilience, but there are two different perspectives on adaptation: the natural sciences and the social sciences (Birkmann et al, 2012, p 295). In natural science, adaptation is often viewed as the development of genetic or behavioural characteristics that enable organisms or systems to cope with environmental changes in order to reproduce and survive. As opposed to this understanding, the social sciences concept often links adaptation with cultural practices allowing societies to survive in the light of a changing environment (Smith and Wandel, 2006, p 283).

In order to manage deep uncertainty, societies must transform from taking a 'predict-and-control' type of approach to more reflexive adaptation practices. 'Deep uncertainty', according to Walker et al (2013), is defined as a level of uncertainty in which there is no option for coming up with calculable numbers (eg for cost–benefit analysis), one cannot differentiate scenarios based on probability and all scenarios are equally possible. Understanding how such a transformation could take place calls for a profound understanding of the present context, the exploration and development of alternative institutions within governance systems, and strategy-making to map out which changes need to be made by which actors.

Furthermore, attention must be paid to the similarities and differences in the local-level cultural and legal administrative systems worldwide, which requires establishing an understanding of governance contexts. This particularly pertains to formal institutions within this context in order to determine to which extent strategies can be applied in specific settings as well as in joint approaches (see Chapter A6).

An important dimension of adaptive capacity in the context of uncertainty is reflexivity, that is, actors being able to respond to unforeseen events spontaneously and effectively. Such reflexive adaptation requires enabling institutional settings. It calls for new approaches to the formal institutional frameworks in governance systems that can enhance their flexibility to adapt to deeply uncertain and dynamic conditions. A key factor here is the extent to which formal institutions support and incentivise the reflexivity of actors and stakeholders through individual and collective social learning (Mostert et al, 2007). Building capacity for reflexive adaptation by creating a dense social context is a critical part of enabling social learning and developing effective responses to climate change (Pelling et al, 2008).

While conventional adaptation often prioritises stabilising these existing frameworks, transformative adaptation requires fundamental changes within governance systems to make them 'change-proof'. Reflexive and transformative adaptation calls for new approaches in the formal institutional frameworks in governance systems that can enhance their flexibility to adapt to deeply uncertain and dynamic conditions. What does this mean, in particular, for spatial planning, being responsible for managing the land and related land uses?

Hallegatte (2009) identified the following five different strategies for dealing with uncertainty, which will be described and elaborated in more detail in the following sections of this chapter:

- No-regret strategies.
- Reversible strategies.
- Safety margin strategies.
- Soft strategies.
- Strategies that reduce decision-making time horizons.

No-regret strategies

In many cases, risk management options are only justifiable under the precondition that the (probabilistic) risk (eg in terms of annual losses with and without the realisation of the management option) is correctly determined. Investments in structural protection systems normally need to be checked against their cost–benefit ratio. Due to the change of climatic and socio-economic framework conditions, this calculation might be wrong.

In other cases, risk management is based on a deterministic approach, which means that the magnitude of an event is just calculated for a predefined return period. On this basis, hazard zones are demarked and may lead to land-use restrictions laid down in spatial plans. Due to an incomplete knowledge about past events or the influence of a changing climate, the extent of the hazard zone might be wrongly determined. In consequence, the need for land-use restrictions might be potentially over- or underestimated by spatial planning.

Both scenarios fundamentally question the economic and legal justifiability of risk management actions. Here, no-regret strategies come into play. Such strategies promise a net benefit irrespective of whether a particular change takes place or not (ie projected climate change) and are consequently, in most cases, acceptable for the current population and not only the future generation, who are potentially negatively affected by changing risks. These no-regret strategies call for risk management options that can be multifunctionally reasoned, for example:

- Urban greening concepts are effective in terms of mitigating urban heat in densely populated city centres and can also be used for storing storm water temporarily in order to unburden sewage systems, but they can also enhance the living quality in a city quarter if they are designed, for example, as parks.
- A decentralised retention of water in the catchment mitigates river flooding but supports the natural water cycle and makes water for agricultural purposes available over the year.

Both examples are addressed by the comprehensive concept 'Rotterdam – Watercity 2035', as illustrated in Figure 84.

Figure 84: Watercity Rotterdam

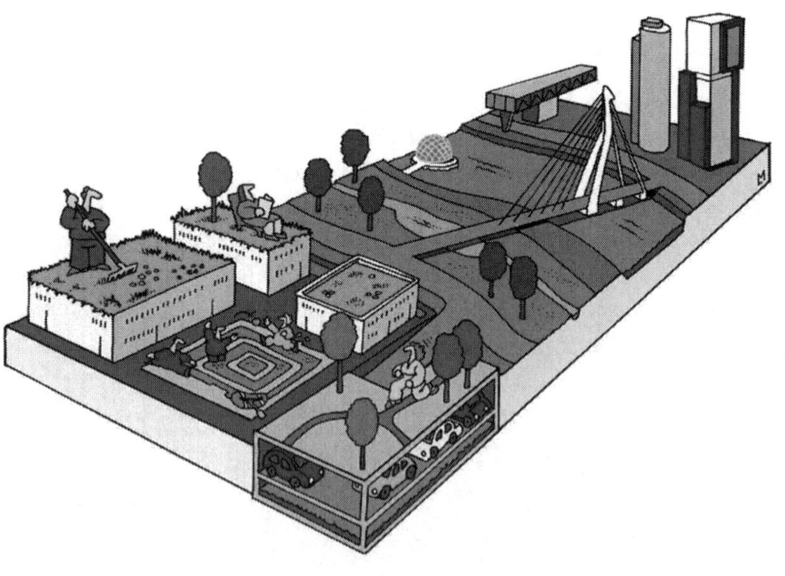

Source: www.rotterdamclimateinitiative.nl

Vertical evacuation concepts have been implemented in various tsunami-prone countries like the US or Japan (Applied Technology Council, 2012). Public cyclone shelters in coastal areas in Bangladesh are constructed within cyclone high-risk areas that are, as paddy fields, economically speaking of high importance for the rural population (Paul, 2011, p 178). These shelters are ordinarily used as mosques, schools or government offices, and promise therefore a net benefit irrespective of whether a severe cyclone takes place or not.

Reversible strategies

Reversible strategies avoid irreversible decisions that cannot be revised after implementation. Here, backcasting comes into play. Backcasting can enable reflection on the challenges presented by the disparities between the 'ideal' reflexive adaptation strategy scenario and that of maintaining a business-as-usual approach. This backcasting allows the consideration of uncertainty and flexibility and serves as an evidence basis for avoiding irreversible, potentially fatal, decisions that run counter to long-term goals.

In this view, scenario development would be a vehicle for social learning, exploring institutional constraints and developing a strategy towards the transformation of the current system. Two scenarios have to be identified (see Figure 85): (1) future impacts if business as usual is maintained (no transformative pathway); and (2) future impacts in a robust adaptation scenario (pursuit of transformative pathway).

These scenarios are determined through a backcasting, or a normative scenario building, approach, with stakeholders providing a focus on

Figure 85: Backcasting

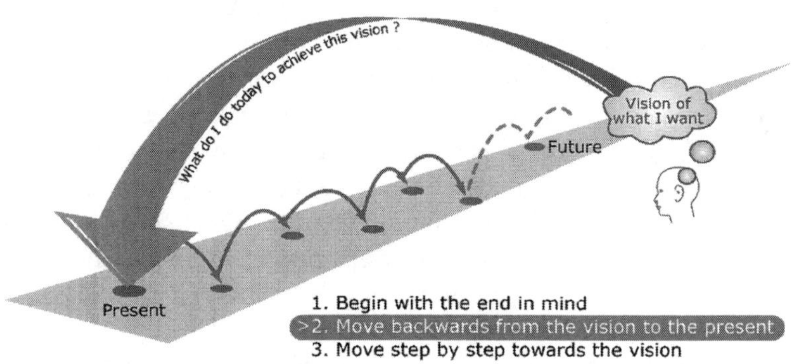

Source: Natural Step (1993)

understanding social structures as opposed to a physical-technical focus more commonly used in backcasting scenario development (Wangel et al, 2013). Existing capacities will have to be compared to that which is needed to achieve the robust scenario. When mismatches occur, a challenge is further identified.

Overcoming these challenges will require solutions that encourage a transformation of the current societal status quo (Robinson, 2003). This requires a strong participatory approach, acknowledging that it provides 'the narrative resources for scenario elaboration (policy learning); while also generating critical self-reflection and preparing the conditions for change (organizational learning). It is in this joint sense that scenario tools can be regarded as "learning machines"' (Berkhout et al, 2002). The Inter-governmental Panel on Climate Change (IPCC) pointed out that such strategies are suitable for disaster risk management: 'Participatory backcasting, which involves local stakeholders in visionary activities related to sustainable development, can also open deliberative opportunities and inclusiveness in decision framing and making' (Field et al, 2012).

Developing strategic planning based on backcasting was recently used in Mozambique in the process of decision-making for flood risk and knowledge management strategies (Mondlane et al, 2013). Another study was undertaken in Austria (Haslauer, 2015) to demonstrate its ability to support sustainable spatial development planning. The model was split into two parts: a model to generate a future scenario and a backwards-running model.

Safety margin strategies

The precautionary principle is of particular importance when a risk management strategy is based on deterministic approaches. This principle is an overarching framework that governs the use of foresight in situations characterised by uncertainty, where there are potentially large costs to both regulatory action and inaction (EEA, 2001). Here, safety margins come into play. Defence systems, as well as hazard zones, are designed for a particular frequency/magnitude function (eg the 1:100 flood). Climate change may lead to an increase in run-off, which means that a flood-prone area for the 1:100 event might have to be enlarged, or in the case of a defined flood zone, the related return period has to be decreased. Due to the inherent uncertainty in climatic models, the change is not exactly calculable.

Under this precondition, a safety margin strategy is based on a worst-case scenario in order to avoid an increase of the risk caused by a changed hazard. Therefore, the vulnerability has to be the key focus: two basic options are possible:

1. an enhancement of existing structural defence systems (eg a seawall, levee or the capacity of a sewage system); or
2. an extension of a hazard zone in order to keep a (potentially enlarged) hazard zone still free of further development.

Such safety margin approaches have already been used in various countries of the developed world (eg The Netherlands, Denmark or Germany) to manage future risks of sea-level rises and for coastal defences and water drainage management (Field et al, 2012, p 354). Moreover, the safety margin concept has been implemented in Japan in order to establish a structural defence system against the so-called 'Level 1' (return period 1:100 years) and 'Level 2' tsunamis (extreme event like in 2011) for predefined return periods (see Chapter A1c in Part A). An interesting example of such a strategy, the so-called 'Klimadeich' (climate-proof coastal defence levee), which was implemented in the German state Schleswig-Holstein, is shown in Figure 86.

Traditionally, the safety level is re-evaluated every 10 years (based on a deterministic approach, levees are designed for the greatest historic event). In 2001, it became obvious that the existing levee (a) does not meet the safety level anymore due to sea-level rises. Thus, the levee was retrofitted (b). The new design already considered a safety margin of about 50cm in order to be prepared for the foreseeable sea-level rises. However, there is enormous uncertainty. The sea level might rise even more depending on the scenario used and progress in understanding and modelling the effects of climate change. Nonetheless, an additional investment in coastal protection (and a related extension of storm surge hazard zones with related land-use restrictions for land-owners)

Figure 86: The 'Klimadeich' concept

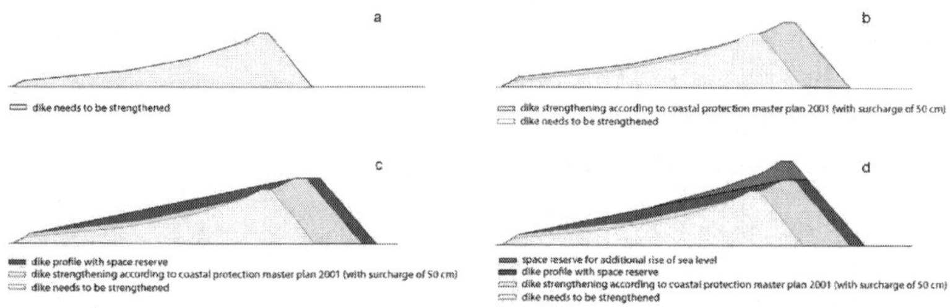

Source: http://www.schleswig-holstein.de/UmweltLandwirtschaft/DE/WasserMeer/09_KuestenschutzHaefen/06_Bemessungsverfahren/ein_node.html can hardly be justified. In order to be able to obtain unforeseeable developments and new knowledge, just the width of the top of the levee was enlarged from 2.5m to 5m (c) for just 10–20% of the construction costs of a traditional approach. As soon as new knowledge about the extent of the further rise of the sea level becomes available, a decision can be taken to increase the height of the levee (d). The estimated costs for the heightening are just 10% of those of a traditional approach.

Soft governance strategies

Governance strategies avoid cost-intensive technical solutions (ie investments in structural protection systems like levees) whose necessity might be questionable under uncertainty and often cause enormous maintenance costs. The key focus of soft strategies lies on discourse-based decisions on the analysis and management of (uncertain) risks. Here, the justification of actions and consensus about thresholds for acceptable risks and response actions becomes important (Greiving and Glade, 2013). The local village recovery plans in Indonesia (see Chapter A2a in Part A) are a good example of such a community-based, consensual governance strategy.

The implementation of a further strategy is described in the following through the example of the district of Cologne, Germany. On behalf of the Federal Institute for Research on Building, Urban Affairs and Spatial Development (Bundesinstitut für Bau-, Stadt- und Raumforschung [BBSR]), a new strategy for analysing and managing disaster risks in regional planning was developed by a scientific consortium of which the author was a part. Moreover, different departments and their experts from the district authority were represented in the network (regional planning, water management, geological experts, natural protection and emission control). The Cologne region is shown by in Figure 87 (which, at the same time, indicates basic information about land use and flood hazard, in blue).

The scientists of the consortium developed and proposed methodologies for the assessment and aggregation of the various components influencing the disaster risks caused by various spatially relevant hazards (river floods, earthquakes, major-accident hazards) that the region is threatened by. These proposals were discussed, if necessary modified and finally agreed by the experts from district authority in the network. The network as a whole concluded that only those hazards could be managed by spatial planning whose impacts are geographically localisable. This is a precondition for a demarcation of hazard zones for which spatial planning is entitled to define specific regulations for land-

Figure 87: Flood hazard profile of the district of Cologne

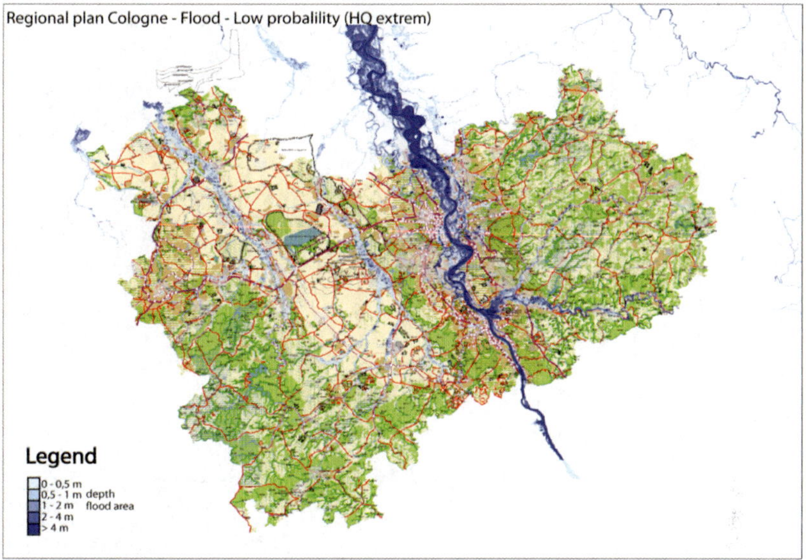

Source: Own figure

use control. Others (eg winter storms or pandemic) affect the entire region ubiquitously. For the aforementioned threats, risk was defined as a function of frequency and magnitude, as shown by Figure 88.

The experts from the authorities supported the scientists in focusing on the most relevant aspects of the sectoral and cross-sectoral assessment. Moreover, they provided impact models, data and, last but not least, their expert knowledge. Normative decisions on thresholds between hazard and vulnerability classes were formulated for each hazard type (floods, earthquakes, major-accident hazards) by the scientists and answered by the authorities. In particular, all land-use types (such as residential areas, industrial areas, different types of public infrastructure and natural protections areas) were consensually classified with regard to their specific susceptibility to the three types of hazards. In doing so, the results of the risk assessment are transparent and acceptable for all involved actors. In this regard, the normative dimension of risk was kept separate from the factual level. Moreover, this approach tackles uncertainty: the network as a whole agreed on the combination of data for both hazard and vulnerability and decided on – from the point of view of the network's member – plausible development pathways for future changes in both the hazard and vulnerability. In each case, the geographical level of analysis was the entire province scale (1:50,000), but areas that are obviously at particular risk were analysed at a scale

Figure 88: Risk index

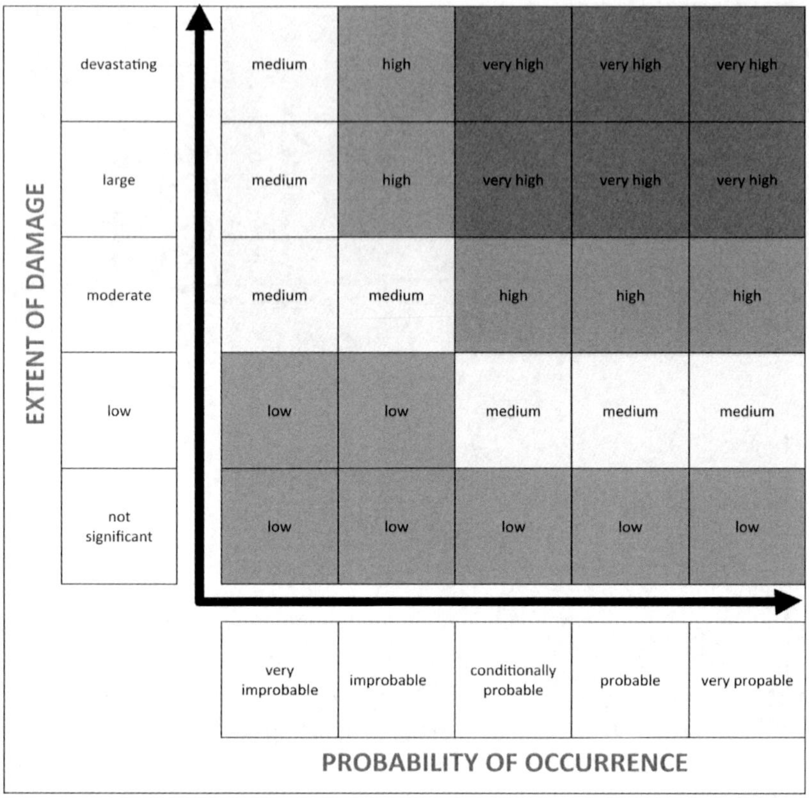

Source: BBK (2010)

of 1:5,000 by differentiating the building and infrastructure stock in a more detailed way.

The result is shown by the example of the flood risk map of the district of Cologne (see Figure 89). This collaborative mapping guarantees an impact on the ground as the results – as such not binding – will be used for the setting up of the new regional plan for the district of Cologne.

Strategies that reduce decision-making time horizons

In general, there is often a conflict in spatial planning between those actors who want to benefit from short-term economic development options (being potentially in conflict with changing risks due to impacts of climate change such as sea-level rises) and others in favour of more long-term, sustainable strategies. However, such strategies are not

Figure 89: Flood risk map of the district of Cologne

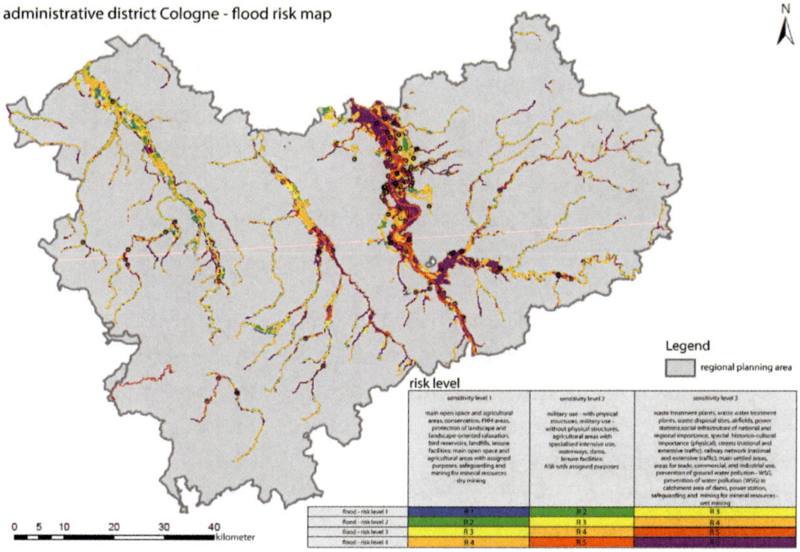

Source: Own figure

necessarily cost-efficient in view of the given uncertainty regarding the aforementioned effects like rises in the sea level. Nonetheless, there might be a consensus between these groups of actors regarding key planning principles, that is, sustainability, or, in more detail, the necessity of disaster risk reduction.

Here, the concept of parametric governance comes into play. In this context, bilateral forms of governance replace traditional top-down command-and-control strategies. This approach aims at governance via parameters or requirements (Greiving et al, 2012). The modality of the achievement of objectives remains in the hands of the given addressees (Cools et al, 2003). In comparison to traditional input-oriented decision-making, this concept offers more flexibility. It is an alternative process to the fragmented management of risks and can be seen as a 'win–win situation' for all included partners, authorities, institutions and so on, especially under circumstances of rapid change and uncertainty. Both objectives and measures are fixed for a limited time period (ie 2–3 years), but can be adjusted over time in order to adapt to new knowledge. Thus, the time horizon of decision-making is rather limited.

Such output-oriented management approaches have been widely used for decades in new public management (Lane, 2000), and also in comprehensive city marketing concepts (Paddison, 1993). Up to now, they have not been common in disaster risk management, though there are several similarities, such as a need for the coordination of the actions of different autonomous actors, common objectives and evaluating the effects of the chosen measures.

However, parametric governance has been implemented in disaster risk management in Switzerland since 2002 (Haering et al, 2002). For the funding period between 2011 and 2015, the central government aimed at supporting the regions in implementing the existing hazard zone maps in regional and local spatial planning, which is a new legal requirement (BAFU, 2011).

The commonly agreed objective (between central government and the regions) is the protection of human life, the environment and human assets against natural hazards. Table 15 explains the functionality of the concept. The elements of the concepts can be adjusted for the next programme period due to the knowledge on changes in frequency/magnitude of natural hazards (more advanced measures in case of an increase of the hazard or less investments/financial support in case of a decline).

Table 15: Elements of agreement on objectives

Performance goal	Performance indicator	Quality indicator	Financial contribution of the central government
Implementation of hazard zone maps	Sum of implemented hazard zone maps	Specific quality criteria (implementation must be in line with state-of-the-art) Contribution to reduction of risk (in %) Cost-efficiency	50% of overall costs

Source: BAFU (2011, p. 125, translated from German into English by author)

Discussion

In Table 16, the advantages, but also limitations, of these five different types of strategies are discussed.

Table 16: Comparative discussion of alternative strategies

Type of strategy	Advantages	Limitations
No-regret strategies	Multifunctionally justifiable and therefore more acceptable for affected population as pure risk mitigation measures Net-benefits in any case	Not possible in case of indispensably needed structural mitigation measures and therefore not combinable with safety margin strategies May fail if the population does not follow the law (informal settlements in hazard-prone areas or public open land)
Reversible strategies	Easily adjustable in case of newly generated knowledge on hazard and/or vulnerability Cost-efficient	Reversible strategies can contain structural measures, but only of a decentralized character (i.e. Strom water storage in small retention ponds), but they are not combinable with safety margin strategies
Safety margin strategies	Adjustment to future changes of risk possible. Can be cost-efficient as long as the measure can be sequentially realized (see example of Germany) Alternative to a retreat. Makes further intensive use (e.g. for settlement purposes) of hazardous areas possible	Call for an open debate on acceptable level of residual risks, but in reality (see examples of Japan and Germany) the level of safety is often just determined by experts Weakened risk awareness behind structural mitigation measures, because people tend to feel safe. Enormous damages if measures fail Call often for extremely costly mega-infrastructures (see example of Japan) and are therefore not implementable in developing countries
Soft governance strategies	Involvement of stakeholders and consensus building foster the acceptability of management option to be taken by the state Make building protection by house owners possible	Consensus is not always achievable (Nimby-phenomenon) In case of a given distrust in public decision-making hardly implementable
Strategies that reduce decision-making time horizons	Easily adjustable in case of newly generated knowledge on hazard and/or vulnerability Foster collaboration of different agencies and can therefore overcome problem of interplay	Cannot function in case of corruption

Source: Own table

Conclusions

Decisions on disaster risk management in general, but particularly planning-based approaches that may limit private property rights, need to be justified. In principle, planners tend to focus on long-term solutions, fixed in stone (or concrete), but their justifiability has already been discredited by a weakened evidence basis. Under a changing environment, and also having in mind the new understanding of what a government stands for, a more reflexive and adaptive practice of disaster risk management is needed. This chapter showed how such a transformation could take place by focusing on the role of spatial planning under uncertainty. Even deep uncertainty, caused by climate and socio-economic change, does not justify hesitation in taking decisions on how to alter risks. Modelling results identify at least a corridor or bandwidth of possible future conditions that help decision-makers to decide whether to follow a medium or worst-case scenario. This decision is, however, a normative one and therefore has to be taken by those who are democratically legitimised and not by experts only.

The main similarity between the different strategies (see section on 'Strategies for dealing with uncertainty' above) is their flexibility, which means that they are adjustable over time and can anticipate new knowledge. Moreover, these strategies are implementable under different legal-administrative environments and risk cultures. Nonetheless, concrete measures must be in line with a given legal framework, but this necessity does not question the usability of the strategy that they are derived from. The strategies are also suitable for the whole bundle of spatially relevant hazards, but are, of course, limited with regard to mega-disasters, for which mitigation measures are hardly possible at all – except a complete retreat, for instance, from tsunami-prone coastal areas – as the Japanese case underlines. It was also shown that communication about risks is crucial irrespective of the presented strategies that are going to be implemented.

References

Applied Technology Council (ed) 2012, *Guidelines for Design of Structures for Vertical Evacuation from Tsunamis*, Second Edition, Online available at: http://www.fema.gov/media-library-data/1426211456953-f02dffee4679d659f62f414639afa806/FEMAP-646_508.pdf

Armaş, J. & Avram, E. 2009, 'Perception of flood risk in Danube Delta, Romania', in *Natural Hazards*, 50(2): pp. 269-287.

BAFU (ed) 2011, *Programmvereinbarungen Schutzbauten und Gefahrengrundlagen. Erläuternder Bericht*, Bern, Bundesamt für Umwelt, Wald und Landschaft.

BBK Bundesamt für Bevölkerungsschutz und Katastrophenhilfe (, ed) 2010, *Methode für die Risikoanalyse im Bevölkerungsschutz*, Bonn.

Berkhout, F., Hertin, J. and Jordan, A. 2002, 'Socio-economic futures in climate change impact assessment: using scenarios as "learning machines"', *Global Environmental Change*, 12(2), pp. 83-95

Birkmann, J., Bach, C. & Vollmer, M. 2012, 'Tools for Resilience Building and Adaptive Spatial Governance – Challenges for Spatial and Urban Planning in Dealing with Vulnerability', in *Raumforschung und Raumordnung,* (70), pp 293-308.

Cools, M., Fürst, D. & Gnest, H. 2003, *Parametrische Steuerung – Operationalisierte Zielvorgaben als neuer Steuerungsmodus in der Raumplanung*, Frankfurt a.M.

Dessai, S. and van de Sluijs, J. 2007, *Uncertainty and Climate Change Adaptation – a Scoping Study*. Utrecht, available from http://dspace.library.uu.nl/bitstream/handle/1874/27160/NWS-E-2007-198.pdf?sequence=1

EC 2009, *A Community approach on the prevention of natural and man-made disasters*. Brussels, 23.2.2009, COM, 82 final.

EC 2014, Directive 2014/52/EU of the European Parliament and of the Council of 16 April 2014, amending Directive 2011/92/EU on the assessment of the effects of certain public and private projects on the environment.

EEA (European Environment Agency) 2001, *Late lessons from early warnings: the precautionary principle 1896-2000*, Environmental issue report No. 22 [Accessed Sept. 15, 2012 from http://www.eea.europa.eu/publications/environmental_issue_report_2001_22].

EEA 2013, *Adaptation in Europe - Addressing risks and opportunities from climate change in the context of socio-economic developments*, EEA report No. 3/2013, Kopenhagen.

Field, C.B., V. Barros, T.F. Stocker, D. Qin, D.J. Dokken, K.L. Ebi, M.D. Mastrandrea, K.J. Mach, G.-K. Plattner, S.K. Allen, M. Tignor, & P.M. Midgley 2012, 'Managing the Risks of Extreme Events and Disasters to Advance Climate Change Adaptation', A Special Report of Working Groups I and II of the Intergovernmental Panel on Climate Change, Cambridge University Press, Cambridge, UK and New York, NY, USA.

Greiving, S. 2009, *Report on Goverscience seminar on inclusive risk governance*, Brussels.

Greiving, S. & Glade, T. 2013, 'Risk governance' in Bobrowsky, P. T. (ed), *Encyclopedia of Natural Hazards*, pp. 804-806.

Greiving, S., Fleischhauer, M. & Wanczura, S. 2006, 'Management of Natural Hazards in Europe: The Role of Spatial Planning in Selected EU Member States' in *Journal of Environmental Planning and Management*; 49(5), pp. 739-757.

Greiving, S., Pratzler-Wanczura, S., Sapountzaki, K., Ferri, F., Grifoni, P., Firus, K., & Xanthopoulos, G. 2012, 'Linking the actors and policies throughout the disaster management cycle by "Agreement on Objectives" – a new output-oriented management approach', *Nat. Hazards Earth Syst. Sci.*, 12, pp. 1085-1107, doi:10.5194/nhess-12-1085-2012

Haering, B., Gsponer, G. & Koch, P. 2002, ‚Effor2 Konzeptbericht – Wirkungsorientierte Subventionspolitik im Rahmen des Waldgesetzes', Umweltmaterialien Nr. 145, Bundesamt für Umwelt, Wald und Landschaft (ed), Zurich.

Hallegatte, S. 2009, 'Strategies to adapt to an uncertain climate change', *Global Environmental Change*, 19(2), pp. 240–247

Haslauer, E. 2015, 'Application of a spatially explicit backcasting model: A case study of sustainable development in Salzburg, Austria' in *Applied Geography*, No. 58, pp. 124-140, http://dx.doi.org/10.1016/j.apgeog.2015.01.018.

Hill, L.J., Sparks, R.S.J & Rougier, J. 2013, 'Risk assessment and uncertainty in natural hazards', in Rougier, J., Hill, L.J. & Sparks, R.S.J (eds), *Risk and uncertainty assessment for natural hazards*, Cambridge University Press, New York, pp. 1-18.

IRGC (International Risk Governance Council) 2005, *White paper on risk governance: Towards an integrative approach*, Geneva.

IRGC (International Risk Governance Council) 2009, *Risk Governance Deficits – An analysis and illustration of the most common deficits in risk governance*, Geneva.

Lane, J.E. 2000, *New Public Management*, Routledge, London.

Löfstedt, R. 2005, *Risk management in post-trust societies*, Houndmills, Basingstoke, Hampshire, New York

Millennium Ecosystem Assessment (MA) 2003, *Ecosystems and Human Well-Being: A Framework for Assessment*, Washington, D.C., Island Press.

Mondlane, A., Hansson, K., Popov, O. & Muianga, X. 2013,' ICT for Social Networking in Flood Risk and Knowledge Management Strategies- An MCDA Approach', in *International Journal of Computer, Control, Quantum and Information Engineering*, Vol. 7, No. 10, 2013, pp. 633-639.

Mostert, E., Pahl-Wostl, C., Rees, Y., Searle, B., Tàbara, D., Tippett, J. 2007, 'Social learning in European river basin management; Barriers and supportive mechanisms from 10 river basins' in *Ecology and Society*, 12(1), p. 19

Natural Step 1993, 'Natural Step-Backcasting', http://www.naturalstep.org/en/backcasting

Paddison, R. 1993, 'City Marketing, Image Reconstruction and Urban Regeneration' in *Urban Stud March*, vol. 30, no. 2, pp. 339-349.

Paul, B. K. 2011, *Environmental hazards and disasters. Contexts, perspectives and management*, Wiley and Blackwell, London.

Pelling, M., High, C., Dearing, J. & Smith, D. 2008, 'Shadow spaces for social learning: a relational understanding of adaptive capacity to climate change within organisations' in *Environment and Planning A*, 40(4), pp. 867-884.

Renn, O., Klinke, A. & van Asselt, M. 2011, 'Coping with Complexity, Uncertainty and Ambiguity in Risk Governance: A Synthesis'. *AMBIO*, vol. 40, pp 231–246.

Robinson, W.I. 2003, *Transnational Conflicts. Central America, Social Change, and Globalization*. London and New York: Verso.

Sapountzaki, K., Wanczura, S., Casertano, G., Greiving, S., Xanthopoulos, G. & Ferrera, F. 2011, 'Disconnected policies and actors and the missing role of spatial planning throughout the risk management cycle', *Natural Hazards* (2 June 2011), pp. 1-30.

Slovic, P. 1999, 'Trust, Emotion, Sex, Politics, and Science: Surveying the Risk-Assessment Battlefield', *Risk Analysis*, 19 (4).

Smith, B. & Wandel, J. 2006, 'Adaptation, adaptive capacity and vulnerability', *Global Environmental Change*, vol. 16, pp. 282-292.

UN 1992, United Nations Conference on Environment & Development Rio de Janerio, Brazil, 3 to 14 June 1992 AGENDA 21.

UN-ISDR (International Strategy for Disaster Reduction) 2005, *Hyogo Framework for Action 2005-2015 - Building the Resilience of Nations and Communities to Disasters*, Extract from the final report of the World Conference on Disaster Reduction (A/CONF.206/6), http://www.unisdr.org/files/1037_hyogoframeworkforactionenglish.pdf

UN-ISDR 2009, *Global Assessment Report on Disaster Risk Reduction*. Geneva, United Nations.

UN-ISDR 2015, *Sendai Framework for Disaster Risk Reduction 2015–2030*, http://www.wcdrr.org/uploads/Sendai_Framework_for_Disaster_Risk_Reduction_2015-2030.pdf

Van Asselt, M. 2005, 'The complex significance of uncertainty in a risk era: logics, manners and strategies in use', *Risk Assessment and Management*, 2/3/4, pp. 125-158.

Walker, W., Harremoes, P., Rotmans, J., Van Der Sluijs, J., Van Asselt, M., Janssen, P. & Krayer von Krauss, M.P. 2003, 'Defining uncertainty: A conceptual basis for uncertainty management in model-based decision support', *Integrated Assessment*, 4(1), pp. 5-17.

Walker, W.E., Haasnoot, M. & Kwakkel, J.H. 2013, 'Adapt or Perish: A Review of Planning Approaches for Adaptation under Deep Uncertainty', in *Sustainability*, vol. 5, pp. 955-979.

Wangel, J., Gustafsson, S. & Svane, Ö. 2013, 'Goal-based socio-technical scenarios: Greening the mobility practices in the Stockholm City District of Bromma, Sweden', in *Futures*, vol. 47, pp. 79-92.

Weichselgartner, J. & Kasperson, R. 2010, 'Barriers in the science-policy-practice interface: Toward a knowledge-action-system in global environmental change research', in *Global Environmental Change* doi:10.1016/j.gloenvcha, 2009.11.006.

Young, O.R. 2002, *The Institutional Dimensions of Environmental Change; Fit, Interplay, and Scale*, Cambridge, Mass., London: The MIT Press.

Young, O.R. (2010), 'Institutional dynamics: Resilience, vulnerability and adaptation in environmental and resource regimes', in *Global Environmental Change*, vol. 20, pp. 378-385.

CONCLUSION

Change-proof cities and regions – an integrated concept for tackling key challenges for spatial development

Stefan Greiving, Kanako Iuchi, Jaroslav Tešliar and Michio Ubaura

Introduction

This final chapter sums up the key findings of both parts of this book: Part A on key insights into current practices of disaster risk mitigation and recovery in five different countries; and Part B on a cross-country analysis of the following five important topics – the transformation of spatial planning after significant disasters; efforts in building spatial resilience after disasters; coordination in building spatial resilience after disasters; participation in rebuilding more resilient space; and spatial planning and uncertainties.

Rebuilding from the Great East Japan Earthquake and the subsequent tsunami in 2011 has been a turning point for spatial planning in *Japan*. There is certainly a nationwide momentum to link risk management and spatial planning. In this context, different relocation strategies were identified: (1) 'Grouped' or 'Scattered'; (2) 'New development' or 'Infill development'; and (3) 'Land-use change' or 'No change'. The reconstruction efforts have made steady progress in the most affected Tohoku region. Local mitigation plans consider the decline of the population as an important framework condition for building back better. That is a very useful approach that should be adopted in other countries faced by a shrinking population.

Learning from the experience in Aceh, *Indonesia*, indicates that the implementation of rehabilitation and reconstruction after disasters should not be limited to physical reconstruction efforts only. Development of social resilience and, in particular, the socialisation and training of spatial planning and disaster risk reduction efforts is also urgently required in other countries. Although the local Village Plans were, in general, considered a success, there were some drawbacks. This was mainly due to the absence of comprehensive planning at the city level and the lack of a sound risk assessment as an evidence basis for decision-making on reconstruction. Moreover, there is still a need for integrating and better coordinating disaster response activities among all relevant stakeholders. In particular, the relation and coordination with the national government level still has to be improved. That is quite relevant for various other countries in the Global South.

The role of spatial planning in the context of both disaster mitigation and recovery is relatively new in the *US*. The recovery process after Hurricane Sandy (which hit New York in 2012) represents the most recent example of recovery and shows how spatial planning and risk reduction can be integrated. There are two central challenges in effective recovery planning: one is balancing the public's interest for speedy action with the need to deliberate carefully about how to spend scarce economic resources; the other is the tension between a desire to simply replace what was lost and a conflicting need to rebuild better and more resiliently. Similar problems occurred in the recovery after many other mega-disasters and therefore need to be addressed by comprehensive disaster risk management strategies.

Also, the *Slovakian* contributions show that a systematic framework of the assessment and mapping of disaster risks is needed in order to create a sound evidence basis for decision-making as to whether to tolerate or alter risks. Geological and hydrological information, such as thematic hazard maps based on 1D and 2D hydrodynamic models, has a very high potential for reducing fatality rates and losses due to natural disasters – as long as decision-makers, in general, and spatial planners, in particular, ground their decisions on this information. This information has to be visualised and communicated in a way that meets the educational background of planners and other stakeholders who are not trained in engineering sciences. From the legislative point of view, there are requirements in the new and modern Building Act that will also include the obligation for spatial planners, as well as all investors, to take the results of a risk assessment fully into account before taking decisions about future land use. Here, the use of the Strategic Environmental Assessment for assessing disaster risks in spatial planning is seen as extremely important – not only for European Union (EU) member states, but also for those outside Europe.

The *German* contributions underline that spatial planning is theoretically able to mitigate disaster risks to a particular extent. However, the quality of planning-based response strategies depends on the quality of risk information. The outcomes of risk assessments have to be tailor-made to the needs of their users, such as planning authorities. There is a growing need to involve all stakeholders from the very beginning of the risk assessment and management process in order to improve the effectiveness of disaster risk response, particularly in already built-up areas. The role of the European Commission for deriving common methodological and procedural standards (and monitoring their implementation in national law) is positively emphasised in regard to both flood risks and major-accident hazards.

Key lessons learned out of the experiences from five different countries

We live in a changing world that is characterised not only by the globalisation of the economy, but also global migration flows. Rapid urbanisation and the rise of mega-cities goes hand in hand with a population decline in many rural areas worldwide. Our climate is changing and adaptation to this change is indispensably required irrespective of whether it is caused by anthropogenic greenhouse gas emissions or by other factors.

All of these ongoing changes influence both natural hazards and the vulnerability of our societies to these hazards. At the same time, these hardly predictable changes question traditional, engineering-based disaster response strategies. Planners in both developed and developing countries have encountered mega-disasters beyond imagination to an extent that they have never experienced before. This is due to an extreme urbanisation and population inflow to urban areas, as well as a deeply interconnected globalised economy leading to increased vulnerability against disasters that can have potentially global impacts. As a result, scientific approaches that rely primarily on statistics and modelling may be very limited in their applicability to emerging types and sizes of possible disasters. Therefore, planners are faced with a situation of approaching mitigation beyond the scientific techniques that they have long been relying on. The environments where planners are placed – whether in developed or developing countries – are thus similar: managing disasters requires a paradigm shift beyond the technical identification of risk as a premier and sole remedy.

From a planner's perspective, however, some progress has been made. New modelling technologies and assessment methodologies have helped to overcome the model-inherent component of uncertainty. Moreover, in the last couple of years, more and more disaster-prone countries – including all five countries addressed in Part A of this book – have made steps forward in establishing planning systems for risk reduction and better linking disaster risk management with land-use planning – in particular, Indonesia, Japan and the US. Unfortunately, these changes are, in most cases, disaster-driven. Advancements in planning systems and assessment methods were always stimulated by mega-disasters, such as the tsunamis in Indonesia and Japan or major floods in Germany and Slovakia. Here, supranational entities like the EU also come into play: the EU launched directives on the management of flood risks and major-accident hazards immediately after mega-events in Europe that had cross-border effects.

Recurrent disasters affecting societies, such as landslides in Slovakia and floods and hurricanes in the US, also play a crucial role in this. The same is true for relocation: this nearly always takes place in the aftermath of disasters. More prospective strategies are needed that anticipate change and understand disaster risk reduction as a key prerequisite for sustainable development. Although planners' technical contribution to proactive disaster mitigation efforts is limited due to other macro-forces (eg the political economy and various inherent systems already established, including administrative and social structures), spatial planners and politicians need to begin to take a step forward to better understand the nature of hazards and risks. In the current system, the severity of risks and hazards is overlooked, and it tends to prioritise short-term economic benefits. There are, however, potentials for synergies between risk mitigation and economic and social development. Examples from different countries have shown that city designs characterised by a network of green and blue structures enhances the quality of living in urban areas by offering space for recreational activities and leisure (see Greiving and Fleischhauer, 2012; Reckien et al, 2013). At the same time, various threats can be absorbed, such as urban heat and urban flooding. Even the coping capacity regarding seismic hazards can be improved as these green structures can serve as locations for safe evacuation areas.

Proposed concept for a step forward to change-proof cities

Chapter B1 on 'Planning systems for risk reduction and issues in pre-disaster implementation' argues that land-use control systems and disaster management systems are often working in parallel, and further effort towards their integration is a must for advancing the way in which spatial planning deals with disasters. Disaster-prone countries establish more and more planning systems for risk reduction and mainstream disaster mitigation. Here, avoiding the usage of hazard-prone areas is considered the first step towards creating a resilient space. Ideally, decisions on land use in these areas should take the specific susceptibility of different land uses into account. However, in every country, mitigation plans are almost neglected prior to disasters. A useful approach would be to inform residents about possible disaster impacts without any spatial control, and about how plan making and plan implementation could contribute to reducing the impact. One example here is the collective relocation programmes in Japan, as well as the flood insurance and buyout programmes that have been used in the US.

Risk assessments as an evidence basis need to focus on the vulnerability of livelihoods; because of the needs of the local population, natural hazards are often not the only and not the most relevant and burning threat for people. Thus, measures should be justified for their multifunctionality (eg vertical evacuation concepts whose facilities are ordinarily used as community centres) in order to create synergies between disaster risk reduction and the standard of living. In the case that structural mitigation measures are suitable (a seawall may negatively affect the access to the coast that is crucial for fishermen), particularly decentralised small structures can be preferable for several reasons: they are more cost-efficient; they can be maintained with local knowledge; and they create jobs for local people.

Chapter B2 on 'Land-use planning after mega-disasters: between disaster prevention and spatial sustainability' points at the fact that while recovery planning almost always considers a spatial component, its practice largely depends on socio-economic and political conditions. Planning processes are also important to make the plan happen on the ground. There is always a need for weighing up between disaster prevention and spatial sustainability. Depending on given political priorities, the one or the other interest may become dominant: the recovery process in Tohoku region (Japan) can be characterised as the type 'setting priority on disaster prevention'; in Aceh (Indonesia), priority was given to 'sustainability of daily life'; the recovery process in New Orleans (US) can be characterised as the type 'balancing disaster prevention and sustainability'; and the land-use planning and implementation process for recovery in Germany and Slovakia can be characterised as a 'fixed' type. In Germany, on-site reconstruction was the basic policy instead of large-scale relocation projects. However, there are potentially some negative drawbacks of such a policy. Germany missed the opportunity to use the given window of time to aggregate the scattered settlements in affected areas in order to create more sustainable land-use structures in shrinking regions.

Chapter B3 on the 'Role of coordination in building spatial resilience after disasters' came up with the following key insight: coordination is influenced by a long tradition of state interventions and cultural traditions and is therefore context-specific. One common observation is the fact that the coordinative unit always depends on the scale of a disaster. However, mega-disasters influence the framework for coordination: there are countries in which after a disaster, recovery/reconstruction agencies were established (ie in Japan, the National Reconstruction Agency; in Indonesia, the Agency for Rehabilitation and Reconstruction; in the US, the Federal Emergency Management Agency; and in Germany, the Federal Office of Civil Protection and Disaster Assistance), which take the lead in the planning process. A further key lesson learned is the need for increased communication between different government entities – both horizontally and vertically. Moreover, communication between different sectors of the governance system (state, private sector and civil society) is regarded as a key factor for resilience. Resident-inclusive processes make them develop risk ownership, which is critical to further implementing mitigation-considered planning (in other words, the plan would not be implemented unless risk is owned by local people). Local knowledge must be taken into consideration for both the assessment and management of risks by national and local public authorities. At the same time, qualified expert knowledge provided by governmental officers or engineering companies is also needed to justify informed local decisions about tolerating or altering risks. The new Sendai Framework for Disaster Risk Reduction makes clear that a sound evidence basis is needed in order to support and improve risk management decisions (Priority 1: 'Understanding Risk'). Here, it is spatial planning, in particular, that urgently requires the specific expertise of sectoral divisions like water boards or geological surveys on a given hazard profile.

Although a fragmentation in responsibilities between the different actors involved is unavoidable due to the necessary specialisation, fragmented risk management strategies are not. In none of the countries studied is there currently a comprehensive management strategy in force in which all actors have agreed on common goals for disaster risk reduction. Response strategies need to be coordinated between comprehensive planning and sectoral planning divisions like water management, environmental planning and emergency response in order to be cost-efficient and effective. This is a key challenge for the future and has recently been outlined by the Sendai Framework for Disaster Risk Reduction (Priority 2: 'Strengthening disaster risk governance to manage disaster risk'), which states: 'Clear vision, plans, competence, guidance and coordination within and across sectors as well as participation of relevant stakeholders are needed.'

Chapter B4 is on 'Residents' participation in rebuilding more resilient space'. The requirements for participation in disaster mitigation and recovery depend on different factors: a community's history of participation should be considered and the scale of the disaster should match the extent of public participation. Moreover, the complexity of the problem needs to be taken into consideration. However, there are always limitations that cannot be overcome by participation: transparency and clarity are required in order to keep (or even rebuild) the trust of the people, but planners need to know and accept that a consensus with regard to an acceptable level of risk and related management actions might not be achievable with everyone. Nonetheless, good risk communication and participation can at least foster a mutual understanding regarding the standpoints of the different groups and their concerns and interests. Furthermore, the limits of risk communication have to be communicated to the population – clarity comes before consensus – doing so helps keep and foster the trust of the people in the institutions and the governance process. Only understood and accepted risk management actions are implementable in practice. This statement is demonstrated by the community-based local Village Plans in Indonesia, which replaced the previous top-down-initiated buffer zone concept that had ignored the economic importance of the coastline for the local population.

Chapter B5 deals with 'Spatial planning and dealing with uncertainties associated with future disasters' and argues that under a changing environment, and also having in mind the new understanding of what a government stands for, a more reflexive and adaptive practice of disaster risk management is needed. Even deep uncertainty does not justify hesitation on how to alter risks. Such decisions are of a normative character and have to be taken by those who are democratically legitimised and not by experts only. They may decide to follow the precautionary principle (and, in consequence, a worst-case scenario) or to tolerate a potential risk in view of given social or economic interests. Both are justifiable but have to be communicated to and discussed with those who are potentially affected by the alternatives. The chapter identifies different strategies for dealing with uncertainty that are flexible and adjustable over time and can anticipate new knowledge about both the hazard and the vulnerability. Moreover, these strategies are implementable under different legal-administrative environments and risk cultures. Ideally, disaster risk management strategies should be mainstreamed into spatial planning and not be seen as standalone strategies. Spatial planning actors are, relative to other actors, in a key position to implement this mainstreaming process as spatial planning requires making decisions about future land use and, in some cases, conducting a weighing-up process that takes into account all relevant concerns and interests – including safety.

Adopting conceptualised findings – the need for a larger emphasis on booming countries

Preparation for unexpected disasters has never been conceptualised in planning. However, such challenges are facing all countries across the globe regardless of their level of economic development or disaster frequency. One of the mainstream approaches in planning continues to be spatial control, which includes keeping hazard-prone areas free of further development and avoiding the usage of high hazard lands. This is considered as a first step towards creating a resilient space. This is particularly relevant for countries where rapid urbanisation is taking place. Here, differentiated decisions on land use in accordance with the specific vulnerability of different types of land use and an adaptation of the buildings to the given hazard profile are often the better risk management strategies, as opposed to a complete avoidance of hazardous areas. However, this traditional planning-based approach of a spatial separation of hazards and vulnerable land uses is also questioned in those countries where the law or 'rational' plans do not fit the urbanisation reality of population agglomeration and hastening development. In particular, informal settlement often takes place in hazard zones – in those areas where a retreat took place after a disaster, as well as in other areas that were never used as a settlement area. Fatally, it is (traditional) risk planning that contributes to the risk – mainly in the cities of the Global South. Their development speed for both population and city expansion are far more rapid than the past development trajectories of what other 'industrialised' countries have experienced.

One of the understandings in international development is that the poor are the hardest hit by disasters, although the economic (and insured) losses are much greater in industrialised countries. However, many mega-disasters in the past (eg Hurricane Mitch in Honduras in 1998 or the devastating earthquake that Haiti was hit by in 2010) have shown that a national economy can be destabilised for decades. The main criterion is not the loss in absolute numbers, but the given capacity to cope. If the dimension of a disaster's impact in relation to national gross domestic product (GDP) exceeds this coping capacity, a society cannot recover on its own (International Strategy for Disaster Reduction and World Bank, 2008). Thus, disaster risk management in the developing world needs to be adjusted to the specifics of these countries and be integrated into comprehensive development strategies.

In this adjustment, local residents negotiating with risk becomes critically important. Despite all of the scientific details in risk assessments, there is no local impact unless the issues are internalised for localised coping strategies. In the end, residents are the ones who have the most detailed knowledge about their place and their ways of sustaining life. Planners and policymakers can integrate this knowledge through guiding discussions, fostering collaboration and sharing information, and working towards improved local ownership in the planning process.

Conclusions for a comprehensive disaster risk management

Most of the aforementioned challenges for spatial planning are quite common in most countries worldwide and are not limited to the management of disasters, but similarly relevant for manifold other areas that are deeply intertwined with risk management: the development of settlements, housing, landscape planning, transport planning and so on. That underlines the need for mainstreaming disaster risk management into spatial planning as an element whose issues have to be weighed up against other concerns and to be integrated in comprehensive urban management strategies. In doing so, the often tremendous consequences of extreme events could be mitigated and mismanagement in the post-disaster recovery phase could be avoided. In this respect, the reader may understand this book as a contribution from the perspective of spatial planning to the necessary global search for clear visions and plans for disaster risk reduction.

References

Greiving, S. & Fleischhauer, M. (2012) National climate change adaptation strategies of European states from a spatial planning and development perspective. *European Planning Studies* Vol. 20, No. 1, January 2012, pp. 27-47.

International Strategy for Disaster Reduction & World Bank (2008) (eds) *Mitigating the Adverse Financial Effects of Natural Hazards. A Study of Disaster Risk Financing Options*. Geneva.

Reckien, D., Flacke, J., Dawsoll .U.-J., Heidrich, O., Olazabal, M., Foley, A., Hamann, J. J.-P., Orru, H., Salvia, M., De Gregoria Hurtado, S., Geneletti, D., Pietrapertosa, F. (2013) Climate change response in Europe: What's the reality? Analysis of adaptation and mitigation plans from 200 urban areas in 11 countries, *Climatic Change* (2014) 122(33): 1-340 ooI10.1007/sI0584-0 13-0989-8

Index

References to figures and tables are shown in *italics*

D

E

F

G

K

L

M

S

T